The Idea of Excellence and Human Enhancement

Adriana Warmbier (Ed.)

The Idea of Excellence and Human Enhancement

Reconsidering the Debate on Transhumanism in Light of Moral Philosophy and Science

Bibliographic Information published by the Deutsche Nationalbibliothek
The Deutsche Nationalbibliothek lists this publication in the
Deutsche Nationalbibliografie; detailed bibliographic data
is available in the internet at http://dnb.d-nb.de.

Library of Congress Cataloging-in-Publication Data
A CIP catalog record for this book has been applied for at the Library of Congress.

This book was published with the support of the Polish National
Science Centre Fund (2012/07/D/HS1/01099).

Cover Design: © Olaf Gloeckler, Atelier Platen, Friedberg
Cover illustration: 'Young Woman Tuning a Lute', painting by Hendrick
Ter Burgghen. © The Fitzwilliam Museum, Cambridge

ISBN 978-3-631-71834-6 (Print)
E-ISBN 978-3-631-76524-1 (E-PDF)
E-ISBN 978-3-631-76525-8 (EPUB)
E-ISBN 978-3-631-76526-5 (MOBI)
DOI 10.3726/b14551

© Peter Lang GmbH
Internationaler Verlag der Wissenschaften
Berlin 2018
All rights reserved.

Peter Lang – Berlin · Bern · Bruxelles · New York ·
Oxford · Warszawa · Wien

All parts of this publication are protected by copyright. Any
utilisation outside the strict limits of the copyright law, without
the permission of the publisher, is forbidden and liable to
prosecution. This applies in particular to reproductions,
translations, microfilming, and storage and processing in
electronic retrieval systems.

This publication has been peer reviewed.

www.peterlang.com

Acknowledgements

This book is the result of a collaborative work of a group of scholars representing different fields in the humanities, social science and science. I would like to express my gratitude for their efforts to write the papers and their contribution to the quality of this volume. I am particularly grateful to my colleagues: Dr Marta Soniewicka, Dr Beata Płonka and Dr Filip Kobiela with whom I am pursuing an interdisciplinary research project on phenomenological and analytic approach to agency, funded by the Polish National Science Centre. I have greatly benefitted from our intense and constructive debates on the topics which are presented in this book. I would like to thank Thomas Douglas for giving permission to reprint his paper in this volume. I am very grateful to reviewers whose useful remarks and suggestions improved the book.

I owe special thanks to Prof. Robert Audi for the meaningful and insightful discussions during a number of seminars that he conducted at the Jagiellonian University. I am extremely grateful to Prof. Richard Holton who read my paper and offered extensive and valuable comments when I was on the Corbridge Trust fellowship of Robinson College in the University of Cambridge.

I would also like to thank Dr Nigel Dower, Prof. Robin Cameron, Dr Elizabeth Shaw, Dr Daniel J. Shaw, Ms Patricia Clarke, Melvin Dalgarno for their helpful remarks, questions and conversations from which I benefitted while visiting King's College of Aberdeen.

I am grateful to Klementyna Chrzanowska and Dr Aeddan Shaw for their help in the linguistic edition of the volume.

The illustration on the jacket – Hendrick ter Brugghen's painting *Young Woman Tuning a Lute*, which I found in the Fitzwilliam Museum – struck me as an apt metaphor for our moral life. The instrument itself presents a challenge for prospective skills. The young musician is tuning the lute – an instrument which is delicate, very sensitive to temperature and humidity, and therefore requires constant adjustments. We may say that human flourishing requires a similar constant, effortful activity of an agent, and much of this volume is concerned with this fundamental issue.

Contents

Introduction 9

**Part I The Development of the Concept of Excellence.
Toward a Debate on Enhancement**

Michał Bizoń
Would Greeks and Romans Endorse Moral Enhancement Through
Genetic Modification? 15

Tomasz Fiedler
Clement of Alexandria: Toward Christian Perfection 49

Karol Wilczyński
The Idea of Human Enhancement and Arabic Neoplatonism ... 67

Jan Kiełbasa
The Perfection of Being and the Perfection of Action. Medieval
Understanding of the Various Dimensions and Limits of Human
Perfection 83

Wojciech Załuski
The Idea of Human Perfection in Modern Philosophy 101

**Part II Enhancement Methods: The Promise and Limitations
of Enhancing Ourselves**

Robert Audi
Moral Philosophy as a Framework for Approaching Human
Enhancement 121

Thomas Douglas
Enhancing Moral Conformity and Enhancing Moral Worth ... 145

Adriana Warmbier
Moral Enhancement. Enhancing Motivational Processes and
Agent-Based Ethics 177

Jacek Jaśtal
Human Flourishing and the Coherence of Knowledge.
An Aristotelian Perspective 199

Wojciech Lewandowski
Intrinsic and Instrumental Values in the Assessment of Human
Enhancement 219

Beata Płonka
Excellence and Its Biological Limitations 237

Robert Poczobut
Transhumanism and Cognitive Science 277

Marta Soniewicka
Human Enhancement: The Question of Fairness and Virtues in
Sport Activities 295

Filip Kobiela
The Brave New Athlete? The Meaning of Perfection in
Contemporary Professional Sport 315

Notes on Contributors 331

Introduction

The aim of this book is to present the debate over the enhancement of the human condition in light of the development of the idea of excellence. Recent advances in neuroscience, genetics, and biotechnology have significantly expanded the range of potential uses of biomedical technologies. Human enhancement means increasing our cognitive, mental, and physical capacities beyond the normal human level. Some argue that biotechnological interventions may even improve our character, strengthen our powers of self-control, and consequently have considerable influence on moral decision making. They claim that enhancements are a moral obligation. The Greeks believed that striving for perfection is an integral part of human flourishing. Yet, since the ancients, our approach to the concept of excellence, and our understanding of self-development and self-fulfilment have changed significantly. This book attempts to answer the question whether one should regard biomedical enhancement as an alternative method for bringing about a better person, or perhaps this goal should rather be achieved through traditional means such as education, training, and voluntary effort in shaping one's own judgments and attitudes. In exploring the origins and the development of the idea of excellence from antiquity to the present-day project of the post-human condition, the book shall afford a lucid account of the practical implications of the ancient and medieval models of human flourishing. No historically influential position in ethics is by itself adequate to serve as the single reliable pattern for self-improvement, and so the most accurate solution is to seek to formulate a view that would capture the best elements of each position. In response to the debate on the promise and perils of bioengineering, which is believed to make the myth of the "self-made man" come true, this book does not only raise the question of whether we should overcome the biological, mental, and social limitations of our condition by biomedical means, but it also attempts to demonstrate the various approaches to the idea of human perfection, each of them based on different assumptions concerning human nature, morality, autonomy, and the nature of human choice. The technical evolution, believed to bring enhanced efficiency, faster speed, and better design, is creating new challenges to our societies in general, and to our ethics in particular, therefore the book does not only discuss the transhumanist project itself, but also presents an alternative model for assessing human development. It does so by turning especially to ancient thought (virtue ethics) and Kantian ethics, attending carefully to their key concepts such as agency, practical reason, *eudaimonia*, moral value, global features and dispositions of character, or the Humanity Formula.

The volume collects fourteen papers, all except one previously unpublished, written by scholars from various countries, representing diverse cultural backgrounds and different fields of research including philosophy, sociology, theology, law, biophysics, and cognitive science. The authors reflect on human enhancement that is both unprejudiced and free of passionate commitment, and present their arguments from a wide range of perspectives.

In Part I, *The Development of the Concept of Excellence. Toward a Debate on Enhancement*, the authors give an account of the idea of human perfection found in various philosophical traditions, such as ancient pagan thought, Christian Antiquity, Arabic Neoplatonism, scholasticism, and modern philosophy. One of the main assumptions that underlie the transhumanist project concerns the very aim of enhancement, which is regarded as being the same as the aim which constitutes the foundations of every ethical theory. Due to this, some argue that the dispute over human enhancement does not pertain to the concept of perfection itself, but refers only to the means of achieving it. However, the diachronic study of the development of the concept of perfection, and its gradual emancipation within anthropological ideas and theories of action in Western philosophy, demonstrates that we do not share an agreed understanding of what human excellence is.

In Part II, *Enhancement Methods: The Promise and Limitations of Enhancing Ourselves*, the authors consider the questions related to the most crucial issues regarding the complex subject of human enhancement while referring to the most recent literature in the field. The opening chapter of this part, written by Robert Audi, provides a vivid account of moral philosophy as a framework for approaching the idea of bioenhancement. The considerations presented here start with a review of four major kinds of ethical theory, namely, virtue ethics, Kantian ethics, utilitarianism, and common-sense intuitionism. Audi's reflections reveal both numerous differences between the four kinds of ethical theory, as well as a certain degree of complementarity between them. Moral philosophy has standardly taken human persons to be its primary concern, and if we are guided by this intuition, then human excellence will be what guides the discussion. Thus, moral philosophy provides an explanation and justification of moral decisions in relation to a framework that takes account of universally respected principles and values. The view put forward by Robert Audi has been thoroughly discussed and developed from different viewpoints presented in four other chapters, which are the fruit of a relevant debate in which most of the contributors to this volume have participated. Since the problem of biomedical enhancement is considered here from various perspectives, this section also addresses three important aspects of the practical dimension of the enhancement debate, namely biology, cognitive science, and sports. The authors discuss the problem of defining the scope of what constitutes physical and mental enhancement, as well as consider the question of whether the results of cognitive science support the transhumanist vision of the future of mankind. The section ends with a reflection on the problem of fairness in sport activities, firstly raised by Michael Sandel in his account of "bionic athletes." The issue calls for a redefinition of the concept of excellence, which requires taking a more balanced view of the roles of individual effort, disciplined training, and enhancement. The present book does not aim to exhaust the complex issue of human enhancement; it rather attempts to offer a profound insight into the assumptions, aims, and consequences connected with transhumanism. Transhumanism is considered here not merely as an interdisciplinary research program, which might entail contentious conception of the human nature and action, but also as a habit of mind which may be of concern to many.

Jagiellonian University   Adriana Warmbier
Krakow, Poland

Part I The Development of the Concept of Excellence. Toward a Debate on Enhancement

Michał Bizoń
Would Greeks and Romans Endorse Moral Enhancement Through Genetic Modification?[1]

Abstract: In this paper I consider ancient Greek notions of character perfection. I begin with a scrutiny of Greek terms for perfection. I argue that there was a close nexus between physical, psychic and ethical traits, as regards perfection of one's nature. Next, I discuss ancient methods of reproductive and population control, and the attitudes of various philosophers to these methods. I focus on the most commonly mentioned method – that of infant exposure. I then discuss four theories of human nature and its perfection: Plato, Aristotle, the Stoics, and the Epicureans. Finally, I consider what could be inferred from these theories regarding enhancement of human nature through genetic modification.

Keywords: Greek philosophy, Plato, Aristotle, the Stoics, Epicurus

1 What could the ancients say on genetics?

A consideration of ancient approaches to genetic enhancement of human nature might appear to be blatantly anachronistic. After all, the ancients did not know the double helix. How could they therefore have taken a stance on the possibility of modifying the human genome?

The ancients did, however, have elaborate theories of human nature and conceptions of its perfection, which played a central role in ancient ethical systems. Theories of human nature were predominantly concerned with psychic traits, but they also included – albeit usually secondarily – physical traits. Ancient philosophers also recognized a close nexus between psychic and physical traits. They were keenly aware that the latter were largely hereditary and accordingly developed various theories of heredity. These theories differed in many ways from the present state of knowledge, e.g. in accepting the heredity of acquired traits. Nevertheless, the ancients were well aware of the fundamental importance of hereditary traits for the shaping of human nature. Few ancients would

---

[1] I would like to thank Prof. Krzysztof Bielawski and Dr. Jakub Filonik for their useful suggestions and help.

be surprised to learn that not only physical, but also psychic traits – including intellectual abilities – are hereditary.

The ancients also acknowledged problems with heredity. If essential human traits are hereditary, what space is left for improvement through education and training? Within the Homeric *ēthos*, excellence of character was almost completely hereditary. The conception that excellence of character may be acquired by training and education originated in some proximity to the advent of democracy. Concomitantly, there arose the controversy over the proportion and importance of hereditary and acquired traits, which appeared for the first time in the late 5th century BCE as part of what is collectively called the *nomos-physis* debate. New conceptions of character formation by paideutic instruction broke the fetters of Homeric hereditary stratification, but also engendered a problem: since noble character traits were not transmitted by heredity, one's offspring could turn out vastly different from the parents. Of course, if acquired traits were inherited, then perfection attained through training would be transmitted as well. Thus, notions of the perfection of nature, on the one hand, and of heredity and training, on the other hand, were closely intertwined.

In this paper I speculate what the views of some ancient philosophers on moral enhancement through genetic modification might have been. Since, by necessity, the discussed philosophers could not hold views on this precise topic, I will adopt the following method of tackling the issue. The point of departure is the question whether, on the ground of the considered ancient theories of human nature, given modern methods of genetic modification of physical and psychic traits, would it have been acceptable and/or advisable to use such methods for the purpose of the enhancement of ethical traits. To answer this question, I discuss four ancient theories of human nature and its perfection, focusing on the nexus between psychic and physical traits. I then suggest to what degree, within these theories of human nature, ethical traits could be enhanced by way of influencing physical and psychic traits. On this basis, I conclude that if within a given theory of human nature ethical traits could be enhanced by means of influencing physical and psychic traits, then it is plausible that on this theory, it would be acceptable and/or advisable to use modern methods of genetic modification for moral enhancement.

I begin with a scrutiny of Greek notions of excellence and argue that they concern simultaneously psychic and physical traits. I then consider ancient attitudes to reproductive and population control, focusing on philosophical arguments for and against certain methods. Next, I turn to physiognomical theories and related philosophical views regarding physical traits with ethical ramifications. I then discuss four theories of human nature, focusing on the nexus between psychic and physical traits. I conclude with a speculative consideration of Greek views on moral enhancement through genetic modification.

2 Greek terms for excellence: the nexus between physical and psychic traits in popular morality

2.1 *Anthrōpinē physis* (ἀνθρωπίνη φύσις)

The adjective *anthrōpinos* (ἀνθρώπινος) is a typical Greek term denoting things specifically human. It is commonly used to qualify several substantives, such as life (βίος) or race (γένος). It is also used independently in the neuter plural (τὰ ἀνθρώπινα) to denote things pertaining to human condition. In philosophical sources, the adjective is often used with *physis* (φύσις) to denote natural traits specific to humans (human nature) and is juxtaposed with divine traits, standardly denoted by the adjective *theios* (θεῖος).

However, it does not follow from this verbal usage that the Greeks universally accepted the existence of a common and unchangeable human nature clearly distinct from divine nature (nor, for that matter, form the nature of animals, hybrids, and monsters). In fact, the boundaries of human nature were anything but stable. The possibility of humans morphing into other beings is a topos in mythology, and no less than two works devoted precisely to this theme have survived, as indicated by their title, *The Metamorphoses*. In various myths, some of the best known described by Ovid in one of these extant *Metamorphoses*, people transform from male to female, from female to male, and from human to god, animal, plant, or monster, usually retaining their personal identity at least to some degree.

In Greek and Roman religion the boundary between human and divine nature was precarious and vague. Although some myths (such as e.g. the one about the birth of Dionysus) suggest that the divide between human and divine nature is great and unsurpassable, it is also common for humans to mingle intimately with the gods and at times become gods, a belief that persisted into Roman times with the occasionally abused institution of deification. Also, Greek and Roman religion explicitly recognized the mixed categories of the demigod (having both divine and mortal attributes) and the hero (descendant of a god and a mortal, who him/herself is mortal), both of whom may have been further promoted to a fully divine status or demoted to the status of a mortal. Plato's injunction to "become like god" (*Teaetetus* 176b) might have sounded to his contemporaries as strenuous, but hardly outlandish.

2.2 Aristeuein (ἀριστεύειν)

A basic tenet of the Homeric hero's moral code was to "always excel and be better than others" (αἰὲν ἀριστεύειν καὶ ὑπείροχον ἔμμεναι ἄλλων, Iliad 6.208).[2] The word rendered as "excel" is *aristeuō* (ἀριστεύω), a verb cognate to the adjective *aristos* (ἄριστος), "best," which is the superlative form of *agathos* (ἀγαθός), "good." Literally, *aristeuō* means "to be best," or "to perform best." Heroes are regularly called *aristoi* (ἄριστοι) – "best" – to signify their superior quality and excellence before others (e.g. Iliad 1.91, 2.82, 2.580, 2.761). Extended epic descriptions of heroic feats were later called *aristeia*, as in the case of the *aristeia* of Diomedes, which comprises much of Scroll 5 of the *Iliad*, or the *aristeia* of Achilles in Scroll 21.

The term *agathos* "commends the most admired type of man"[3] in the Homeric society in all relevant aspects of physical, psychic, and ethical excellence. A hero is the embodiment of all of these three qualities: he excels physically, he is a leader of men, and he has the moral right to do so. On the other end of the spectrum, physical ugliness and moral baseness were also coupled, as can be seen in the description of Thersites in Scroll 2 of the *Iliad*. Thersites, described as an ugly hunchback, challenges and rebukes the heroic leaders of the Achaeans in counsel. He is called *aischistos* (αἴσχιστος), the opposite of *aristos*. This term denotes equally physical deformity and moral baseness. Thersites is on the lowest end of the Homeric social hierarchy, lacking excellence in all three areas: physical, intellectual, and ethical.

According to the Homeric ethical outlook, moral right – or even moral superiority – is essentially associated with physical and intellectual excellence. This composite quality of being *aristos* is measured in terms of success. Intention plays little or no role in the Homeric ethical view,[4] and moral error is hardly distinguished from mistake. The standard of excellence in the Homeric system of values "is used to commend skills, physical gifts, or inherited social advantages. None of these can be attained merely by good intentions."[5] It is difficult to say whether moral excellence is different from – but concomitant with – physical

---

2 An in-depth discussion of the Homeric hero as ἄριστος and the Homeric notion of excellence can be found in: G. Nagy, *The Best of the Achaeans: Concepts of the Hero in Archaic Greek Poetry* (Baltimore: The John Hopkins University Press, 1979).
3 A. Adkins, *Merit and Responsibility. A Study in Greek Values* (Oxford: Clarendon Press, 1960), p. 32.
4 For a detailed discussion of this, see Adkins, *Merit and Responsibility*, pp. 30–60.
5 Adkins, *Merit and Responsibility*, p. 46.

Would Greeks and Romans Endorse Moral Enhancement? 19

and intellectual prowess, or whether these are rather various aspects of the same quality of heroic excellence. It is clear, however, that moral quality follows upon physical and intellectual excellence as their consequence, and cannot be detached from them; it is enhanced and diminished along with them.

2.3 *Kalokagathia* (καλοκἀγαθία)

Another important Greek term is *kalokagathia*. This abstract substantive is derived from the adjective phrase *kalos kagathos* (καλὸς κἀγαθός), meaning literally "beautiful and good." The first term, *kalos* (καλός), denotes simultaneously physical beauty and noble nature. It was also used in an ethical sense, as in "right, noble action." The second term, *agathos* (ἀγαθός), is also endowed with many meanings. In the most general sense, it means "good." The many meanings of this word are well seen in its different comparative and superlative forms: *agathos* may mean "morally good," "physically strong," or "apt" and "fitting."

In the phrase *kalos kagathos* the qualities of moral and physical excellence are gradually becoming differentiated, but they are still coupled together and closely intertwined. In fact, it is still difficult to separate them completely. Werner Jaeger described this aristocratic notion of excellence as "the chivalrous ideal of the complete human personality, harmonious in mind and body, foursquare in battle and speech, song and action."[6] In texts from the 5th century BCE, *kalos kagathos* seems to denote primarily the qualities of a gentleman (e.g. Herodotus, *Histories* 1.30, Aristophanes, *Knights* 184–6) and sometimes the qualities of the good citizen, such as bravery and prowess in battle (Thucydides, *The Peloponnesian War* 4.40, cf. Xenophon, *Hellenica* 5.3.9; *Anabasis* 4.1.19). These usages were undoubtedly conditioned in part by the predominant meaning of the term *kalos*, which clearly – although never exclusively – signified physical beauty and was a standard term of commendation of young boys, written alongside their depictions on dedicatory vases.[7]

In the 4th century BCE, however, as the term *kalon* (the neuter form of *kalos*) increasingly came to be used in a moral sense, the compound *kalos kagathos*, retaining its ambiguous nature, acquired a stronger moral meaning (cf. Demosthenes, *Letter* 1.16; Xenophon, *Anabasis* 2.1.9; *Memorabilia* 1.2.23; Plato, *Apology* 21d). For both Plato and Aristotle, *kalokagathia* meant the "sum

---

6 W. Jaeger, *Paideia: the Ideals of Greek Culture*, Vol. 1, trans. G. Highet (Oxford: Basil Blackwell, 1965), p. 62.
7 For a detailed discussion on this custom and the meaning of καλός in this context, see K. Dover, *Greek Homosexuality* (Harvard: Harvard University Press, 1978).

total of all the special virtues," and therefore signified at once the moral, intellectual, and even physical perfection of the whole person (see Aristotle, *Eudemian Ethics*, VIII 15; cf. *Nicomachean Ethics* 1124a4; 1179b10). The *Magna Moralia* of Pseudo-Aristotle shows this shift in the usage of *kalos kagathos* particularly well:

> There is a phrase, then, which is not badly used of the perfectly good man, namely, "nobility and goodness" (καλοκἀγαθία). For "he is noble and good" (καλὸς κἀγαθός), they say, when a man is perfectly virtuous. For it is in the case of virtue that they use the expression "noble and good;" for instance, they say that the just man is noble and good, the brave man, the temperate, and generally in the case of the virtues. (*Magna Moralia* 1207b21-27, trans. Ross)

2.4 *Aretē* (ἀρετή)

The Homeric term for heroic excellence, *aristos* (ἄριστος), is cognate with *aretē* (ἀρετή), perhaps the most fundamental general Greek term for excellence. The term is difficult to translate. It could be rendered as "virtue," "fitness," "perfection," or "excellence." It was applied not only to humans, but gods (*Iliad*, 9.498), animals (Herodotus, *Histories* 3.88), and even inanimate objects to denote outstanding performance in accordance with an ideal of perfection accepted for the given kind of entity (cf. Plato, *Hippias Minor* 369a, ff). *Aretē* "is the achievement of some perfection" (Aristotle, *Metaphysics*, 1021b20, cf. *Nicomachean Ethics*, 1106a15). In Plato's *Protagoras*, the title protagonist promises his would-be apprentice that from him he can learn *aretē*, and "every day progress towards the better" (318a).

While in the Homeric epics *aretē* denoted mainly valor, especially of heroic warriors and gods (cf. *Iliad* 20.411; 15.642; *Odyssey* 2.206), philosophers use this term more often to denote moral virtue (cf. e.g. Heraclitus fr. 112; Democritus fr. 179; Gorgias fr. 6). Nevertheless, this term still retained its more general meaning of "fitness" and continued to be applied to human and non-human subjects equally (cf. Herodotus, *Histories*, 8.92; 4.198; Thucydides, *The Peloponnesian War* 1.2). For Socrates, *aretē* became the highest aim of human endeavors, nearly synonymous with "happiness" (εὐδαιμονία, cf. Plato, *Apology* 29de, 36c). *Aretē* is for him the spring of all other excellences and all goods, including external ones (*Apology* 30b). Plato also makes heavy use of the term, developing the old tradition of the four cardinal virtues, and making them the complementary, necessary, and sufficient components of a harmonious soul and *kalokagathia* (cf. *Republic* 427e, ff; 442bd). Aristotle employs *aretē* as a technical ethical term, and draws a distinction between ethical and intellectual virtues (*Nicomachean Ethics* 2.5–7, 6 passim). However, *aretē* was for him merely a necessary, not a sufficient, condition for happiness (εὐδαιμονία) which required also other components, notably an adequate level of physical well-being. According to Aristotle, happiness

> needs the external goods as well. [...] and there are some things the lack of which takes the lustre from happiness, as good birth, goodly children, beauty; for the man who is very ugly in appearance or ill-born or solitary and childless is not very likely to be happy, and perhaps a man would be still less likely if he had thoroughly bad children or friends or had lost good children or friends by death. (*Nicomachean Ethics* 1099a15-17, cf. 1099b11, trans. Ross)

2.5 Phyē (φυή), euphyia (εὐφυία), euphyēs (εὐφυής)

Finally, there are terms pertaining primarily to positive qualities of the body yet used also in contexts bearing a moral meaning. It is clear that health and physical wellness, as well as physical beauty, were among the qualities most highly valued by the Greeks. This is well illustrated by an "old drinking song" cited by Socrates in Plato's *Gorgias*. In the song "singers enumerate the goods of life, first health, beauty next, thirdly [...] wealth honestly obtained" (451e, trans. Jowett). Socrates, of course, argues that virtue of the soul should take precedence before health and beauty. Nevertheless, he does not question the high value of the latter. A similar sentiment is attributed to Thales in Diogenes Laërtius: "What man is happy? 'He who has a healthy body, a resourceful mind and a docile nature'" (*Lives of Eminent Philosophers*, 1.37). That Socrates himself considered the well-being of the body as an important good can be seen in *Crito* 47e, where he argues that a life with a "corrupted body" is not worth living (still less, of course, a life with a corrupted soul). As shown in the earlier mentioned fragment (*Nicomachean Ethics* 1099a15–17), Aristotle also held the view that physical well-being is a necessary condition of complete happiness.

Physical well-being and bodily beauty were thus associated with ethical quality. This association was expressed in a number of terms, one of which was *phyē* (φυή), a noun cognate with the verb *phyō* (φύω), "to grow," "to develop spontaneously," and possibly with the noun *physis* (φύσις). *Phyē* generally denotes "the totality of the natural world," and "growth" (of any kind), but it is often used to denote "fine growth," or "noble stature." In Homer, it is used in this meaning exclusively in reference to the human form (*Iliad*, 22.370; 3.208; *Odyssey* 6.152). This meaning is also found in later authors, such as Hesiod (*Works and Days* 129), Bacchylides (5.168), and Theocritus (22.160). Another – relatively rare – term is *euphyia* (εὐφυία), which can be found, among others, in Aristotle and Plutarch. Its more common form is the adjective *euphyēs* (εὐφυής) and it.is occasionally used in a general sense to denote a person with a "good natural disposition" (Xenophon, *Memorabilia* 1.6.13; Aristotle, *Nicomachean Ethics* 1114b8; Theophrastus, *Characters* 29.4). It may, however, also be used with the meaning "genius" or "natural cleverness" (Isocrates 7.49; Aristotle, *Politics* 1455a32; cf. Plato, *Republic* 401c). This latter instance is particularly striking, since it refers to a psychic, intellectual quality by means of a term originally pertaining to the body and its well-being.

The above-described attitude to body and health persisted well into late antiquity. Wellness of the body, mind, and soul remained closely associated with each other and, consequently, with ethical quality. Two centuries into the Christian era, the poet Juvenal sang that "one ought to pray for a healthy mind in a healthy body" ("orandum est ut sit mens sana in corpore sano", *Satires* 10.356). It would appear that in pagan antiquity physical, psychic, and ethical excellence were never clearly distinguished, still less dissociated from each other.

3 Greek and Roman practices for reproductive and population control and the attitudes of philosophers to these practices

The high value attributed to physical fitness, as well as the perceived nexus between physical and psychic features, conditioned ancient attitudes towards reproductive and population control. A brief scrutiny of philosophical views on reproductive and population control methods will demonstrate the importance attached by their authors to physical fitness. This will serve as a point of departure for the analysis of philosophical views on the nexus of physical and psychic traits.

The most common methods used in Greece and Rome were contraception, abortion, and infant exposure. There exists an ample body of literature on this subject.[8] Although estimates are very approximate and tentative, it is safe to

---

8   Y. Charbit and A. Virmani, "The Platonic City: History and Utopia," *Population*, Vol. 57, No. 2 (2002), pp. 207–235; R. H. Feen, "Keeping the Balance: Ancient Greek Philosophical Concerns with Population and Environment," *Population and Environment*, Vol. 17, No. 6 (1996), pp. 447–458; M. Golden, "Demography and the Exposure of Girls at Athens," *Phoenix*, Vol. 35, No. 4 (1981), pp. 316–331; W.V. Harris, "Child-Exposure in the Roman Empire," *The Journal of Roman Studies*, Vol. 84 (1994), pp. 1–22; W. Ingalls, "Demography and Dowries: Perspectives on Female Infanticide in Classical Greece," *Phoenix* Vol. 56, No. 3–4 (2002), pp. 246–254; P. Kraeger, "Aristotle and Open Population Thinking," *Population and Development Review*, Vol. 34, No. 4 (2008), pp. 599–629; C. Patterson, "'Not Worth the Rearing': The Causes of Infant Exposure in Ancient Greece," *Transactions of the American Philological Association*, Vol.

say that these practices constituted a common aspect of everyday life. Various methods of reproductive control are frequently mentioned in literature and receive divergent verdicts, more often expressing a casual or mildly approving rather than dismissive attitude.

Philosophers seldom discuss contraception and abortion. This could be read either as indifference to these issues or their acceptance as common practice. Somewhat more common are remarks on infant exposure.[9] It might be tempting to make the *a fortiori* inference that since only the more drastic practice of exposure receives due attention, the relatively milder methods of contraception and abortion were accepted as a matter of fact. As the example of Aristotle shows, however, such an inference is illicit. It is possible for an ancient philosopher to endorse exposure and condemn – under certain conditions – such practices as abortion.

In the *Republic*, Plato advocates a strict system of reproductive and population control. This involves an elaborate method of selecting couples depending on their fitness. It is not said explicitly whether this fitness regards physical, intellectual, and/or ethical, but it is reasonable to assume that all of them are important. Furthermore, selective breeding is to be employed, apparently, through the practice of exposure:

> "It follows from our former admissions, I said, that the best men must cohabit with the best women in as many cases as possible and the worst with the worst in the fewest, and that the offspring of the one must be reared and that of the other not, if the flock is to be as perfect as possible". (*Republic* 459de)

This passage might be taken as a clear indication that Plato advocates infant exposure for the purpose of promoting preferred physical, intellectual, and ethical features. However, a caveat must be made: the mentioned regulation concerns

---

115 (1985), pp. 103–123; H. D. Rankin, "Plato's Eugenic ΕΥΦΗΜΙΑ and ΑΠΟΘΕΣΙΣ in Republic, Book V," *Hermes*, Vol. 93, No. 4, (1965), pp. 407–420; J. M. Riddle, J.W. Este and J. C. Russel, "Ever Since Eve... Birth Control in the Ancient World," *Archaeology*, Vol. 47, No. 2 (1994), pp. 29–35; L.P. Wilkinson, "Classical Approaches to Population and Family Planning," *Population and Development Review*, Vol. 4, No. 3 (1978), pp. 439–455; C. Gardner, "The Remnants of the Family: The Role of Women and Eugenics in Republic V," *History of Philosophy Quarterly*, Vol. 17, No. 3 (2000), pp. 217–235; J. M. Bryant, *Moral Codes and Social Structure in Ancient Greece: A Sociology of Greek Ethics from Homer to the Epicureans and Stoics*. New York: State University of New York Press, 1996.

9 Cf. John M. Rist, *Human Value: A Study in Ancient Philosophical Ethics* (Leiden: Brill, 1982), p. 9.

particularly Kallipolis, i.e. the ideal polity described in the *Republic*. The status of this polity is far from evident. As Kallipolis is explicitly introduced as an analogue to the soul and serves an instrumental function for analyzing justice in an individual man, it is possible that it was never meant as a political entity. On such a reading, it would not be permitted to extrapolate the institutions and regulations of the Kallipolis to physical political entities.

We may, however, compare the cited passage with another mention of exposure, from the *Theaetetus*. In this dialogue Socrates elaborates at length on the metaphor of thought as mental offspring. Being the son of a midwife, he inherited his mother's skills, which he employs to facilitate intellectual rather than bodily labour. He incites his discussant to give himself to this maieutic power:

> "Apply, then, to me, remembering that I am the son of a midwife and have myself a midwife's gifts, and do your best to answer the questions I ask as I ask them. And if, when I have examined any of the things you say, it should prove that I think it is a mere image and not real, and therefore quietly take it from you and throw it away, do not be angry as women are when they are deprived of their first offspring" (*Theaetetus* 151bc).

It is clear that this passage does not advocate the exposure of human infants, but uses the practice as a metaphor for the selection of arguments and conclusions. However, Plato does use the example of the practice of exposure in a casual way, which suggests both that it was a common feature of life and that he was not opposed to it enough to not use it in this casual way. On the basis of these two passages we can conclude that there is some – albeit tentative – evidence that Plato might have advocated infant exposure, and little reason to believe that he would oppose the practice.[10]

Aristotle is perhaps most often used – and abused – in discussions concerning ancient attitudes towards abortion and infant exposure. A key passage that is often cited can be found in *Politics* 1335b19–26:

> "As to exposing or rearing the children born, let there be a law that no deformed child shall be reared; but on the ground of number of children, if the regular customs hinder any of those born being exposed, there must be a limit fixed to the procreation of offspring, and if any people have a child as a result of intercourse in contravention of these regulations, abortion must be practised on it before it has developed sensation and life;

---

10 One may compare this casual attitude with that of Isocrates, a prominent opponent of Plato. In his *Panathenaicus* 12.122 Isocrates numbers the exposure of infants among serious crimes practiced in legendary times by the enemies of Athens, with which he favorably – and unrealistically – compares contemporary Athens. Cf. also Aristotle's criticism of the lack of population control as far as the number of begotten children go in *Politics* 1265a38, ff.

for the line between lawful and unlawful (ὅσιον) abortion will be marked by the fact of having sensation and being alive" (trans. Rackham).

This rather laconic statement concerning abortion has received a sizable body of literature.[11] It will suffice for the present discussion to note two caveats regarding its interpretation. The first concerns the contentious term *hosion* (ὅσιον), sometimes translated as "holy" (e.g. by Richard Kraut), which is highly misleading and may spark an unnecessary and misguided discussion. Whereas *hosion* may mean "holy" in the sense of "specially sanctioned," it is often contrasted with "divinely sanctioned," expressed standardly by a different term, *hieros*, ἱερός. In such contexts, *hosion* denotes things sanctioned by human laws, albeit perhaps more strictly upheld than normally. In the present passage there is no reference to the divine sphere and it is more precise to render *hosion* as "lawful" or a similar term denoting human legislation.

Secondly, it is sometimes claimed that knowledge of modern embryology would induce Aristotle to restrict his tentative acceptance of abortion.[12] In fact, his estimate of forty days for male and ninety days for female embryos is clearly open to revision in view of modern science. However, it must be stressed that Aristotle indicates as the boundary for lawful abortion the development of sensual perception – *aisthesis* (αἴσθεσις) – and not merely life, *dzōē* (ζωή). Therefore, it is possible that at least cases such as anencephalia could, on this theory, lead to a more permissive rather than restrictive view. In general, it is more plausible that if Aristotle had known modern science (in particular, the much later than 40 days development of a central nervous system required for sense perception), he would not have been inclined to restrict his views on abortion.

Another statement of Aristotle relevant for the present topic regards infant exposure. The philosopher proposes a regulation based on physical fitness which

---

11 See, e.g.: M. Lu, "Aristotle on Abortion and Infanticide," *International Philosophical Quarterly*, Vol. 53, No. 1, Issue 209 (2013), pp. 47–62; P. Drum, "Hylomorphism and Abortion," *Australian Journal of Professional and Applied Ethics*, Vol. 2 (2000), pp. 71–74; P. Drum, "Certain Errors Regarding Aristotle and Abortion," *Ethics Education*, Vol. 14 (2008), pp. 38–39.

12 P. L.P. Simpson, *A Philosophical Commentary on the Politics of Aristotle* (Chapel Hill: The University of North Carolina Press, 1998), p. 247, n. 82; D. Dobbs, "Natural Right and the Problem of Aristotle's Defense of Slavery," *The Journal of Politics*, Vol. 56 (1994), pp. 70–71, n. 2; M. Lu, "Abortion and Virtue Ethics," in: *Persons, Moral Worth, and Embryos*, ed. S. Napier. (Dordrecht, Netherlands: Springer, 2011), p. 122, n. 23; Cf. M. Lu, "Aristotle on Abortion and Infanticide," *International Philosophical Quarterly*, Vol. 53, No. 1, Issue 209 (2013).

appears to apply only to physical traits. However, as will be argued below, Aristotle recognized a nexus between physical and psychic traits. His views on how intellectual and character traits are reflected in physical features are scattered among his writings (they were later reworked and expanded in a treatise on physiognomy attributed to him). A regulation concerning the selection of physically fit offspring provides some – although weak – ground for inferring that Aristotle would also regulate the selection of offspring on the ground of psychic (especially intellectual) fitness.

On the basis of a telling passage in Epictetus, it is argued that Epicurus spoke against having a family and also advocated the exposure of infants.[13] This highly biased report shows the position of Epicurus as rather harsh, advising abandoning children – at least in the case of the wise man – not on the ground of physical deformity (which is not mentioned), but for the mundane reason of the child being a potential burden. Perhaps the Stoic Epictetus exaggerated Epicurus' original statement for polemical purpose. However, it would still be plausible that Epicurus voiced some favorable opinions on the issue of infant exposure, which are no longer extant.

This same passage shows that Epictetus' personal opinion on the issue was decidedly unfavorable. On this point he shared the view of his teacher, Musonius

---

13 "Even Epicurus perceives that we are by nature social, but having once placed our good in the husk he is no longer able to say anything else. For on the other hand lie strongly maintains this, that we ought not to admire nor to accept anything which is detached from the nature of good; and he is right in maintaining this. How then are we [suspicious], if we have no natural affection to our children? Why do you advise the wise man not to bring up children? Why are you afraid that he may thus fall into trouble? For does he fall into trouble on account of the mouse which is nurtured in the house? What does he care if a little mouse in the house makes lamentation to him? But Epicurus knows that if once a child is born, it is no longer in our power not to love it nor care about it. For this reason, Epicurus says, that a man who has any sense also does not engage in political matters; for he knows what a man must do who is engaged in such things; for indeed, if you intend to behave among men as you would among a swarm of flies, what hinders you? But Epicurus, who knows this, ventures to say that we should not bring up children. But a sheep does not desert its own offspring, nor yet a wolf; and shall a man desert his child? What do you mean? that we should be as silly as sheep? but not even do they desert their offspring: or as savage as wolves, but not even do wolves desert their young. Well, who would follow your advice, if he saw his child weeping after falling on the ground? For my part I think that even if your mother and your father had been told by an oracle, that you would say what you have said, they would not have cast you away" *Diatribes* 1.23.

Rufus.[14] Using the common Stoic strategy of appealing to the example of animals, to which humans are compared, he argues against exposure from the assumption that it is natural for humans to show affection for their children. Presumably, this would follow from the Stoic tenet that the basic natural instinct is for self-preservation, and some further assumptions along the lines that children are, to some degree, a part of the parent.

However, condemnation of infant exposure was by no means a view universally accepted in the Stoic tradition. Another prominent representative of the school and contemporary of Musonius, Seneca the Younger, displays a rather casual attitude to the practice:

> "Mad dogs we knock on the head; the fierce and savage ox we slay; sickly sheep we put to the knife to keep them from infecting the flock; unnatural progeny we destroy; we drown even children who at birth are weakly and abnormal. Yet it is not anger but reason that separates the harmful from the sound" (*On Anger* I.15.2).

It may be noted that in the passage cited above, Epictetus does not mention the case of deformed offspring, with which Seneca is explicitly preoccupied.

The views on methods of reproductive and population control varied among philosophers. Important for the present topic are the grounds they put forward for endorsing or condemning such practices. Both Plato and Aristotle endorsed, at least in some cases, the practice of infant exposure for the reason of eliminating from the community offspring with inferior physical and possibly also psychic traits. The same reason was given by Seneca the Younger. Musonius Rufus and Epictetus spoke against infant exposure, but did not address the case of deformed offspring. *Prima facie*, neither of the cited passages suggests that the practice of exposure could have been advocated with the aim of selecting individuals with superior psychic and/or ethical traits. After all, how would such capacities be assessed in a new-born infant (or prenatally, in the case of abortion)? As it turns out, there are at least two possibilities. Firstly, psychic traits could be perceived as inherited, and so the offspring of parents displaying preferred traits might be privileged. It is likely that this would be the view of Plato, as evidenced in his Myth of the Metals, discussed below. Secondly, a nexus between physical and psychic traits could be assumed, allowing some level of inference regarding the

---

14 "But what seems to me very terrible is that some who do not even have poverty as an excuse but are prosperous and even wealthy none the less have the effrontery not to rear later-born offspring in order that those born earlier may inherit greater wealth [...]. So that their children may have a greater share of their father's goods, they destroy their children's brothers." Musonius Rufus, fr. 15 Lutz, in: W.V. Harris, "Child-Exposure in the Roman Empire," *The Journal of Roman Studies*, Vol. 84 (1994), p. 7.

latter on the basis of the former. This was the view of Aristotle and some of his imitators, who developed the study of physiognomy, as well as of several Stoics.

4 Philosophers on the nexus between physical and psychic traits

There are no less than three extant ancient treatises on physiognomy, the study of how psychic traits are manifested in physical features. The authors of two are not known. The oldest of these was in antiquity attributed to Aristotle, while the other was written by a Latin author in the 3rd or 4th c. CE. The third text was written by a certain Polemo (2nd c. CE). There are also numerous anecdotes involving the practice of physiognomy, such as the story of a certain Zopyrus, who was said to have inferred from Socrates' unbecoming features that the philosopher was by nature stolid and lustful. In fact, it was quite a common view that one's character and psychic traits – including ethical ones – can be inferred from physical appearance and demeanor. This acknowledgment of a nexus between physical and psychic traits does not in itself indicate whether the former or the latter were perceived as the cause of this congruence (or, in fact, whether any causal connection was postulated). The sources point to a spectrum of views. Let us consider some examples that are relevant to the topic of enhancing psychic traits by influencing physical ones.

In the *Parmenides*, Plato uses language suggestive of a nexus between psychic and physical traits. Describing Parmenides, the philosopher from Elea, Plato appears to connect physical beauty and ethical worthiness: "Parmenides was already quite old, very grey-haired, handsome and noble to look at (καλὸν δὲ κἀγαθὸν τὴν ὄψιν), and around sixty five years of age" (*Parmenides* 127b). The phrase *kalos kagathos*, discussed above, is here used in reference to Parmenides' physical appearance. It is a pregnant term, denoting simultaneously Parmenides' becoming stature, implying his high intellectual acumen, and suggesting an upright ethical character. Significantly, the psychic and ethical traits are implied to manifest themselves in physical features. Such reasoning will be the hallmark of later physiognomical theories.

In the *Prior Analytics*, Aristotle claims that "it is possible to infer character from features." On its own this statement does not settle the issue of causality. However, Aristotle immediately adds that "the body and the soul are changed together by the natural affections." This indicates that he recognized a causal connection not so much between physical and psychic traits, as between certain external factors and both psychic and physical qualities.[15] Aristotle also

---

15 "It is possible to infer character from features, if it is granted that the body and the soul are changed together by the natural affections: I say "natural", for though perhaps

attributes different characters to ethnic groups, and implies that climate is the responsible factor (*Politics* 1327b20–33).[16]

The Stoics recognized a particularly close nexus between physical and psychic traits. Posidonius of Apamea inferred from a statement made by Zeno of Citium regarding desirable external features for well-bred youths that certain physical features may determine one's character.[17] Chrysippus claimed in the second book of his *On the End* that ethical qualities can be manifested physically:

"The following is a sufficient basis for us to be able to say that goods and evils are perceptible: for not only are passions perceptible, along with their species, such as grief, and fear, and the like, but so also is it possible to perceive theft and adultery and similar;

---

by learning music a man has made some change in his soul, this is not one of those affections natural to us; rather I refer to passions and desires when I speak of natural emotions. If then this were granted and also that for each change there is a corresponding sign, and we could state the affection and sign proper to each kind of animal, we shall be able to infer character from features" (Aristotle, Prior Analytics 2.27, Trans. A. J. Jenkinson).

16 Cf. The *Physiognomy* of Adamantius the Sophist: "If there are people in whom the Greek and Ionic race has been preserved pure, these are rather large men, fairly broad, upright, firm, rather fair-skinned and blond; the mixture of their flesh is moderate and rather firm, their legs are straight, extremities well-formed; the head is of a medium size and round, neck robust, hair rather blond, fairly soft and lightly curly; the face is square, lips are thin, nose is straight, eyes are moist, dark blue, intense, and there is much light in them; for the Greeks have the best eyes among all nations" (Adamantius the Sophist, *Physiognomy* II.32).

17 Zeno of Citium seems to sketch a beautiful and lovely picture of a young man, and this is how he sculpts him: "Let his face be pure, his brow not relaxed, his eyes neither goggling nor squinting, his neck not stretched, the parts of his body not languid, but poised, with good tension [...]. Let his gestures and movements give no hope to the licentious, and let modesty and manliness bloom upon his appearance" (Clement of Alexandria, *Paedagogus* III.11=SVF I, 246). "Posidonius links this argument to the evident facts of physiognomy. It is true of both animals and men that broad-breasted and hotter types are always more spirited by nature, whereas broad-hipped and colder types are more cowardly. Men also differ considerably in character as far as cowardice and courage, and love of pleasure and love of toil are concerned, according to the places in which they live, for the emotional movements of the soul always follow the condition of the body, and this condition is greatly influenced by the mixture of the environment. He also says that the blood in animals differs in being hot or cold, thick or fine, and in many other respects, too. Aristotle has discussed these things most thoroughly" (Posidonius, fr. 169.84–96). (Edelstein and Kidd, cyt. za. Knuuttila & Sihvola 2013: 629).

and, in general, folly and cowardice and many other vies; not only joy and good works and many other right actions, but also wisdom and courage and the rest of the virtues" (SVF iii.85).

According to some Stoics, the ethical quality of the sage can be perceived in his or her external features (*eidos*, SVF i.204). Cleanthes, the second head of the school, argued for the heredity of not only physical features (*corporis lineamentis*) but also of psychic traits (*animae notis*, SVF i.518): "Not only, says Cleanthes, do we become like our parents in body, but also psychologically, in our passions (*pathē*), characters (*ēthē*), and dispositions (*diatheseis*)" (SVF i.518, from Nemesius, *On the Nature of Man* 2, 20. 14–17 Morani). A similar view was attributed by Cicero to Panetius: "Panaetius has it [...] that souls are born, as our similarity to our parents makes clear – a similarity which appears in our characters (*in ingeniis*), not only our bodies" (fr. 83 van Straaten=Cicero, *Tusculan Disputations* 1.79).[18] It was common, both in Peripatetic and Stoic traditions, to differentiate psychic traits in relation to gender.[19]

From this brief scrutiny of physiognomical passages we can infer some conclusions regarding views held in antiquity on the nexus of physical and psychic traits. Many authors, among them Plato, Aristotle, and various Stoics recognized that psychic traits, including ethical ones, may manifest themselves in one's physical appearance. Moreover, Aristotle accepted that – at least in some cases – external conditions may shape one's physical traits, simultaneously influencing one's character, including cognitive and ethical traits. Posidonius held the view that features of character may be determined by one's physical constitution.

---

18 Cf. the view of Chrysippus reported by Plutarch: "Chrysippus adduces as proof that the soul has come into existence, and comes into existence later [sc. than the body], the fact that children become like their parents both in their manner (*tropos*) and their character (*ēthos*)" (SVF ii.806=Plutarch, *Stoic Self-Contradictions* 1053cd).

19 "The masculine character is vehement, easily provoked to action, never resentful, generous, straight, cannot be duped or won through cunning or plot, eager to win through its own merit, magnanimous. The feminine character is clever, prone to anger, resentful, merciless and envious, easily tired of work, docile, cunning, bitter, reckless, and cowardly" (Anonymus Latinus, *Physiognomonia* 4). Cf.: "Merciful men are delicate, with pale skin and glossy eyes; the top of their nose is wrinkled, and they are constantly shedding tears. These same men are fond of women, beget female children, and in character they are erotically inclined, have a good memory, and are talented and hot [...]. The wise, cowardly and modest type tends to be merciful, whereas the ignorant and shameless type is without mercy" (Pseudo-Aristotle, *Physiognomy* 3, 808a33–808b2).

5 Philosophers on the possibility of enhancing psychic intellectual and ethical traits by somatic or external factors

We have seen that in ancient Greece and Rome the selection of preferred physical traits in infants was a fairly commonly accepted practice. But what about psychic – intellectual and ethical – traits? Many philosophers – including Plato, Aristotle, and the Stoics – accepted some nexus between physical and psychic traits. It might be tempting to infer from this that their endorsement of control of physical traits would naturally extend to psychic traits as well. However, the relevant passages, cited above, are insufficient to make this inference.

It is uncontentious that Greek philosophers almost universally[20] advocated the enhancement and perfection of human nature and character. Perfection was an explicit goal of most ancient eudaimonistic ethical systems. Moreover, perfection was usually understood in terms of enhancing cognitive, ethical, and more generally psychic traits (physical traits were of secondary importance). However, it certainly does not follow – nor does it preclude – that for ancient philosophers it would be viable and/or desirable to attain the goal of perfecting human nature through the method of influencing physical traits. For the most part, perfection was to be attained through paideutic and ascetic means. What more can be said about ancient philosophical views regarding the possibility of enhancing psychic traits by external and/or bodily means? Let us approach this question by first considering some prominent ancient theories of human nature and the methods of its perfection. I consider in turn Plato, Aristotle, the Stoics, and the Epicureans.

On Plato's tripartite psychology, justice consists in each soul component "doing what belongs to it." This constitutes a harmonious unity of the soul, in which each component possesses its particular virtue (*Republic* 442be). Despite having introduced two non-rational components of the soul, Plato appears to retain in broad outline the Socratic premise that virtue is a kind of knowledge. Accordingly, he devises a complex educational curriculum for the future guardians of the Kallipolis (*Republic* 535a-540a) which, by imparting physical and intellectual training on a very demanding level, aims at imparting virtue to the guardians. Therefore, virtue can, at least to a rudimentary degree, be acquired through involuntary and external factors.

The Myth of the Metals from *Republic* III, 415ac, *pace* its allegorical function, suggests that Plato held the view that human character and intellectual traits are

---

20 Epicureans and, for different reasons, Cynics are – as I argue – an exception.

innately diverse and, for the most part, hereditary.[21] This view is the basis for an elaborate mechanism of selective breeding of the guardians of the "city in words" (*Republic* 459d-461e, 546ae). The inherited psychic traits (the so-called golden, silver, and bronze souls) are necessary prerequisites for wisdom and virtue. To become a guardian or philosopher-king, one must supplement it with the long and arduous education described in *Re*public VI-VII. However, hereditary traits are a necessary precondition of education and therefore also of virtue.

Aristotle characterizes virtue as "a state of character concerned with choice, lying in a mean, i.e. the mean relative to us, this being determined by a rational principle, and by that principle by which the man of practical wisdom would determine it" (*Nicomachean Ethics* 1106b35–1107a1, trans. Ross). Excellency of character, i.e. being virtuous, is achieved by regularly performing virtuous acts (1103a14–25). This view seems to be encumbered with a fatal flaw; namely, if one needs to perform virtuous acts in order to be virtuous, but only the virtuous perform virtuous acts, how then does one become virtuous in the first place? This view of virtue seems to debar the very possibility of moral development.

Aristotle is aware of this problem and solves it by postulating an ingenuous theory of ethical development. In analogy to the learning of crafts, one begins by performing the correct (i.e. virtuous) action unreflectively, without the appropriate disposition. Over time, however, once the agent comes to understand the propriety of the action, this disposition is developed. It is only then that the agent becomes virtuous in the strict sense, since virtue requires that both the action is appropriate for the given situation and that it is performed for the appropriate reason (Ross 1995: 203).

---

21 "The god who made you mixed some gold into those who are adequately equipped to rule, because they are most valuable. He put silver in those who are auxiliaries and iron and bronze in the farmers and other craftsmen. For the most part you will produce children like yourselves, but, because you are all related, a silver child will occasionally be born from a golden parent, and vice versa, and all the other from each other. So the first and most important command from the god to the rulers is that there is nothing that they must guar better or watch more carefully than the mixture of metals in the souls of the next generation. If an offspring of theirs should be found to have a mixture of iron or bronze, they must not pity him in any way, but give him the rank appropriate to his nature and drive him out to join the craftsmen and farmers. But if an offspring of these people is found to have a mixture of gold or silver, they will honor him and take him p to join the guardians or the auxiliaries, for there is an oracle which will say that the city will be ruined if it ever has an iron or a bronze guardian" (415ac).cf. *Laws* 783de-ff.

Since virtue is not an unreflective performance of an appropriate action, it cannot be something like a feeling or emotion. This is because these are passive reactions to stimuli, while Aristotle has further conditions for virtue, namely that they must involve rational choice (*prohairesis*).[22] Virtue is therefore not only the exercise of a disposition, even a reflective one: it is a rational choice made on the basis of the disposition (*Nicomachean Ethics* 1143b18–1145ab). Choice is a "deliberate desire of things in our own power" (*Nicomachean Ethics* 1112a18–1113a14) or, as Aristotle laconically characterizes it, it is "either desireful reason or reasonable desire, and that sort of origin of action is a man" (*Nicomachean Ethics* 1139a4). Thus, ethical action is by definition voluntary.

Virtue may be the outcome of rational choice, but it may be argued that on Aristotle's theory this choice is determined by one's character. As Michael Frede argues,

"[...] Aristotle does not have a notion of a will. One's willing, one's desire of reason, is a direct function of one's cognitive state, of what reason takes to be a good thing to do. One's nonrational desire is a direct function of the state of the nonrational part of the soul. One acts either on a rational desire, a willing, or on a nonrational desire, an appetite" (Frede 2011: 24).

Every virtuous and vicious action consists in a rational choice. However, this does not mean that one is free to choose to engage or not to engage in a particular action. Rather, choice is determined by one's character, which in turn is the result of a long (in fact, life-long) series of choices, education, and training. If one is virtuous, one cannot but choose in a virtuous way. Conversely, if one has a bad character, or is an acratic, one cannot but choose to act in an evil way. This is expressed in the definition of virtue as a habit (albeit realized through choice). As Frede concludes: "For Aristotle a good life is not a matter of a free will but of hard work and hard thought, always presupposing the proper realization of human nature in the individual, and a good upbringing" (Frede 2011: 30). If this is so, it could be argued that enhancement of one's character, habits, and non-rational traits would contribute to ethical enhancement.

The Stoics, being monists, defined the basic stuff of all things as *pneuma*, which formed particular beings by varying fields of tension (*tonos*). They also

---

22 It could also be possible that virtue is a capacity. This would however, on Aristotle's technical terminology, make it not stable and persistent enough. It is therefore defined as a disposition, i.e. a capacity which has become ingrained in one's character through repeated, lengthy performance (*Nicomachean Ethics* 1105b19–1106a13; 1106a14–1107a2).

adhered to the view that there is a strict analogy and correspondence between the *pneuma* of the individual human soul and the *pneuma* of the cosmos, i.e. the *logos* (see, e.g. Sextus Empiricus *Adversus Mathematicos* IX.104–10, Seneca, *Epistulae* 92.30).[23] *Pneuma* determines one's nature, which in turn defines the ultimate goal (*telos*) of life's natural development.[24] The pivotal notion is *oikeiōsis*, a process of reaching a full realization of one's rational nature. The starting point is the innate impulse for self-preservation (cf. Seneca, *Epistulae* 121)[25]. This is at first directed at physical survival, and is the same in humans and other animals. Upon the development of reason, man becomes different from non-rational animals. What becomes then the proper object of preservation is one's rational nature.

There were many alternative formulations of the Stoic *telos*, but in all of them the central notion was nature, and the goal of life was the realization of one's nature to the fullest extent. This was simultaneously the goal of ethical life, as virtue was for the Stoics synonymous with living according to nature. The way to do this was by bringing one's nature in agreement with cosmic nature, since only in a narrow sense they conceived nature as the constitution one is born with and which one develops over the span of life. In a wider sense, human nature is for

---

23 The human soul, being a special kind of highly condensed *pneuma*, determines one's human nature. The notion of nature was central to Stoic psychology and ethics. According to Diogenes, Laërtius Zeno of Citium, the founder of the school, wrote no less than two treaties on the topic: *On the life according to nature*, and *On impulse* also titled *On human nature* (VII.4, cf. VII.87).

24 "Zeno defined the goal thus: 'living in agreement'. This means living according to a single and consonant rational principle, since those who live in conflict are unhappy. Those who came after him made further distinctions and expressed it thus: 'living in agreement with nature', supposing that Zeno's formulation was an incomplete predicate. For Cleanthes, who first inherited [the leadership of] his school, added 'with nature' and defined it thus: 'the goal is living in agreement with nature'. Chrysippus wanted to make this clearer and expressed it in this way: 'to live according to experience of the things which happen by nature'. And Diogenes: 'to be reasonable in the selection and rejection of natural things'. And Archedemus: 'to live completing all the appropriate acts'. And Antipater: 'to live invariably selecting natural things and rejecting unnatural things'. He often defined it thus as well: 'invariably and unswervingly to do everything in one's power for the attainment of the principal natural things'" (Arius Didymus in Stobaeus II 75.11–76.15).

25 This theory is opposed to the Epicurean view that the first impulse upon birth (of all creatures, not only humans) is towards pleasure and away from pain. See Diogenes Laertius VII 85–6 cf. Cicero *De finibus bonorum malorumque* III.16–17; Inwood and Donini 2002: 677.

the Stoics a part of cosmic nature (cf. Seneca, *Epistulae* 92.30). It is the goal of life to assimilate one's nature to this cosmic nature, which is tantamount to reaching a state of perfection, of a full development of one's nature. It might therefore be said that for the Stoics everyone is born with a natural kernel, which should be subsequently developed over the course of life in order to reach its full and perfect form.[26]

Thus, according to the Stoics, ethical life consists in a process of developing one's nature from an innate kernel towards a perfected state in which one lives in full coherence with cosmic nature, being properly its part. A state of a full development of one's nature is therefore "what is complete according to nature for a rational being qua rational" (Diogenes Laertius VII 95, cf. Cicero *De finibus bonorum malorumque* III 33). This state is synonymous with virtue and characterizes the sage.

Moreover, for the Stoics, who followed the Socratic tradition, virtue is a kind of knowledge, especially knowledge about what is good and evil. Accordingly, virtuous actions consisted in doing everything for the right reason, i.e. for the good. In order to become virtuous, therefore, one has to acquire knowledge of the good and evil, and then train oneself to always act upon this knowledge, and not some ulterior motive, such as material gain or pleasure.

Here enters the key Stoic notion of assent (*synkatathesis*). The development of character proceeds by assenting to the right motivations, which are opinions of reason (*to hēgemonikon*, i.e. ruling aspect of the soul), resulting from sense-perception (*phantasia*). The result of assent feeds back to the *hēgemonikon*, resulting in new opinions being created, which in turn determine future assents. Therefore, one develops one's rational character by building on previous assents – on good ones for the better, and on bad ones for the worse. The outcome of the

---

26 "Thus Zeno first, in his book *On the Nature of Man*, said that the goal was to live in agreement with nature, which is to live according to virtue. For nature leads us to virtue. And similarly Cleanthes in *On Pleasure* and Posidonius and Hecaton in their books *On the Goal*. Again, 'to live according to virtue' is equivalent to living according to the experience of events which occur by nature, as Chrysippus says in book one of his *On Goals*. For our natures are parts of the nature of the universe. Therefore the goal becomes 'to live consistently with nature', i.e. according to one's own nature and that of the universe, doing nothing which is forbidden by the common law, which is right reason, penetrating all things, being the same as Zeus who is the leader of the administration of things. And this itself is the virtue of the happy man and a smooth flow of life, whenever all things are done according to the harmony of the divinity in each of us with the will of the administrator of the universe" (Diogenes Laertius VII 87–8).

process depends therefore on one's initial character and the chain of assents. It is strictly up to the agent what character he or she develops. Nevertheless, this process is deterministic in the sense that one's assents depend on character, and one cannot assent contrary to the current state of character.

For Epicurus, the ultimate goal of life is *eudaimonia* conceived as a particular kind of pleasure: *aponia*, or freedom from toil, and *ataraxia*, or freedom from anxiety.[27] Attainment of this state is the Epicurean idea of human perfection. As Epicurus says in the *Letter to Menoeceus*, having reached it "you will live amongst men like a god" (*Letter to Menoeceus* 135). To reach this state one needs to format one's character in a specific way through learning a set of basic teachings, and then practicing them in life. These are summarized in the so-called *Tetrapharmakos*: "God presents no fears, death no cause for alarm; it is easy to procure what is good; it is also easy to endure what is evil."[28]

According to Epicurus, the first two tenets of the *Tetrapharmakos* can be inferred from his atomistic physics. According to Epicurus' atomism, the basic component of things are indivisible entities, moving through a void, from time to time diverted from their trajectories by a random force called the swerve (cf. *Letter to Menoeceus* 125).[29] This includes the soul, which is composed of atoms and is destroyed upon death, once they dissipate in different directions. Regarding the gods, Epicurus argues against their involvement in human matters on the basis of a version of the *unde malum* problem. Therefore, knowledge is a *sine qua non* for a happy, perfect, and good life of *ataraxia* and *aponia*.[30]

---

27 "We declare pleasure to be the beginning and end of the blessed life; for we have recognized pleasure as the first and natural good, and from this we start in every choice and avoidance, and this we make our goal, using feeling as the canon by which we judge every good" (Epicurus, *Letter to Menoeceus* 128). Cf. Cicero, *De Finibus Bonorum Malorumque* I 30.
28 Erler and Schofield 2002: 645. *Tetrapharmakos* means literally "fourfold remedy." Epicurus and his followers often speak of philosophy in the language of medicine. Philosophy is a therapy, which expels unwanted and detrimental passion from the soul.
29 Incidentally, the swerve (gr. παρέγκλισις, lat. *clinamen*) is used by Epicurus to justify that men have free will, in fact, radical free will. This view is in direct opposition to the determinism of the Stoics. For Epicurus, man is completely free in that nothing determines his decisions, not even character or natural dispositions (which, of course, may nevertheless influence these decisions).
30 "There is no way to dispel being afraid about matters of supreme importance if someone does not know what the nature of the universe is, but is anxious about some of the things retailed in myths. Hence without natural philosophy it is impossible to secure the purity of our pleasures" (*Kyriai Doxai* 12). It is important to note, however, that Epicurus did not consider the knowledge required for understanding the first two

These two tenets are necessary, but not sufficient conditions for attaining a happy life of *ataraxia* and *aponia*. For this, the remaining two tenets, namely that "it is easy to procure what is good; it is also easy to endure what is evil," are required. Whereas the first two tenets can be arrived at by theoretical reason alone, the remaining two require not only theoretical understanding, but also practical judgment. They belong therefore to the sphere of *phronēsis* (i.e. practical, as opposed to theoretical, reason, cf. *Letter to Menoeceus* 132).

Physics has taught us that certain things which pertain predominantly to the pleasures and pains of the body are not to be feared, as they help us mainly in attaining *aponia*. However, pleasure and pain can also pertain to the mind. Obtaining the former and being free from the latter is tantamount to *ataraxia*. To reach this state it is not sufficient to be free from physical pain: one must also be free from anxiety, which may arise, for instance, as a result of a desire for something unattainable or pernicious. Such desires are, according to Epicurus, unnatural. Natural desires, on the other hand, are those of a prudent person (*Letter to Menoeceus* 129–32). These include bodily desires (for food, drink, and shelter), but also the desire for (mental) *ataraxia*.

It is of the utmost importance to exstirpate unnatural desires, as they are the primal cause of anxiety or disturbance, i.e. mental pain. According to Epicurus, unnatural desires are a peculiar feature of humans. They are absent from the animal soul (both in animals and in man's own animal part). They originate in the mind, when it is directed at things not within man's proper nature which is presumably possible because of man's freedom, absent in animals. In animals bodily desires are naturally directed at what is required by their proper natures. In humans, however, bodily desires may be inflated by the mind, when it is guided by what Epicurus terms "empty opinion," and may indeed go to infinity (cf. *Kyriai Doxai* 15, 19).

Therefore, no amount of satisfaction of bodily desires can secure *ataraxia*. This must be done by reason, which should limit desires to their natural

---

tenets to be very vast or difficult. Indeed, it is obtainable by practically everyone, and does not require any highly technical studies. "Hoist your sail, he told Pythocles, and flee from every form of paideia" (Diogenes Laertius X 5). All of it he rejected in the belief that such studies are not necessary to reach that state of happiness in which men can scarcely be distinguished from gods. Epicurus thought the only thing needed to make someone an Epicurean was knowledge and acceptance of his basic teachings, as these are summarized in the *Tetrapharmakos*. In this way it is possible even for ordinary people to attain happiness without previous education (*paideia*). The key to happiness is the inner attitude (*diathesis*) of which we ourselves are masters.

bounds: "Neither drinking-parties nor continual revelry nor the enjoyment of boys and girls, or fish and all that a lavish table can offer, can provide a life of pleasure – only sober reason" (*Letter to Menoeceus* 132).In order to liberate oneself from unnatural desires, one must first come to know one's nature, i.e. what desires are natural for humans, and then train one's character to desire only those things.

Natural desires are twofold: they may pertain to the body or the mind. Regarding the body, it is only natural to desire that which is in accordance with the natural constitution of man. It is therefore natural to desire health and self-preservation. It is not natural to desire that which is physically unattainable for man, like immortality. Regarding the mind, it is natural to desire pleasure and avoid pain. It is therefore not natural to desire greater pleasures which might involve a certain admixture of pain.

It may finally be noted that, on the ground of Epicurus' ethics, the natural prescription for happiness is effectively a recipe for self-sufficiency (*Letter to Menoeceus* 131). Self-sufficiency is to be regarded not as a regime of self-denial, but as freedom:

> "For these things to which I aspire are not trivial. It is a question of things which make my condition resemble that of the gods and which prove to me as a human being that, despite mortality, I do not take second place to that nature which is immortal and blessed. For whilst I am alive, I have pleasure to the same degree as the gods" (Diogenes of Oenoanda, fr. 125–6, ed. Smith).

6 Philosophers on genetic enhancement of psychic and ethical traits

In this and the following final section I gather the results obtained above and draw some conclusions regarding speculative ancient philosophical attitudes to genetic enhancement. This part of the paper borders on, what I would call, philosophical fiction. We can no more than speculate how ancient philosophers would accommodate modern knowledge of genetic editing in their theories of human nature and its perfection. On the basis of the results obtained above, I consider how modern genetic enhancement methods fit with the four discussed theories of human nature. I consider in turn three aspects of the question: the enhancement of physical, psychic (cognitive), and ethical traits.

As I have argued, in what could be called Greek popular morality, physical, psychic, and ethical perfection were closely associated. This is particularly evident in older texts, such as the Homeric epics, the lyric poets, but also in authors from the 5th and 4th century, such as Herodotus and Xenophon. However, this

Would Greeks and Romans Endorse Moral Enhancement? 39

association is equally evident in the thoughts of philosophers from Plato to the Stoics, who – following Socrates – tried rather to dissociate ethical virtue from bodily and external, agent-independent factors. Even if ethical perfection was clearly distinguished from such factors, it was still closely related to intellectual perfection, as might be expected from proponents of intellectualistic notions of virtue.

It is uncontentious in my view that enhancement of physical traits is acceptable and desirable within all four philosophies discussed above. Accordingly, I do not consider this aspect of the question in detail. In the *Republic*, Plato suggests the exposure of unfit guardian offspring. Aristotle advocates legal compulsion for the exposure of unfit children. Although there are ancient voices disapproving of infant exposure – notably the Stoics Musonius Rufus and Epictetus – they do not utterly condemn the practice, but rather its abuses, notably the exposure of healthy unwanted infants that are considered a burden. Other Stoics, such as Seneca the Younger, advocate exposure for the explicit reason of eliminating physical unfitness. On this basis, it is plausible that authors advocating exposure for the reason of eliminating physical unfitness would also endorse modern methods of prenatal diagnosis and genetic enhancement of physical traits. Let us now consider in more detail the possibility of enhancing cognitive and ethical traits.

In the *Republic*, Plato explicitly advocates methods of reproductive control in order to promote preferred physical and psychic traits. In light of Plato's intellectualistic ethics, the possibility of enhancing ethical character should be treated as strictly correlated with enhancement of cognitive traits. Ethical quality is directly dependent on the possession of wisdom. There is no reason to think that Plato would reject the possibility of genetic enhancement of cognitive capacity. Given his intellectualistic ethics, this would be tantamount to at least some degree of ethical enhancement, as it would make the person cleverer (not wiser!), and therefore more capable of attaining wisdom.

Moreover, according to the tripartite psychology, all three soul-components have their proper virtue: wisdom (*sophia*) for reason, courage (*andreia*) for the emotive component, and moderation (*sophrosynē*) for the appetitive component.[31] Justice is the psychic state in which all three soul-components possess their proper virtues and cooperate harmoniously. Plato advises the selection of

---

31 Strictly speaking, moderation is a virtue of all three soul-components and consists in the component's doing what belongs to it. However, it is particularly associated with the appetitive component.

guardians not only according to their physical fitness and cognitive capacity (the quality of the highest soul-component), but also according to the appropriate quality of the two lower soul-components. In particular, the guardians ought to be neither too timid nor too aggressive (*Republic* 386a-389a, 416ac). This is to be achieved by an elaborate and arduous education (416b, ff). However, given modern methods, desired levels of timidity and aggression could be achieved by genetic engineering. There is no evidence that Plato would reject such a possibility outright.

All this being said, it is unlikely that Plato would consider genetic enhancement as sufficient for the attainment of virtue. Despite the eugenic schemes of the *Republic*, it is impossible, on the basis of Platonic moral psychology, to engineer an ethical person. This follows from the all-important role of self-examination, poignantly expressed by Socrates in the *Apology*: "the unexamined life is not worth living for a man" (ὁ δὲ ἀνεξέταστος βίος οὐ βιωτὸ ςἀνθρώπῳ, *Apology* 38a). The examined life consists in constant elenctic testing of one's opinions and making pre-meditated, voluntary decisions on the basis of one's best knowledge. Virtuous life thus consists in voluntary action based on knowledge and choice. Involuntary factors, such as molding character traits by education or genetic enhancement can constitute only concomitant – but never decisive – components of virtuous life.

The central role of rational choice for Platonic ethics is well demonstrated in the Myth of Er from Book 10 of the *Republic*. In the text, disembodied souls are described as choosing their next life. The choice is made autonomously, without external influence (*Republic* 617e). The quality of one's next life, which could be read as a metaphor of one's ethical quality in the present life, is dependent on one's character and level of wisdom. The reason for souls making terribly misguided choices leading to a life of vice is "participating in virtue by habit and not by philosophy" (*Republic* 619d). Habit is the domain of involuntary character traits, such as the virtues of two lower soul-components, emotions and appetite, and also cognitive capacity, as distinguished from its exercise. These can all be enhanced by involuntary methods, such as paideutic training and – today – by genetic modification. According to Plato, these are necessary, but not sufficient conditions of virtue. Moreover, relying, as regards virtuous life, exclusively on these involuntary factors (and their enhancement) would not only be a mistake, but would lead to the opposite of the desired goal, namely to a life of vice. The only way to virtue is through a life of rational self-examination and conscious, rational choice.

A similar result can, in my view, be argued for in the case of Aristotle. In Aristotelian ethics the central and necessary component of virtuous life is choice (*prohairesis*). Although virtue is defined as a kind of habit (*hexis*), this comprises only its part. No action can be deemed virtuous (or vicious) if it is not the outcome of rational choice. Virtue is therefore defined as habit pertaining to choice (*hexis prohairetikē*). Thus, in Aristotelian ethics genetic modification of character traits cannot have a decisive contribution to the attainment of virtue.

Indeed, it could be argued that within Aristotelian ethics genetic engineering of ethical traits is a contradiction in terms. The sphere of the ethical concerns by definition that which is voluntary. Only voluntary action can be the subject of praise and blame, and thus only such action is, strictly speaking, ethical. One cannot be assessed ethically on the ground of involuntary traits and actions. If Aristotle were aware of modern scientific studies, according to which certain character traits that he would consider ethical (e.g. aggression or sociability) might be determined by inherent and hereditary factors – thus involuntary – he would likely rather discount them as ethical, than deem their genetic modification to be ethical enhancement. This is not to say, of course, that he would not endorse the modification of such traits. This procedure, however, would not constitute ethical enhancement.

Nevertheless, although choice is necessary for virtue, Aristotle does say that external goods are needed for happiness. Goods such as health, physical fitness, and material wealth are the raw matter of virtue. Although their possession does not constitute being virtuous, it aids the performance of virtuous actions; conversely, their lack may hinder one from acting in such a way. It is therefore not unreasonable to assume that Aristotle would have endorsed genetic modification, if not for the attainment of virtue itself, then at least as a way to eliminate the obstacles that hinder or prevent one from acting in a virtuous way.

A final note concerns Aristotle's notion of the natural slave. The defining form for humans is rationality, i.e. the possession of reason. However, reason comes in degrees and one may be endowed with a higher or lower cognitive capacity. Aristotle recognizes the possibility of having such a low level of cognitive capacity that one is unfit to care for oneself, including one's ethical life. Such people are not self-sufficient, and thus they do not belong to themselves – they are entirely dependent for their physical and ethical livelihood on intellectually superior individuals, who are equal and free citizens of a political community. The non-self-sufficient people are by definition natural slaves, i.e. by nature they do not belong to themselves but to others, and this being their natural state makes the existence of slavery something that is in their best physical and ethical interest.

Given modern methods of cognitive enhancement it is imaginable, on the ground of Aristotelian psychology and ethics, that a natural slave could be enhanced to the level of a self-sufficient citizen. This would constitute a change of the slave's nature, as it would involve modification of his/her substantive form. As reason is the decisive – but certainly not the only – factor responsible for ethical quality, such modification would constitute ethical enhancement. Moreover, this reasoning could just as well be turned around. Since one's facility for rational choice and virtue depends on the one's cognitive capacity, then genetic modification of the latter could, on the ground of Aristotelian ethics, also be used for the engineering of a race of intellectually inferior natural slaves.

The matter is different in deterministic Stoic ethics, which rejects both psychic tripartition and the notion of rational choice, introducing in its place the notion of assent. To be sure, as for Plato and Aristotle, for the Stoics ethics is a theory of the perfection of human nature. Enhancement of human nature is therefore not only allowed, but is in fact an ethical imperative. Whether this could involve genetic modification is a different matter. To answer this question we have to inquire what constituted, on the basis of Stoic ethics, virtuous life.

According to the Stoics, what makes one human is one's rational soul. The Stoics subscribed to Socratic intellectualist ethics, according to which virtue is a kind of knowledge. Moreover, they rejected the view that anything but knowledge might influence one's ethical decisions. Even one's passions, e.g. emotions and appetites, are beliefs of a kind. They can, and should, be extirpated by appropriate education and training. Thus, one's intellectual development is a direct cause of ethical growth. There is no unequivocal evidence that the Stoics would reject the possibility of enhancing cognitive capacity. Given their intellectualist ethics, this would constitute at least some level of ethical enhancement. Would such enhancement constitute a merely concomitant component of virtue – as in Plato and Aristotle, who introduced further the notion rational choice – or would it be possible on the ground of Stoic ethics to fully engineer an ethical person (the ideal Wise Man) by purely involuntary, deterministic methods?

Deterministic and naturalistic Stoic moral psychology appears to be more conducive to this possibility than the dualistic psychologies of Plato and Aristotle. On the latter theories, ethical life – although strongly rooted in bodily factors – was ultimately the domain of the immaterial soul. Ethical traits could therefore be at best influenced, but not determined by somatic means, such as genetic enhancement. For the Stoics, the soul is a body; it is therefore more plausible that they would acknowledge the possibility of its enhancement by bodily means. Moreover, in contrast to Plato and Aristotle, for the Stoics assent is not a choice between alternatives which, although directed by one's character, could turn out otherwise. Stoic assent is fully determined by one's character, which is ultimately of a bodily nature. This being said, it must be stressed that assent is nevertheless a conscious act, not an instinctive reaction. As such, it cannot be induced by any other factor than reason and rational deliberation. On the whole, however, Stoic moral psychology and ethics appear to be most congenial for accommodating modern methods of genetic ethical enhancement.

The least likely candidate for endorsing genetic enhancement of psychic and ethical traits is Epicurus. This follows from the relatively humble requirements for happiness in hedonistic ethics. For Epicurus neither extensive knowledge nor elaborate education and training are necessary for a happy life. As for the former, all that is needed is rudimentary knowledge of physics sufficient for extirpating fear of death and the gods. As for the latter, education might be just as beneficial as pernicious – by causing anxiety – while virtue is only instrumentally conducive to happiness and does not constitute the focus of ethical life.

Epicurus does consider unnatural desires as a serious hindrance in attaining the happy state of *aponia* and *ataraxia*. As these desires consist, for the large part in physical urges, they could be extirpated by genetic means. It is therefore at least possible that Epicurus would welcome such a possibility, e.g. for curbing unnatural sexual drives, and this would constitute genetic enhancement of some ethical traits. Moreover, in the case of intellectually impaired people, some level of cognitive enhancement might be desirable, in order to facilitate understanding physical knowledge required in the first two precepts of the *Tetrapharmacos*. Incidentally, it might be asked whether Epicurus would modify – perhaps expand – the scope of this physical knowledge, having access to the findings of modern science.

However, Epicurus explicitly states in the two latter precepts of the *Tetrapharmacos* that *aponia* and *ataraxia* are easy to attain and consist in developing a proper mental attitude, rather than in elaborate ethical training. Given therefore a rudimentary level of physical well-being and sufficient cognitive capacity, no further enhancement of nature is necessary. This includes ethical nature, as virtue is of limited importance for happiness. Moreover, it must be borne in mind that Epicurus subscribed to a very strong notion of freedom. Accordingly, in his view, one's mental life could not be determined by involuntary means and, in fact, it did not need to be: psychic disturbance, being a kind of false opinion could and ought to be removed by easily attainable intellectual persuasion and training.

7 The Greeks and genetic enhancement of human nature

There is some evidence that the Greeks would have been more sympathetic than not to the possibility of genetic enhancement of physical traits, had such technology been available to them. Greek culture was fiercely agonistic and placed a high value on individual excellence. This spirit of competitiveness is well seen not only in the sport culture of the Greeks, but in theatre (which was a form of competition), and in almost all other areas of creativity. As testified by Plato's *Hippias Minor*, during the Olympic Games poets and literates of all sorts competed, alongside athletes, with their various literary compositions. The ultimate aim of Greek life was individual perfection, "to always excel and be better than others" (*Iliad* 6.208). Given the modern methods of genetic modification of cognitive and character traits, it is highly likely that the Greeks would use them for individual perfection and the achievement of supremacy over competitors rather than to attain civic equity.

The Greek competitive attitude can also be well seen in philosophical theories of human nature and its perfection, particularly in the stratification of ethical traits. The imaginary "state in words" of Plato's *Republic* consists of three strata of citizens, segregated according to their degree of physical, psychic, and ethical excellence. According to Aristotle, a political community consists of free and equal citizens and slaves (natural and unnatural), who are discriminated according to their cognitive capacities influencing their ethical quality. According to the Stoics, humanity constitutes a continuous spectrum of diverse characters, at the high end of which stands the ideal – and hardly attainable – Wise Man. Perhaps the most egalitarian ethical outlook was advanced by the Epicureans, which followed from their less stringent requirements for the ethical goal of happiness.

Moreover, neither popular morality nor the philosophers held the body to be "sacred" or untouchable. For the Greeks bizarre hybrids were a cultural commonplace, and the spawning of human-animal hybrids was considered a very real possibility. Also, bodily fitness and beauty were things of prime importance in Greek culture in general and were also highly stressed in philosophical doctrines (with the odd exception of the Cynics). The evidence for that can be seen in the ideal canon of the human body developed by Praxiteles, which reflects the prevalent Greek strive towards physical perfection. Thus, it is likely that most Greeks would endorse physical enhancement (e.g. elimination of physical defects, genetic diseases, perhaps even enhancement of stamina and physical skills, *vide* the praise of "super-men" by Callicles) – as long as it would not hinder psychic faculties, which were universally considered more important than those of the body.

The only part of man which the ancient Greeks might have considered "sacred" (a notion which had a very different meaning then than now) was the rational soul, seen as immaterial and immortal (although only by some authors). This is not to say, however, that the rational soul could not have been altered. In fact, according to most schools it was an ethical imperative to assiduously develop and perfect one's rational soul by *paideia* and philosophy (with the exception of the Epicureans, who advocated a much milder degree of education). All Greek ethical systems placed human nature in the center of their attention and its perfection was their central tenet. It could be argued that ancient philosophers would endorse genetic enhancement in accordance with their respective notions of human nature. Yet, Greek ethical systems did not delimit nature rigorously, apart from the ubiquitous defining feature of rationality. Because of this Greek ideas of the perfection of human nature could be equally broad and malleable. It is, therefore, not surprising that philosophers set no less an aim for ethical life than for the quester to become like a god (cf. Plato, *Theaetetus* 176b, Aristotle, *Nicomachean Ethics* 1077b30-34, Epicurean doctrine in Diogenes of Oenoanda, fr. 125-6, and the Stoic analogy between human *pneuma* and divine *logos*).

References

Adkins, A. *Merit and Responsibility. A Study in Greek Values.* Oxford: Clarendon Press, 1960.

Bryant, J. M. *Moral Codes and Social Structure in Ancient Greece: A Sociology of Greek Ethics from Homer to the Epicureans and Stoics.* New York: State University of New York Press, 1996.

Charbit, Y. and A. Virmani. "The Platonic City: History and Utopia," *Population*, Vol. 57, No. 2 (2002), pp. 207-235.

Dobbs, D. "Natural Right and the Problem of Aristotle's Defense of Slavery," *The Journal of Politics*, Vol. 56 (1994), pp. 70-71, n. 2.

Dover, K. *Greek Homosexuality.* Cambridge, Massachussets: Harvard University Press, 1978.

Drum, P. "Hylomorphism and Abortion," *Australian Journal of Professional and Applied Ethics*, Vol. 2 (2000), pp. 71-74.

Drum, P. "Certain Errors Regarding Aristotle and Abortion," *Ethics Education*, Vol. 14 (2008), pp. 38-39.

Erler, M. and M. Schofield., "Epicurean ethics." In: *The Cambridge History of Hellenistic Philosophy*, ed. K. Algra, J. Barnes, J. Mansfeld and M. Schofield. Cambridge: Cambridge University Press, 2002.

Feen, R. H. "Keeping the Balance: Ancient Greek Philosophical Concerns with Population and Environment," *Population and Environment*, Vol. 17, No. 6 (1996), pp. 447-458.

Frede, M. *A Free Will. Origins of the Notion in Ancient Thought*, ed. A. A. Long. Berkeley, Los Angeles, London: University of California Press, 2011.

Gardner, C. "The Remnants of the Family: The Role of Women and Eugenics in Republic V," *History of Philosophy Quarterly*, Vol. 17, No. 3 (2000), pp. 217–235.

Golden, M. "Demography and the Exposure of Girls at Athens," *Phoenix*, Vol. 35, No. 4 (1981), pp. 316–331.

Harris, W. V. "Child-Exposure in the Roman Empire," *The Journal of Roman Studies*, Vol. 84 (1994), pp. 1–22.

Ingalls, W. "Demography and Dowries: Perspectives on Female Infanticide in Classical Greece," *Phoenix* Vol. 56, No. 3–4 (2002), pp. 246–254.

Inwood, B. and P. Donini., "Stoic ethics." In: *The Cambridge History of Hellenistic Philosophy*, eds. K. Algra, J. Barnes, J. Mansfeld and M. Schofield. Cambridge: Cambridge University Press, 2002.

Jaeger, W. *Paideia: the Ideals of Greek Culture*. Vol. 1. Trans. G. Highet. Oxford: Basil Blackwell, 1965.

Knuuttila S. and J. Sihvola. *Sourcebook for the History of the Philosophy of Mind: Philosophical Psychology from Plato to Kant*. Dordrecht: Springer Science & Business Media, 2014.

Kraeger, P. "Aristotle and Open Population Thinking," *Population and Development Review*, Vol. 34, No. 4 (2008), pp. 599–629.

Lu, M. "Aristotle on Abortion and Infanticide," *International Philosophical Quarterly*, Vol. 53, No. 1, Issue 209 (2013), pp. 47–62.

Lu, M. "Abortion and Virtue Ethics," in: *Persons, Moral Worth, and Embryos*, ed. S. Napier. (Dordrecht, Netherlands: Springer, 2011), p. 122, n. 23.

Nagy, G. *The Best of the Achaeans: Concepts of the Hero in Archaic Greek Poetry*. Baltimore: The John Hopkins University Press, 1979.

Patterson, C. "'Not Worth the Rearing': The Causes of Infant Exposure in Ancient Greece," *Transactions of the American Philological Association*, Vol. 115 (1985), pp.103–123.

Rankin, H.D. "Plato's Eugenic ΕΥΦΗΜΙΑ and ΑΠΟΘΕΣΙΣ in Republic, Book V," *Hermes*, Vol. 93, No. 4, (1965), pp. 407–420.

Riddle, J.M., J.W. Este and J. C. Russel, "Ever Since Eve... Birth Control in the Ancient World," *Archaeology*, Vol. 47, No. 2 (1994), pp. 29–35.

Rist, J.M. *Human Value: A Study in Ancient Philosophical Ethics* (Leiden: Brill, 1982), p. 9.

Ross, W. D. *Aristotle*. London, New York: Routledge, 1995.

Simpson, P. L. P. *A Philosophical Commentary on the Politics of Aristotle* (Chapel Hill: The University of North Carolina Press, 1998), p. 247, n. 82.

Wilkinson, L.P. "Classical Approaches to Population and Family Planning," *Population and Development Review*, Vol. 4, No. 3(1978), pp. 439–455.

Tomasz Fiedler

Clement of Alexandria: Toward Christian Perfection

Abstract: This paper[1] is a summary exposition of a Christian vision of human perfection as described by an early Christian writer, Clement of Alexandria. I present his thought in two steps. First, I analyze Clement's anthropology, setting forth his understanding of human being, his nature, origin, and goal. Secondly, I demonstrate how this understanding underlies a way of life that would lead to the achievement of human perfection. I conclude with some remarks on the relationship between Clement's Greek cultural background and his Christian faith that clearly influences the shape of his vision. Clement's vision of man is inscribed within the biblical scheme: creation – fall – salvation. Created as a being both earthly and heavenly, meant to achieve maturity so as to fully participate in the divine life, man rejects the plan of God and interrupts the process of his becoming. The result is a state of degradation of his genuinely human features and capacities, of alienation from his true self and true life. Salvation appears as a restoration of the human ability to relate to God, to know the truth of his being and purpose. It is a resumption of the process of human growth, sustained by divine grace, but requiring human cooperation, leading towards the perfection of human nature. This dynamic, theoretical vision of man clearly requires completion with a practical counterpart. Thus, Clement offers a detailed picture of a way of life that responds to the need for human collaboration in order to achieve the final goal of man. It regulates Christian life in every particular, so as to shape correct behavior, heal the soul from its passions and attachments, and finally enable man to share in the divine life for which he was created.

Keywords: Christianity, Philosophy, Way of life, Asceticism, Perfection, Knowledge

Clement of Alexandria (150–215) is one of the most fascinating figures of early Christianity. According to tradition, he was born in Athens in a pagan family. As a philosopher, he travelled across the Mediterranean looking for a teacher that could lead him on the way toward wisdom. His quest came to an end in Alexandria, where he encountered Christianity. It appeared to him as the answer to his questions and, therefore, as the true philosophy: an expression of true wisdom, and the way to attain it. Christian faith revealed to him the mystery of man and new horizons of human growth. Complemented by his knowledge of

---

1 This paper is a reworked excerpt from my S.T.L. thesis, defended at Boston College, School of Theology and Ministry (2012).

Greek philosophy, it gave rise to a project of Christian perfection, combining a vision of man with a way of life leading him to his ultimate goal.

The major work of Clement is the trilogy consisting of the *Protrepticus*, the *Paedagogus* and the *Stromata*. The latter presents a detailed portrait of a "true gnostic" – the perfect Christian, alongside Clement's reflections on faith, knowledge, and relationship between Christianity and culture. The first two, to which I will refer in this paper, present Clement's vision of man and his model of the Christian way of life. Additionally, the texts also discuss the subjects of ignorance and knowledge, divine grace, and human moral effort, which are woven into this vision and form a complex whole.

1 The origin and goal of man

Clement's vision of man is inscribed within the biblical scheme of creation – fall – salvation. This movement from the original state of man, characterized by a "communion with heaven", to a condition of ignorance and degradation of human nature, and the return to man's true, heavenly, and divine dignity, pervades the whole work of Clement.[2] In particular, the *Protrepticus* presents the drama of fallen humanity and the perspective of salvation, revealing the true nature of mankind – its heavenly origin and goal. Man appears throughout these pages as a being which is both earthly and heavenly, suspended between these two realms, free to determine the direction of his becoming.

1.1 Creation

Man is created in the image and likeness of God – this biblical notion is in the center of Clement's anthropology, revealing the fundamental truth about man as a being intrinsically related to God by virtue of the very fabric of his nature.[3] This essential relationship to God, expressed by the "image and likeness", seems to be understood by Clement in two distinct, yet complementary ways.

Clement sees the image of God in the human mind (νοῦς). The creation "in the image" is mediated by the divine Logos who is properly called "the image of God", being "the true son of the Νοῦς, archetypal light of the light." Next, the image of the Logos is the "true man" or – more specifically – "the νοῦς in man." It

---

2   See Clement of Alexandria, *Protrepticus*, 25,3–27,3 (*Le Protreptique*, Sources Chrétiennes 2bis, Paris: Les Editions du Cerf, 2004) for a synthetic presentation of this pattern.

3   Gen 1:26; cf. Clement of Alexandia, *Protrepticus*, 98,4; cf. Clement of Alexandria, *Paedagogus*, I 9,1; I 98,2 (*Le Pédagogue I*, Sources Chrétiennes 70, Paris: Les Editions du Cerf, 2008).

is because of his mind, says Clement, that "man is said to have been made in the image and likeness of God, being assimilated to the divine Logos by the intelligence of heart and thus reasonable (λογικός)."[4] Man is therefore an image of the image – his affinity with the Logos, reposing in his intellectual faculty (νοῦς), is what constitutes his kinship with God, the original Νοῦς.

Another way in which Clement understands the image is with regard to man's ability to enter into a personal relationship with God. This ability is due to the special way in which God created man. While He made all the other creatures by His order, simply calling them into being, He fashioned man with His own hands and breathed into him something of His own. This breath of God, which is something divine in man, is a charm on account of which God loves man.[5] Furthermore, if God loves man much more than all the other creatures, it is not only because man is the most beautiful among creatures, but also because he is the living being which loves God (φιλόθεον ζῷον).[6] This reciprocal love between God and man, made possible by the special creative act "in the image", is at the same time what brings this act to perfection, to the fulfillment of "image and likeness."[7]

These two ways of understanding creation "in the image and likeness" can actually be seen as two aspects of one and the same reality – man's capacity to relate to God. This capacity for relationship presupposes the capacity to know God, where knowledge of God is understood not as "objective", but as personal, that is, knowledge that includes relationship, a familiarity with an "other" based on intimacy of an interpersonal encounter. At this point we can observe that the νοῦς, which is the image of God in man, must not be understood in a rationalist sense, but rather as a capacity of perception of the divine reality described by Clement as Νοῦς. This νοῦς is thus this divine spark in man that makes it possible for him to know his like, the faculty to perceive similar through similar.

The last important aspect of the creation of man is the fact that this creation is not yet completed – man is a being *in fieri*, with an open perspective of becoming, of attaining his perfection. To describe this reality, Clement makes a distinction between "image" and "likeness" of the creative act of God, the first being given, and the latter to be achieved.[8] Referring to the all-embracing activity

---

4   Clement of Alexandria, *Protrepticus*, 98,4.
5   Cf. Clement of Alexandria, *Paedagogus*, I 7,1–3.
6   Cf. Clement of Alexandria, *Paedagogus*, I 63,1.
7   Cf. Clement of Alexandria, *Paedagogus*, I 9,1.
8   The distinction is based on the biblical text itself: while Gen 1:26 speaks of the project of creation of man "in our image and likeness," the following verse states that "God created man in His image" only (Gen 1:27).

of Christ who creates, redeems, and guides man toward salvation, Clement states that His goal is to "transform the earthborn man into a holy and heavenly one, so that the word of God: 'Let us make man in our image and likeness' may be completely fulfilled." It is because only Christ realized this word fully – as the image of the Father perfectly like Him – while other men are intended as images only, not yet possessing the likeness to the archetype.[9]

1.2 Fall

The fall is the reality that follows creation and stands in opposition to it. It shows how man, a heavenly being created in the image of God and called to live in communion with Him, by a free act of his will disobeys his Creator and chooses the opposite course, thus interrupting the process of becoming the man that God's creative plan intended him to be. Referring to the biblical story of the fall in *Genesis*, Clement says that the first man in paradise was yet a child of God. Then, succumbing to pleasure, he was misled by his desires and diverted from his course.[10]

The consequence of the fall, which is seen by Clement as an act of disobedience toward God and the Logos, is the deprivation of λόγος – meaning deprivation of reason, but also of the pattern of becoming in the likeness of God through the mediation of the Logos. Thus, from λογικός ("reasonable" or "Logos-like") man becomes ἄλογος. For this reason he is likened to beasts (and not to God), becoming just an irrational animal delivered to his desires. Further still, inasmuch as he commits sin, man himself becomes worse and so dead to God. The moral disorder following sin means therefore for man a separation from God as the source of life and – in ultimate consequence – death.[11]

Underlying this reality of sin and moral disorder in man is the problem of ignorance, which is seen by Clement as both the cause and the effect of sin. In fact, Clement opposes virtue, which is "a disposition of a soul made harmonious with the Logos in the entire life", to sin, which is "any fault committed by reason of ignorance of the Logos."[12] Ignorance of the Logos consists here in a failure to harmonize man's conduct with Him, and must be seen as opposed to the knowledge of God, that is, familiarity with God acquired through an intimate relationship that transforms the entire person. Man, created as a child, a yet imperfect

---

9 Cf. Clement of Alexandria, *Paedagogus*, I 98,2–3.
10 Cf. Clement of Alexandria, *Protrepticus*, 111,1.
11 Cf. Clement of Alexandria, *Paedagogus*, I 101,3–102,1; II 100,1.
12 Clement of Alexandria, *Paedagogus,* I 101,2.

image of the Logos, did not have this knowledge in the beginning. He was supposed to acquire it gradually, as a result of his becoming familiar with the Logos through the imitation of Him, thus achieving likeness to God through the kinship of virtue, that is, being made harmonious with the Logos in his way of life. This process, however, was interrupted by man who chose pleasure instead of the way of growing in virtue, thus diverting from his original course and remaining in ignorance about the Logos.

This choice of man was encouraged by his ignorance or misperception of the true good – being unfamiliar with the Logos, failing to live according to his heavenly nature and to realize his heavenly dignity, he was overcome by his earthly nature that prevailed. Desires stirred by pleasure turned man's attention away from his Creator and from the way of becoming in His likeness toward the more immediately accessible reality: the material creation and the rewards that it promised. At the same time, the choice itself was an act of disobedience toward God who had indicated man the way of achieving his perfection, that is, likeness to God, knowledge of Him and of man's heavenly nature and goal. As man refused to walk this way, he both broke his relationship with God, failing to put faith in Him, and doomed himself to remain in the state of ignorance, living not according to the heavenly, but to his earthly nature. These two realities – ignorance of God and disobedience toward Him – are seen by Clement as two interrelated causes of the fall.[13]

As an unexpected result, man found himself enchained by his sin as a slave, not a master of creation,[14] attached to the earth without being able to raise his sight to heaven. The state of ignorance was further aggravated inasmuch as man limited his horizon to the earthly existence: "The innate, original communion of men with heaven was obscured by ignorance", and eventually "the erroneous conceptions, diverting from the straight path and truly destructive, led man, the heavenly plant, away from the heavenly conduct, knocked him down to the earth, and moved him to attach himself to the earthly creatures."[15] The fallen man, as depicted by Clement on the pages of the *Protrepticus*, presents a truly desperate image of a creature who failed to follow the path of maturation in its heavenly nature, and lost the capacity to live according to it, becoming totally absorbed in the earthly pursuits.

---

13 Cf. Clement of Alexandria, *Paedagogus*, I 101,1–3.
14 Cf. Clement of Alexandria, *Paedagogus*, II 9,2 (*Le Pédagogue II*, Sources Chrétiennes 108, Paris: Les Editions du Cerf, 1991).
15 Clement of Alexandria, *Protrepticus*, 25,3–4.

1.3 Salvation

Clement explains the reality of salvation in contrast to that of the fall: to the movement downwards, resulting in the loss of heaven, attachment to the earth, and degradation of man, is opposed the movement upwards, and a perspective of return to heaven and restoration of humanity to its original state is laid open. Among many oppositions employed by Clement on the pages of the *Protrepticus* (death-life, earth-heaven, fall-restoration, etc.), an important place is given to that of darkness-light, error-truth, and ignorance-knowledge. Since ignorance is both the cause and the most grievous effect of the fall, alienating man from his true, heavenly nature and limiting his horizon to his earthly existence, removing ignorance is what is most necessary in order to enable man to recognize his true nature, turn his mind and heart away from the material creation, and thus reorient his whole life toward God.

This goal is accomplished by Christ through His Incarnation. Clement says: "the Word of God became man, so that you may learn from man how man can become God."[16] Christ invites men to the recognition of the truth, to the knowledge of God, which is proper for man "made for familiarity with God", "born for the contemplation of heaven."[17] He restores the faculty of seeing, gives light that enables to know God and man, illuminates, and sharpens the eye of the soul. Men can thus remove ignorance and darkness that, like a mist, impeded their sight, and contemplate the true God.[18] The light of Christ, shining in the heart of man, makes "the rays of knowledge rise and so makes manifest and radiant the hidden inner man, the disciple of light, kinsman and coheir of Christ."[19] Clement affirms thus that the knowledge of the true nature of man and his relationship to God does not consist in learning some external truth, but rather in recognizing, thanks to divine illumination, the inner truth that man carries within himself. Accordingly, Clement will say that man has no better means than himself to walk on the way toward the knowledge of God.[20] Hence "the greatest of all knowledge is to know oneself, since if one knows oneself, he will know God, and knowing God, he will be made like God."[21]

---

16 *Clement of Alexandria, Protrepticus*, 8,4.
17 Cf. Clement of Alexandria, *Protrepticus*, 85,3; 100,2–3.
18 Cf. Clement of Alexandria, *Protrepticus*, 113,2–114,2.
19 Clement of Alexandria, *Protrepticus*, 115,4.
20 Cf. Clement of Alexandria, *Paedagogus*, II 1,3.
21 *Clement of Alexandria, Paedagogus*, III 1,1 (*Le Pédagogue III*, Sources Chrétiennes 158, Paris: Les Editions du Cerf, 2008).

The Incarnation of Christ is in the center of Clement's vision of man: It manifests God's love for man and His eternal desire to save humanity,[22] the intimate relationship that unites the two and becomes the privileged means of salvation, including the call to the imitation of Christ as the way to achieve the final perfection of man. Salvation appears as a restoration of man to his proper condition, renovation of creation in what constitutes the heavenly nature of man, the image of God in him – his capacity to perceive, to know God and to enter into relationship with Him. Christ is thus revealed in the role of a double mediator – the archetype for creation of man, and the One who introduces man into a new, filial relation with the Father. The knowledge of God is thereby presented in its two aspects – vision and communion, contemplation of God and reception of the grace of adoption as a child of the Father and coheir of the Son.[23] This knowledge comes through the awareness of man's true identity in his relation to God, that is, of his heavenly origin and goal.

Salvation is thus bringing the creation of mankind to perfection, achieving the likeness to God. In this way man – a being *in fieri*, created as a child – reaches his maturity, becomes what he was meant to become in the initial plan of God. It is not a mere return to the primitive, childlike state of humanity in the paradise, but coming closer to its final, adult condition. The new life given by Christ is greater than the lost paradise: "he who fell from the paradise, receives a greater prize of his obedience, that is, heaven."[24] Man redeemed by Christ not only recovers his λόγος, but also receives the Logos, that is, the knowledge of God. Being an image yet without resemblance, man is corrected by Christ after the archetype, so that he may become like Him as well.[25] Being inhabited by the Logos, man has the form of the Logos; he becomes like God, or simply, he becomes God. Clement recapitulates thus the mystery of man: "God is in man and man is God."[26]

Clement's notion of salvation is therefore presented as raising human existence to a new, entirely different level, whereby man is introduced into the divine life of the Trinity. Man is inhabited by God, and through this most intimate familiarity with Him he receives the true knowledge of God, he himself being made God. The central place, again, is given to Christ, with a particular emphasis on the concept of the Logos as a mediator who effects salvation by uniting humanity

---

22 Cf. Clement of Alexandria, *Protrepticus*, 116,1; 87,3.
23 Cf. Clement of Alexandria, *Protrepticus*, 82,5–7.
24 Clement of Alexandria, *Protrepticus*, 111,3.
25 Cf. Clement of Alexandria, *Protrepticus*, 120,3–4.
26 Cf. Clement of Alexandria, *Paedagogus*, III 1,5–2,1.

with God. Salvation is thus truly a completion of the work of creation, a final transformation of man into someone he had not been in the beginning.

However, this transformation of man does not take place without his cooperation: he has his part of responsibility to assume in this process. The account of salvation set forth by Clement is thoroughly Christocentric as it underlines the preeminence of Christ and His divine initiative in the salvation of man. At the same time, Clement is fully aware that without his consent, man will remain outside the scope of the saving action of Christ: Man is free to accept or to refuse the gift of grace, choose obedience to Christ or remain in a fallen state. Just as man was free to disobey God in paradise, man is also free to refuse to be saved.[27]

But the consent is not enough – man is called to active cooperation with God. Christ regenerates man and makes him a temple of God – man has to take the responsibility for what he has become: "Purify this temple, abandon to wind and to fire pleasures and amusements as ephemeral flowers, cultivate the fruits of temperance, consecrate yourself to God as the first fruits, so as not to be only the work of God, but also the grace of God."[28] Having become the people of God and learned their true dignity, Christians have to live accordingly and exercise themselves in walking in the newness of life.[29] Made fellow citizens of God by grace, they have to become now His imitators.[30] The likeness to God cannot be realized without the cooperation of man: "It is fitting that we love in return Him who leads us lovingly to the best life, that we live according to the ordinances of His will (...); thus we can accomplish by resemblance the works of the Pedagogue, so that the words 'in the image and likeness' may be fulfilled."[31] The realization of the divine plan of salvation, expressed by the "likeness to God" or by the "divinization", depends on man's effort to pursue here on earth the heavenly way of life and follow the traces of God.[32]

The above passages let us understand that for Clement salvation is not a single event that happens once and for all, but a process of transformation of human nature. The divine grace marks the beginning of this process and remains active throughout, but it has to be received and, as it were, "interiorized" by a corresponding way of life. Christ is thus at the same time God bestowing grace, and man giving example of the way of life that enables its fruitful work within us.

---

27 Cf. *Clement of Alexandria, Protrepticus*, 95,1–2.
28 Cf. Clement of Alexandria, *Protrepticus*, 117,4–5.
29 Cf. Clement of Alexandria, *Protrepticus*, 59,3.
30 Cf. Clement of Alexandria, *Protrepticus*, 117,1.
31 Clement of Alexandria, *Paedagogus*, I 9,1.
32 Cf. Clement of Alexandria, *Paedagogus*, I 98,2–3.

Clement articulates this reality in one sentence: "the Lord comes to our help as man and as God: as God he remits our sins, as man he teaches us not to sin."[33] The theme of the imitation of Christ as a way to achieve likeness to God becomes thus essential to the vision of Clement. Similarly, the Christian way of life he sets forth appears as a direct consequence of his anthropology: the vision of man as heavenly being called to realize the perfection of his nature demands a way of conduct that makes this goal reality.

2 The way to achieve human perfection

2.1 Vision calling for practice

According to Clement's vision of man, salvation, which means perfection of human being, demands cooperation on the part of man, that is, imitation of Christ that will result in harmonization of human life with Him, achieving a balance between the earthly and the heavenly parts of his nature, coming to a deeper knowledge of himself and of God, and thus in attaining the fullness of life according to the truth of his nature. The principles of Christian conduct presented by Clement in the *Paedagogus* form a project of a way of life which leads man to the accomplishment of this goal. In this way the anthropological vision of Clement receives its practical realization, being applied to details of everyday life in order that the entire Christian existence may be an imitation of Christ leading to salvation, a raising of human life from earthly to a heavenly dimension. Christianity is thus for Clement an inseparable unity of vision and practice that it inspires, a revelation of truth that has to be lived out.

This need for a proper way of life has a deeper justification for Clement. Since the fall was a result of ignorance, that is, of insufficient acquaintance with the Logos, salvation can only be achieved through acquiring this familiarity with the archetype of man, which alone can give him the knowledge of truth about his heavenly nature. This knowledge cannot be attained by a mere notional learning, an assent to a vision of salvation – it comes as a result of the imitation of Christ, as an experience that involves and transforms the entire man, leading him to a discovery and growth of his true, inner self made in the image of God. Accordingly, Clement says: "The lie is not rejected by a simple comparison with the truth, but it is by the practice of truth that it is expelled and banished."[34] Although the way of salvation may seem austere and difficult for people who

---

33 Clement of Alexandria, *Paedagogus*, I 7,1.
34 Clement of Alexandria, *Protrepticus*, 77,3.

are spontaneously attracted to the pleasures of life, those who pursue the truth should not be afraid of being turned away from the way of wisdom by many pleasing appearances: once they get to know the truth, they will "reject the trumpery of custom just as children grown adult throw away their toys."[35]

This passage shows us well what Clement understands by acquiring the knowledge of truth and becoming a mature human being – it is a condition in which the perception of the nature of his being and the true good, directs all desire of man toward God. Such a man is no longer susceptible to being overcome by an illusion of earthly pleasures which, contrasted with the truth he now clearly sees, lose all their appeal for him. Now we can also understand better in what sense sin is a result of ignorance. Someone who does not undertake the effort to pursue the way of truth and wisdom that alone can give him the experiential knowledge of his highest good – the union with God – will spontaneously turn to the goods that are more easily perceived and accessible, although illusory, because turning man away from his real goal and leading to death. Consequently, the way of life proposed by Clement has as its immediate objectives to detach man from earthly pleasures, check his spontaneous inclinations toward illusory goods, and teach him instead self-control and discipline that will enable him to walk the more demanding way of truth.

2.2 Pedagogy of the logos

The way toward acquiring the knowledge of God and the likeness to God is presented by Clement as the pedagogy of the Logos, a joint activity of Christ and of man. It has a threefold purpose. First, through precepts and examples it shapes a correct moral behavior, a way of life centered on the imitation of Christ, leading to the formation of virtue. Second, it heals a sick and weak soul from its passions, which results in detachment from the transitory world and its illusory pleasures. Third, through this purification of the soul, it enables it to receive a deeper knowledge of man and of God.[36]

For Clement, Christ is God who became man in order to teach man how man can become God.[37] Jesus gives the model of the true life. It is through the imitation of His human life that man can bring to perfection His image in him. The Logos has the authority to show the way and to educate – men learn to imitate His virtues in order to be made like God through the kinship of virtue. Clement

---

35 Cf. Clement of Alexandria, *Protrepticus*, 109,1–3.
36 Cf. Clement of Alexandria, *Paedagogus*, I 1,2–3,3.
37 Cf. Clement of Alexandria, *Protrepticus*, 8,4.

says that "just as there is a way of life proper to philosophers, rhetors or wrestlers, so there is a noble disposition turned toward good, coming from the pedagogy of Christ", characterizing all aspects of Christian life.[38] Christians should therefore live according to the indications of the truth, loving the Pedagogue and His precepts, harmonizing their life and their actions with Him, and so live the true life.[39] Clement thus affirms that "salvation is the following of Christ"[40] and describes the Christian as "the one who follows the Logos."[41]

The second objective of the pedagogy of the Logos, achieved through practice of the Christian way of life, is the healing of passions and sinful desires. For this purpose the Logos gives us "spiritual remedies, exposes mischiefs, shows the causes of passions, cuts the roots of unreasonable desires, recommending to abstain from them, and giving to the sick all the antidotes of salvation."[42] His pedagogy consists of education and sanctification: the first is directed at guiding human moral effort, the latter is the work of God's sanctifying grace in man. Its vehicles are the sacraments: baptism that removes sins and illuminates the eye of the soul,[43] and the Eucharist, a "spiritual nourishment",[44] whereby, Clement says "we put Savior in our hearts in order to destroy the passions of our flesh"[45]. Thus "both body and soul are sanctified."[46] The sacraments are seen by Clement as means by which God effects the sanctification of man. They sustain the efforts of man in pursuing his heavenly goal and make them fruitful.

In the process of healing passions, the Logos "constrains the appetency (ὀρέξεις) to bring fruit, not to be lost in desires (ἐπιθυμεῖν)."[47] Clement distinguishes different kinds of desire.[48] Its basic form is ὄρεξις – meaning "appetency", a general ability to desire. It is this term he uses when he speaks about the desire for eternal life or desire for truth.[49] Another term is ἐπιθυμία, which is desire turned

---

38 Cf. Clement of Alexandria, *Paedagogus*, I 98,1–99,2.
39 Cf. Clement of Alexandria, *Paedagogus*, I 100,3.
40 Clement of Alexandria, *Paedagogus*, I 27,1.
41 Clement of Alexandria, *Paedagogus*, II 110,2.
42 Clement of Alexandria, *Paedagogus*, I 100,1.
43 Cf. Clement of Alexandria, *Paedagogus*, I 25,1–32,4.
44 Cf. Clement of Alexandria, *Paedagogus*, I 50,3.
45 Clement of Alexandria, *Paedagogus*, 43,1.
46 Clement of Alexandria, *Paedagogus*, II 20,1.
47 Clement of Alexandria, *Paedagogus*, I 66,5.
48 This question is treated in detail by David Hunter in "The Language of Desire: Clement of Alexandria's Transformation of Ascetic Discourse," *Semeia*, Vol. 57 (1992), pp. 95–111.
49 See e.g., Clement of Alexandria, *Paedagogus*, I 1,1; 1,3; 102,3.

toward pleasure. The goal of Christian pedagogy, as we see in the last quotation above, is not the suppression of desire, but its proper orientation. When desire is turned toward eternal life and God, it becomes the necessary basis of faith and of Christian life. Only when it turns toward the material creation and its pleasures, diverting man from his heavenly goal, does it become something negative and in need of healing, that is, of reorientation toward the true good. The whole process of healing of passions, therefore, wants to detach man from the earth and to direct his desire and steps toward heaven.

Once this is accomplished and a person following the Christian way of life is established in virtue and freed from passions, the ultimate goal of the pedagogy of the Logos is also realized: preparation for the knowledge of God. In fact, Clement defines pedagogy as the "education for the knowledge of truth, (...) indication of a straight path of truth for the contemplation of God."[50] Asserting that the purpose of man is to contemplate the divine, Clement invites us to contemplate human nature as well,[51] for the one who

> acquires the knowledge of what happens in man according to his nature, will not pursue the exterior things, but will purify what is proper to man, the eye of the soul, cleansing also his flesh. The one who, becoming pure, died to those things on account of which he was yet dust, what better means than himself has he to walk on the way toward the knowledge of God?[52]

The knowledge of self leads therefore to the knowledge of God and to the likeness to God,[53] if one previously purifies oneself, that is, becomes free from earthly attachments and pursuits that impeded his sight, and so is able to discover and perfect his true, heavenly nature. In the words of Tatakis, Christian self-knowledge rises to the knowledge of God when "the Christian discovers the divine side of the human being, the fact that it is created after the image and likeness of God, and consequently demands that this divine side dominate life and fill it with its light."[54]

The Christian way of life originates thus in the pedagogy of Christ who shows "the model of the true philosophy."[55] Man is educated and sanctified by the Pedagogue as a child of God, made a citizen of heaven, he receives on high the

---

50 Clement of Alexandria, *Paedagogus*, I 53,3–54,1.
51 Cf. Clement of Alexandria, *Paedagogus*, I 100,3.
52 Clement of Alexandria, *Paedagogus*, II 1,2–3.
53 Cf. Clement of Alexandria, *Paedagogus*, III 1,1.
54 Basil N. Tatakis, *Christian Philosophy in the Patristic and Byzantine Tradition*, trans. G. D. Dragas (Rollinsford, NH: Orthodox Research Institute, 2007), p. 42.
55 Cf. Clement of Alexandria, *Paedagogus*, II 117,4.

Father whom he learns to know on earth, and so is transformed by the Logos.[56] The goal of the pedagogy of the Logos is to perfect the creation of man and lead him to the knowledge of God, as Clement affirms quoting St. Paul: "He revealed the object of our quest saying: 'until we all attain to the unity of faith and to the knowledge of God, to a perfect man.'"[57] It is again expressed in the final prayer with which Clement concludes the *Paedagogus*: "Grant us, who follow your precepts, to perfect the likeness of the image and to experience the goodness of God."[58]

We can see here a vital relationship between knowledge and human moral choices. "To know God" is for Clement equivalent to "be saved" – it is one of the ways in which he describes the state of the final perfection of man reaching union with God. Knowledge is thus the goal of the process of perfection of man and, at the same time, the thread of the entire vision. Its lack is a cause of sin and moral degradation; moral purification is a pre-condition for acquiring it; its possession prevents man from sinning. The progress of the Christian on the way to union with God is thus characterized by this dynamic interplay: moral purification – knowledge – moral perfection, in a kind of spiral ascent.

2.3 Principles of Christian life

Clement opens the second book of the *Paedagogus* saying: "Keeping our eyes fixed on our goal, and choosing the Scriptures useful for the practical part of the pedagogy, we have to sketch summarily what must be in all the life the one who is called a Christian."[59] Realization of the heavenly nature of man becomes thus the idea regulating Christian life in all its details. Accordingly, Clement fills the second and the third book of his work with often meticulous recommendations concerning all aspects of daily life: food, drink, sleep, clothing, entertainment, wealth, marriage, sexual life, etc. The analysis of this practical part of Clement's thought lets us discover the logic underlying his project of the Christian way of life and allows to point to five general principles of his asceticism.

(1) First, Clement acknowledges the reality of the earthly, fleshly aspect of man with its demands as an integral part of human nature which, according to the Christian faith, is meant to participate in its final, heavenly and divine condition. Consequently, Clement does not disparage the flesh nor downplay

---

56 Cf. Clement of Alexandria, *Paedagogus*, III 99,1–2.
57 Eph 4:13, as quoted by Clement in Clement of Alexandria, *Paedagogus*, I 18,3.
58 Clement of Alexandria, *Paedagogus*, III 101,1.
59 Clement of Alexandria, *Paedagogus*, II 1,1.

its needs, opposing it to the soul, but sees its inherent place and role in the Christian struggle for perfection, perceiving man as a unified whole. Thus, he makes space for bodily needs – food, drink, sleep, rest, etc. – in their natural function of sustaining life. Because of this concern for the body, Clement is often more moderate in his ascetic rules than most philosophical schools of his time.[60]

(2) At the same time, however, Clement clearly sees that human existence is not limited to the earth and has to transcend itself in order to reach its true, heavenly goal. Our earthly life is ephemeral and transitory – the true source of our being is God who alone can make us partakers of His immortal life. Consequently, Clement is never tired of opposing the temporary earthly existence, with its illusory goods and concerns, to the eternal life and its enduring treasures as the only ones that really exist and can satisfy human desire for life. Through this contrast Clement reminds us of our heavenly nature and goal and invites us to live in this world as travellers, free from attachments and anxieties, mindful of our true, heavenly homeland.

(3) The two sides of the human being – earthly and heavenly – are to be kept, therefore, in a proper balance. Bodily needs are to be satisfied in the limits of necessary, without absorbing all human attention, or becoming a means of seeking fleshly pleasure. What is natural and necessary for man should not be suppressed, but granted according to proper measure and time, and used on purpose, as help to achieve the higher good and subordinated to this goal, without becoming an end in itself. Clement adopts a few basic rules that express this intent: that of the golden mean, sufficiency, utility, frugality, and simplicity. Being satisfied with what is necessary, rejecting the superfluous, and reducing one's needs, become a hallmark of a life of interior freedom, independent of human opinion and of self-delusion, always mindful of its goal.

(4) Essential to maintain this harmony between the two sides of human nature is self-control and the mastering of natural impulses. They have to be kept in the limits of the necessary, so as not to give rise to desires seeking sensual enjoyments. Satiety should be avoided because it opens way to indulge in pleasures. Christians should be on their guard against everything that can stir passions and result in base thoughts and actions, diverting them from their heavenly goal. It is necessary, therefore, to control one's senses, not to pay excessive attention to the body, and to avoid purposeless pleasure, not

---

60 See the remark of Marrou: *Le Pédagogue II* (SC 108), p. 156, n. 6.

related to the necessities of life. The desire of man has to be directed toward his true good and lose all taste for earthly enjoyments, so as to support him on his way to heaven.

(5) What is needed for the proper orientation of desire is an attitude of attentiveness, of staying awake and alert, mindful of the heavenly reality. It is necessary not to be heavy with food or wine which lessens attention and capacity for reasoning, darkens the soul, makes one forgetful of the real life and limits horizons to the earthly reality, resulting in attachments and stirring of passions. The Christian has to stay sober and watchful, constantly turning his mind to God in prayer, being awake for His presence and desirous for the knowledge of God. He cannot quench this desire for truth and the spirit of quest in sensuality, but must maintain its flame always bright. He has to keep his soul sensitive, not overwhelmed by improper, sensual impressions or by futile occupations that dissipate energy in idleness.

The asceticism of Clement, characterized by these five principles, is clearly dependent on his vision of man, being a practical means by which Christians can achieve their goal that they come to know through revelation and follow in faith, with the help of divine grace. It is meant to help the followers of Christ achieve an attitude of interior freedom and detachment from the earth, and to direct all their desires and activities toward the realization of their heavenly nature. In the words of Tatakis, it is "an effective method of salvation."[61] The *Paedagogus* is thus the earliest known document that offers a practical manual of Christian life, combining in a consistent and elaborate way theological vision with everyday life.

3 Conclusion: Clement and his Greek heritage

The biblical inspiration constitutes the most fundamental layer of Clement's thought: it is the basis of his anthropological vision and of the consequent project of the Christianway of life. Man is for Clement primarily a heavenly being, whose origin and goal are defined with respect to God. Accordingly, the true identity of man is seen in the divine side of his being: in his inner self, in his capacity for knowledge of God and for communion with God. Consequently, the way of life proposed by Clement is based on the same pattern: it is focused on developing the heavenly aspect of human nature while detaching man from the earth.

---

61 Cf. Tatakis, *Christian Philosophy in the Patristic and Byzantine Tradition*, p. 31.

At the same time, alongside this biblical vision we find many elements coming from the Greek background of Clement. It is true first of all for his asceticism. Many of the practical instructions he gives in the *Paedagogus* are identical with ascetical practices of ancient philosophical schools, which Pierre Hadot calls "spiritual exercises" constituting the practical part of philosophy.[62] Many authors examining the relationship between Clement and ancient philosophical schools pointed to such similarities, and indicated elements which were assimilated by Clement.

Describing Christianity as philosophy and as a way of life, Clement refers to the ancient model of philosophy that, as explained by Hadot, distinguished two poles of philosophical activity: "On the one hand, the choice and practice of a way of life; on the other, a philosophical discourse which is at the same time an integral part of this way of life, and renders explicit the theoretical presuppositions implicated in this way of life."[63] Accordingly, when Clement presents Christianity as philosophy, he refers to these two elements: Discourse, which for him is the vision of man and his relationship with God based on the biblical revelation, and a way of life, which is a means of perfecting the human nature in order to achieve its goal set forth by this vision.

We can thus observe how Greek culture informs Clement's categories of thought, his way of approaching the question of human existence and human growth. However, his assimilation of Greek philosophy is not an indiscriminate choice of material. Clement is very careful and self-conscious in examining his cultural background, understanding the elements of his culture from a new perspective of Christian revelation. Including them in his vision of man and his relationship with God, he often effects a shift in their meaning and purpose. In this way, as it was pointed out by Hadot, the philosophical spiritual exercises became part of a broader ensemble of specifically Christian practices.[64] As a result, the entire ascetic endeavor, with its different stages and practices, acquires a new meaning – it becomes a way of ascent to union with the Triune God, helping man realize his heavenly goal.

This interaction between Christianity and culture does not occur between abstract entities or in some abstract medium. It is Clement himself who is the channel of this influence. He encounters Christianity as a Greek philosopher formed by his culture and, consequently, comes to understand it in a way

---

62 Cf. Pierre Hadot, *What is Ancient Philosophy?*, trans. M. Chase (Cambridge, Massachusetts: Harvard University Press, 2002), pp. 66–67.
63 Hadot, *What is Ancient Philosophy?*, p. 76.
64 Cf. Hadot, *What is Ancient Philosophy?*, p. 248.

determined by the conceptual tools available to this culture. As a philosopher seeking truth and wisdom he meets Christ whom he recognizes as the Wisdom and the Truth, the One who gives ultimate answers to his existential quest. But the questions and aspirations that put Clement on this quest (or at least the way of asking them) remain Greek and determine the way in which he presents the answer that he has found.

We can conclude, therefore, that Clement does give Christianity a Greek shape. But at the same time we have to recognize that it is the Gospel that remains the organizing principle of this synthesis. The biblical vision of man and his relationship with God, seen in the perspective of the history of salvation, constitutes the frame of Clement's thought. The elements of Greek culture are included within this frame after a careful discernment and with a clear purpose of serving the Christians in arriving at their goal as it is revealed by Christ.

References

Clement of Alexandria, *Protrepticus: Le Protreptique*. Introduction, translation and notes C. Mondésert. Sources Chrétiennes 2bis. Paris: Les Editions du Cerf, 2004 (2nd edition with corrections).

Clement of Alexandria, *Paedagogus: Le Pédagogue I*. Introduction and notes, H.-I. Marrou, Trans. M. Harl. Sources Chrétiennes 70 (Paris: Les Editions du Cerf, 2008) (reprint).

Clement of Alexandria, *Paedagogus: Le Pédagogue II*. Trans. C. Mondésert, notes H.-I. Marrou. Sources Chrétiennes 108 (Paris: Les Editions du Cerf, 1991) (2nd edition reviewed and corrected).

Clement of Alexandria, *Paedagogus: Le Pédagogue III*. Trans. C. Mondésert and C. Matray, notes H.-I. Marrou. Sources Chrétiennes 158. (Paris: Les Editions du Cerf, 2008) (reprint).

Hadot, Pierre. *What is Ancient Philosophy?* Trans. M. Chase. Cambridge, Massachusetts: Harvard University Press, 2002.

Hunter, David G. "The Language of Desire: Clement of Alexandria's Transformation of Ascetic Discourse." *Semeia*, Vol. 57, 1992, pp. 95–111.

Tatakis, Basil N. *Christian Philosophy in the Patristic and Byzantine Tradition*. Trans. G. D. Dragas. Rollinsford, New Hampshire: Orthodox Research Institute, 2007.

Karol Wilczyński

The Idea of Human Enhancement and Arabic Neoplatonism[1]

Abstract: The subject of the present paper is the idea of human enhancement as described by Neoplatonic philosophers, especially those of Arabic origin. A short presentation of two important figures (Al-Kindī and Al-Fārābī) from the beginning of Arabic Neoplatonism is followed by an analysis of the usage of this idea in a fragment of Avicenna's *The Metaphysics of the Healing*. Although the above mentioned thinkers used only Arabic, they were able to introduce and develop concepts known to Greek philosophers such as Plato, Aristotle, or Plotinus. These concepts were crucial for the development of the doctrine and anthropology of the late Arabic Neoplatonism. For Avicenna and other Arabic philosophers, the very idea of human development was situated within the intellectual faculty (principle) of the human soul. They generally considered human improvement as a great idea to pursue, but also as problematic and based merely on an approximation. Avicenna's argumentation and his description of the idea of human improvement had a fundamental significance for understanding human excellence in medieval philosophy.

Keywords: Neoplatonism, Arabic philosophy, Al-Farabi, Avicenna, Human perfection

Neoplatonism, the tradition initiated by Plotinus (204–270), seems to be one of the more underestimated philosophical schools in the Western tradition.[2] The arguments of Neoplatonic thinkers play a minor role in contemporary discussion about the idea of human enhancement, and even though among representatives of Neoplatonism we can count Plotinus Porphyry, Proclus, or Iamblichus,[3] it appears that their theories are no longer an object of interest for modern thinkers. Neoplatonic writers saved the antique thought from being forgotten by establishing numerous schools in which Plato's and Aristotle's works were read up to 8th–9th century in their original language, and handing on this tradition to the Arabic schools of Baghdad and later Andalusia. Owing to their

---

1 I would like to express my gratitude to my supervisor, Prof. Jan Kiełbasa (Associate Professor at the Jagiellonian University in Krakow), and Prof. Peter Adamson (Ludwig-Maximilians-Universität) for the inspiring comments.
2 Cf. L. P. Gerson, "Foreword," in: *Neoplatonism*, R. T. Wallis (London: Duckworth, 1995).
3 Who in turn influenced the shaping of the ideas of such authors as Pseudo-Dionisus, Al-Fārābī or Maimonides.

translations into Arabic language, the texts of Aristotle and also (but to a lesser extent) Plotinus could be later recovered in the Latin West.

The aim of the following paper is to briefly present Neoplatonic anthropology with a particular emphasis on its view of humans and human enhancement in Neoplatonic Arabic philosophy, which not only took on the subject of the perfection of man but also discussed it in depth. The first part of the text will present the main features of Neoplatonic thought which contributed to development of the idea of human enhancement. The second part, after a short presentation of two figures (Al-Kindī and Al-Fārābī) from the beginning of Arabic Neoplatonism, will analyze the usage of this idea in the passage from Avicenna's *The Metaphysics of the Healing*, which is one of the most important writings of Arabic Neoplatonism. All together, the paper aims at demonstrating the arguments as well as determining the stand of the representatives of early Arabic philosophy in the discussion about human enhancement.

At this point it is worth mentioning that the essential nature of Neoplatonic philosophy and its impressive legacy (it is estimated that the Neoplatonic texts created between 2nd and 6th century numbered up to 15,000 pages)[4] are particularly unique because of the cultural conditions in which they came to life. Regardless of the different religious backgrounds of the thinkers (Neoplatonism connected not only the polytheistic, but also the three biggest monotheistic religions – Christianity, Judaism, and Islam), their idea of excellence relates to God as the highest and most perfect being. Because of that, when the aforementioned writers studied the idea of human enhancement or the idea of excellence, they tried to base their theories on entirely different premises and had an entirely different approach that the one represented by present day scholars. I am convinced, however, that this does not exclude the possibility of presenting their arguments in the light of the present-day debate.

1 Greek Neoplatonism

Plotinus was the intellectual heir of Plato and Aristotle, at least in terms of the language used in his greatest work *Enneads*. It is from them that he adopted three ideas of what we may translate as "excellent" or "perfect": τέλειος, κάλλιστον/βέλτιστον, and ἀρετή. Moreover, Plotinus also discussed human enhancement within four areas which had been previously taken up by many Greek philosophers, defining it as something that allows us to strive for ageless

---

4   Gerson, "Foreword".

bodies (c.f. *Enneads* 1.5.6, 1.6.1), perfection (*purification*) of souls (c.f. 5.1.1), supreme efficiency (c.f. 4.6.3), and – of course – a better education of one's children (c.f. 3.1.5 and 18). All those topics, which were systematically discussed by Plato in his dialogues, appear in the writings of Plotinus and other Neoplatonic thinkers.

The theme of reaching "excellence" or "perfection" of the human person was examined by Aristotle, and his definitions influenced Plotinus deeply. The idea of τέλειος, developed by Aristotle, is connected to τέλος, meaning "aim", "goal", or "end". In *Metaphysics* Aristotle defined what he wanted to express by τέλειος:

> It means: (...) (b) That which, in respect of goodness or virtues [gr. ἀρεταί], cannot be surpassed in its genus; (...) (c) And goodness is a kind of perfection [gr. τελείωσις]. For each thing, and every substance, is perfect when, and only when, in respect of the form of its peculiar excellence, it lacks no particle of its natural magnitude. (d) Things which have attained their end, if their end is good, are called 'perfect'; for they are perfect by attaining their end. (*Metaphysics* 1021b)

Especially the latter was decisive for Neoplatonic thought. Plotinus believed that the human τέλος is not here, in this world, but is connected to the Intellect, which is the best representation of the One, the Beauty. Therefore, as long as human soul is "chained" by sensation and matter, it remains in a "tomb", it is not achieving its proper goal, τέλος (cf. *Enneads* 4.8.4): "The central moral truth that Platonist philosophy reveals (or claims to) is that our true life and our true good consist in a constant effort of ascension from our physical embodiment."[5] This is because only in the realm of the Intellect may a human person fulfill their aim.

This basic Neoplatonic premise was connected with the idea of κάλλιστον/βέλτιστον – i.e., what is perfect, the best, in its proper sense.[6] Aristotle, who also influenced Neoplatonic philosophy in this matter, defined κάλλιστον in the following way:

> Those who suppose (...) that perfect beauty and goodness [gr. τὸ κάλλιστον καὶ ἄριστον] do not exist in the beginning (...) are mistaken in their views. For seed comes from prior creatures which are perfect, and that which is first is not the seed, but what is perfect [gr. τὸ τέλειον]. (...) It is evident (...) that there is some substance which is eternal and immovable and separate from the sensible; and it has also been shown that this substance can have no magnitude, but is impartible and indivisible (...); and moreover that it is unaffected [gr. ἀπαθής] and unalterable; for all the other kinds of motion are posterior

---

5 J. Cooper, *Pursuits of Wisdom. Six Ways of Life in Ancient Philosophy from Socrates to Plotinus* (Princeton: Princeton University Press, 2013), p. 382.
6 Cf. R. T. Wallis, *Neoplatonism..*.

to spatial motion. Thus it is clear why this substance has these attributes. (*Metaphysics* 1072b–1073a, cf. also *Nicomachean Ethics* 1152b)[7]

There are three factors that determine whether we can call something "the best", κάλλιστον: duration (*eternal, immovable*), identity (*separate from the sensible, indivisible*), and internal power or intensity (*unaffected, unalterable*) (cf. *Enneads* 5.1). Plotinus made use of this definition and often referred to the Sun as an example of what is "the best", as something which is immovable, separated from the Earth and also cannot be affected. Of course, this was only a metaphor which allowed Plotinus to explain real beauty: "The beauty [gr. κάλλιστον] that is in the sense-realm is an index of the nobleness of the Intellectual sphere, displaying its power and its goodness alike: and all things are for ever linked" (*Enneads* 4.8.6).

What is "the best" in a human being? For Plotinus, by accepting the presuppositions described above, it was easy to decide that it is the intellectual faculty, and that the aim of human life can be fulfilled only through the use of intellect – there is no other way: "to set oneself above Intellect is to fall away from it" (*Enneads* 2.9.9). How can then one fulfill one's aim? Does it mean that every human person should know more? Or does it mean that it is necessary to build a philosophical system which would lead to a better vision of the Cosmos? For Plotinus the answer is not that simple. Of course, he claims that the intellect is something which all humans share (cf. O'Brien 1964: 24–30). He also states that the faculty of intellect is a device to make people closer to "the Intellectual sphere" and to attain excellence (gr. ἀρετή),[8] which is another obvious reference to Aristotelian ethics on "proper function". Moreover, he adds that man's proper function is intellect, shared neither with animals nor plants, and that using one's intellect in a proper way is the most excellent and the best action possible. But how is it possible, in the Neoplatonic perspective, that through this action one may attain the "Intellectual sphere"? How is it possible to tear oneself away from what is movable, material, and divisible?

Plotinus writes the following: "And towards the Intellectual-Principle what is our relation? By this I mean, not that faculty in the soul which is one of the emanations from the Intellectual Principle, but The Intellectual-Principle itself. And how do we possess the Divinity? In that the Divinity is contained in the

---

7 As seen here, Aristotle's definition was not original but was influenced by many thinkers who preceded him, especially Plato and Parmenides.
8 C. D'Ancona, "Greek into Arabic: Neoplatonism in Translation," in: *The Cambridge Companion to Arabic Philosophy*, ed. P. Adamson and R. Taylor (Cambridge: Cambridge University Press, 2005), p. 11.

Intellectual-Principle and Authentic-Existence" (*Enneads* 1.1.8). Thus, thanks to the faculty of intellect we are able to grasp the Divine and to try to "enhance" ourselves.[9] As Plotinus puts it, a human person resembles the cosmological order and so do the virtues:

> Virtues in the Soul run in a sequence correspondent to that existing in the over-world, that is among their exemplars in the Intellectual-Principle. In the Supreme, Intellection constitutes Knowledge and Wisdom; self-concentration is Sophrosyne; Its proper Act is Its Dutifulness; Its Immateriality, by which It remains inviolate within Itself is the equivalent of Fortitude. In the Soul, the direction of vision towards the Intellectual Principle is Wisdom and Prudence, soul-virtues not appropriate to the Supreme, where Thinker and Thought are identical. All the other virtues have similar correspondences. (*Enneads* 1.2.7)

In the above analogy, one may find an answer on how to excel and enhance one's intellect.[10] What is important is that it is not possible to "purify" the soul exercising only one virtue – one needs to attain all virtues: "And if the term of purification is the production of a pure being, then the purification of the Soul must produce all the virtues; if any are lacking, then not one of them is perfect".[11] Being closer to the Intellectual Sphere is therefore rather a kind of spiritual exercise, being engaged in self-discipline, than gaining knowledge of what is sensible, mutable, and far from the Beauty.

The three concepts that were shortly described above (τέλειος, κάλλιστον/βέλτιστον and ἀρετή) were crucial for the development of the Neoplatonic doctrine and the Neoplatonic understanding of the idea of human enhancement. There is no doubt that Plotinus' anthropology was also influenced by Aristotle's theory of the soul as an entelechy (gr. *perfection, completeness*) of the body, which in this case was rejected and widely criticized.[12] The critique was carried out mainly by Plotinus and Proclus, and had a fundamental significance

---

9   J. Rist, *Plotinus: The Road to Reality* (Cambridge: Cambridge University Press, 1967), p. 299.
10 Cf. G. Leroux, "Human Freedom in the Thought of Plotinus," in: *The Cambridge Companion to Plotinus*, ed. L. P. Gerson (Cambridge: Cambridge University Press, 1996), pp. 293–294.
11 Leroux, "Human Freedom in the Thought of Plotinus," pp. 293–294; c.f. also Simplicius, *in Caelo* (*Commentary on Aristotle's on the Heavens*), ed. J.L. Heiberg (Berlin: Reimer, 1894), pp. 165, 7–9; H. J. Blumenthal, *Plotinus' Psychology: his Doctrines of the Embodied Soul* (The Hague: Martinus Nijhoff, 1971), p. 38.
12 Cf. R. Wisnovsky, *Avicenna's Metaphysics in Context* (New York: Cornell University Press, 2003), pp. 79–98.

for the understanding of the notion of human excellence. Neoplatonic thinkers shared the old Platonic view of the body, being that it is the worst, or least valuable, part of human entity. *Enneads* also recognize the soul as "a slave of rational forming principles" the source of which is in the body (cf. 4.4.12). Those "principles" (gr. λόγοις) are connected with a desire that arises in the human soul through sensory cognition. The act of choosing them leads to an inner struggle and – in the case of the lack of strong will – to the fall of the soul (*ibidem*). What is more, Plotinus made it clear that those bodily "principles" exert violence on the soul and are accidental to it (4.8.1). Plotinus, however, could not accept the Aristotelian view of the soul as entelechy. He rejected the idea of the body as something which may be used to attain perfection since it was just a frame or instrument for what is the best in humans – the faculty of intellect: "Hence the Soul may feel sorrow and pain and every other affection that belongs to the body; and from this again will spring desire, the Soul seeking the mending of its instrument" (*Enneads*, 1.1.3). Can therefore body enhancement play any important role for Neoplatonic thinkers?

The answer is "yes", and "no". "No", because the body has nothing to do with the Intellectual Sphere. After death, as it was in Plato's writings, the human soul will become separated from the body. But, on the other hand, one needs to examine Matter in a proper way to discover its imperfection, formlessness, and – actually – non-existence (*Enneads* 1.8.3). A "perfect human person" would know what is the part of her intellect, and what is not: "The body has acquired life, it is the body that will acquire, with life, sensation and the affections coming by sensation. Desire therefore will belong to the body, as the objects of desire are to be enjoyed by the body. And also a feeling of fear will belong to the body alone; for it is the body's doom to fail of its joys and to perish" (*Enneads* 1.1.4). In Neoplatonic anthropology every act of the body is determined in advance and is usually related to the fall of the soul. Also the bodily and material desires are not connected with the choice of will, but are simply its weakness. The human soul is forced to adopt the "stigma of coincidence" (cf. 6.8.14). Plotinus simply acknowledges that the fall of the soul is "voluntary, but with no reflection" (4.3.13, see also 17–18). Because of that what stays in the power of will is compliance with divine laws, and control of what is from the body. Apart from rejecting the idea of the soul as entelechy, Neoplatonic thinkers wanted to point out that what is free in man can be found in the universal emanationism: what we can choose by ourselves and what is determined.[13]

---

13 Cf. Wallis, *Neoplatonism*, pp. 74–77.

The responsibility, or the role, of the soul lies merely in what is related to virtue – it has in fact no relation to the body. Therefore, according to Plotinus, the bigger the virtue, the closer a certain being is to the divine Unity – which also concerns the human sphere:

> Who is then a Sage? In him surely the higher part of the soul is working. It does not suffice to perfect virtue to have only this Spirit [equivalent in all men] as a cooperator in life: the acting force in the Sage is the Intellective Principle [the diviner phase of the human Soul] which therefore is itself his presiding spirit or is guided by a presiding spirit of its own, no other than the very Divinity. (*Enneads* 3.4.6)

In the same way, it might seem that a man – by means of the decision to take on a discipline or a spiritual exercise – can get closer to Excellence. However, I believe this is not what Plotinus means, because, as he stresses, "the Soul is so constituted that its life-history and its general tendency will answer not merely to its own nature but also to the conditions among which it acts" (*Ibidem*). In other words, by necessity, humans are tied to their bodies, combining what is "lion-like" with what is "variegated" (I.1.7) and despite their efforts they will never liberate themselves from their corporeality and will never attain "pure freedom". It is worth mentioning that freedom, for Plotinus, means being free of any physical determination, and identifying oneself only with the soul and not material body.

Here we need to deal with a paradox – human enhancement may be fulfilled in a fully determined world, in necessity. Hence follows reserving freedom exclusively for the intellect and also the rejection of Aristotle's anthropology by Plotinus. The soul cannot perfect something that by necessity is placed at the bottom and cannot rise up. The soul's aim is self-purification and detachment from the body. Yet what is the scope of the process of purification? How much may one *detach* oneself from passion, body, and carnal desires? To what degree is it possible for the human soul to be perfected, i.e., be more connected to intellect? What is important here is that this very topic was brought up in Proclus' *Elements of Theology* (p. 194-195) which – in a manner akin to the *Enneads* – were paraphrased by a Syriac or Arabic translator and passed on to Al-Kindī and other researchers of the Baghdad school.[14] Proclus repeatedly comes back to that problem, for example in the *Commentary to Timaeus* (e.i. II.216).[15] This is where

---

14 P. Adamson, "The Kindian Tradition. The Structure of Philosophy in Arabic Neoplatonism," in: *The Libraries of the Neoplatonists*, ed. C. d'Ancona, (Leiden: Brill, 2007), p. 352.
15 cf. D. T. Runia, M. Share, *Proclus: Commentary on Plato's Timaeus*, Vol II, B. 2: *Proclus on the Causes of the Cosmos and its Creation*, trans. and notes. D. Runia and M. Share (Cambridge: Cambridge University Press, 2008), p. 52.

the question of relation of the body and soul was posed: If the soul sins by choice, then how does it stay "pure"? How can the soul detach itself from what comes from the body? The answer offered by Proclus is an acknowledgement that as a result of wrong choices the soul "gets dirty" and falls. The consequences of that are (1) a tighter connection of the body and soul, and (2) the soul losing a stable connection with the mind (gr. *nous*). It is worth mentioning that this is most probably where the emergence of the traditional pagan ceremonies and rituals within philosophy of Iamblichus came from.[16]

2 Arabic Neoplatonism

In this shape, the problem became transferred to the arena of Arabic philosophy.[17] As Cristina d'Ancona puts it: "the fact that Plotinus' *Enneads* and Proclus' *Elements of Theology* were among the first works translated into Arabic had long-term consequences for the entire development of *falsafa*."[18] It was defined for the first time in the so-called *Theology* of Pseudo-Aristotle, an Arabic paraphrase of Plotinus' *Enneads*[19]: "Human was defined by divine Plato (...), as something that is nothing other than soul using its body properly. During that the most noble and divine soul uses body in a different sense, i.e., through the animal [part] of the soul [arab. *al-nafs al-haywāniyya*]" (Ps.-Aristotle, *Theology*, X, 72). What is worth noticing is that many Syrian and Arabic Neoplatonic thinkers read *Enneads* in the following way: the sin – the fall – is in the power of man, but the soul itself does not choose it – it is being deceived (cf. *Enneads* 4.8.8).

Coming back to the problem of excellence, it should be pointed out that the idea of human enhancement does not have many equivalents when it comes to Arabic terminology. As demonstrated by Robert Wisnovsky,

unlike late-antique Greek thinkers (...), Avicenna [as well as other Arabic thinkers] was helped along by the early Greek-Arabic translations, and particularly by the translations of the *Metaphysics* and the *Uthūlūjiyā* [i.e., *Theology* of Pseudo-Aristotle], which, by using the same Arabic root – *t-m-m* [ar. تامّ means 'complete', 'perfect'] – to render the Greek *entelekheia*, *teleiotês* and *telos*, did part of the job for him.[20]

---

16 R. E. Witt, "Iamblichus as a Forerunner of Julian," in: *De Jamblique a Proclus*, ed. Heinrich Dörrie (Vandoevres, Geneve: Fondation Hardt, 1974).
17 Cf. A. Ivry, *Al-Kindi's Metaphysics* (Albany: State University of New York Press, 1974), p. 19.
18 Falsafa is the Arabic term for philosophy. D'Ancona, "Greek into Arabic," p. 21.
19 Cf. P. Adamson, *The Arabic Plotinus. A Study of the Theology of Aristotle and Related Texts* (London: Duckworth, 2002).
20 Wisnovsky, *Avicenna's Metaphysics in Context*, p. 113.

Despite the semantic narrowing of the topic of the idea of excellence, philosophers such as Al-Kindī (800–870), Al-Fārābi (880–950), and Ibn-Sinā (lat. Avicenna, 980–1037) developed a very interesting concept of human excellence using only Arabic, a language that was unfamiliar to philosophy at that time. In this part of the article I will only loosely outline the ideas of the first two philosophers. I will dedicate its larger part to the analysis of a short passage of the most important book of Ibn-Sinā: *The Book of Healing: Metaphysics* (ar. *Al-Ilāhiyyāt*) in which the issue of coming back to the original, best condition of the soul is considered (*Al-Ilāhiyyāt* IX.7).

Unfortunately, the majority of the works of Al-Kindī have not survived to the present times, or are still hidden in private libraries somewhere in the East. Thus, only a few remarks on ethics or anthropology are available and none of them are devoted to the topic of perfection or happiness. There is, however, one book of Al-Kindī, which may shed light on his opinion about human enhancement, showing by whom he was influenced: "Yet it is impossible that someone obtain all that he wants, or be safe from losing all that he loves, since stability and constancy do not exist in the world of coming to be and passing away, in which we live. Necessarily, stability and constancy only exist in the world of the intellect, which we can contemplate" (Al-Kindī, *On Dispelling Sorrows* I, 2). What are the conclusions? Al-Kindī puts them straightforwardly: "When we contemplate objects of desire which bring people pleasure in the intellect in a certain stability in the soul's form for the duration of its allotted time and its ability to produce both something similar to it and something which protects it from pain and affords it rest, then we take them in the best way as the need may be" (*Ibidem* II.2). As it was pointed out by numerous researchers, Al-Kindī was in fact the first thinker to formulate numerous philosophical problems in Arabic, which was completely unfamiliar to philosophy at that time. Al-Kindī knew extensive fragments of part IV and part V of *Enneads* (as well as a short fragment of VI) in the form of paraphrases – Pseudo-Aristotle's *Theology* mentioned earlier – formulated in the Bagdad school in the 8th – 9th century by a Syrian researcher.

There is no doubt that Al-Kindī was the intellectual heir to Plotinus. However, it was Al-Fārābi who developed a more widely recognized and more mature theory of the ideal state. Contrary to a popular belief, however, Al-Fārābi's political theory was based on metaphysical and anthropological foundations. The "Second Teacher" was drawn from well-prepared metaphysical and religious assumptions – based probably on *The Republic and Laws* by Plato – where the principles govern not only a perfect political organism, but also every single human being.

Al-Fārābī, considered by some researchers as the founder of Arabic Neoplatonism,[21] was the author of several works on political philosophy, ethics, and – particularly – human happiness (which was considered as an innovation in his times). What, according to al-Fārābī, does it mean to be perfect? He explains it in his most important work *Book on the Opinions of Citizens of the Perfect State* (ar. *Ārā' Ahl al-Madīnah al-Fāḍilah*): "Beauty, perfection and brilliance of every entity is to exist in its most excellent state of existence and to attain its ultimate perfection" (82.17). Of course, it is the First Entity, which is the best and most beautiful (82.20). Therefore, it is the apprehension of Its Beauty and Perfection, which lets other entities become perfected and enhanced. Al-Fārābī used here Aristotelian language and a Neoplatonic framework to describe the real value of man, which is the true happiness (cf. *Ārā'*, 112.1).[22] As in the case of his Greek predecessors, Al-Fārābī used the concept of four classical virtues to describe what is a perfect human person. Moreover, he stressed that perfection may be obtained only by attaining all of them. One may not be partially virtuous, partially complete: "The Complete (ar. *tamm*) is that, beyond which exists anything, that could pertain to it" (cf. *Ārā'*, 60.14).

What is important, as in case of other Neoplatonic thinkers, was that Al-Fārābī's view of human perfection was truly intellectualistic: to progress on the path of happiness means to attain virtues by knowing God, Allah (*Ārā'*, 84.11). One should use his intellect, as well as body and passions, to discover the similarities between the world and the Perfect. That is why Al-Fārābī was interested in logic: reasoning properly and having a correct sense of what are the principles led to good actions. This conclusion had an impact on Al-Fārābī's political philosophy and his view on the ideal, or perfect, state. According to him, a perfect state should be led by an ideal ruler who is acquainted with the principles of the world and intellect. Being an absolute leader and ruling the whole state is possible only with the help of philosophers and sages. Logic and philosophy are indispensable in providing him with the best decisions and securing the happiness of all citizens. The truths acquired by the rules of logic are not only safer, but they also purify the soul and sanctify it,[23] and place man closer to God.[24] These ideas of

---

21 M. Fakhry, *History of Islamic Philosophy* (New York: Columbia University Press, 1983), p. 147.
22 Cf. also I. R. Netton, *Al-Fārābī and his School* (Surrey: Curzon, 1999).
23 P. Adamson, "The Kindian Tradition", p. 370; Cf. also Ivry, *Al-Kindī's Metaphysics*, p. 28–32.
24 This is in fact one of the most commonly accepted statements in the world of Islam. Even today, in the majority of Muslim communities, mentally disabled persons are considered as saints – it is believed that their intellect stays ahead of their owners on the

Arabic thinkers, especially of Al-Fārābī, were essential for the development of one of the greatest Neoplatonic philosophers – Ibn Sīnā. I would like to take one of the chapters of his most important book *Al-Ilāhiyyāt* as an example of the use of the Neoplatonic framework in shaping the idea of human enhancement.

Unlike many of his Neoplatonic predecessors, Ibn Sīnā, being more Aristotelian, conceived the soul as the best part (and perfection – ar. *kamāl*) of a human person,[25] which may still may be perfected. He employed the Neoplatonic hierarchy to explain how the soul's perfection works in practice, and define what it means for a soul to be truly perfect: immortal, separate from imperfect body, and self-sufficiently knowing (and, therefore, choosing) what is the best. We may find those practical considerations in a short chapter entitled *Concerning "the return"* (*Al-Ilāhiyyāt* IX.7). Ibn Sīnā tried to describe there "the states of the human souls when they separate from their bodies and to what state they will come to be" (IX.7.1).[26] After confirming that "true religion" is the source of knowledge about those states, he explained by the method "of reason and demonstrative syllogism" what consists of the happiness "which belongs to souls" (IX.7.3).

What is interesting, is that Ibn Sīnā used the language of pleasures (IX.7.4) and perfection (IX.7.5 – ar. *kamāl*) to describe the state of the best, i.e., happy soul. This is why he claimed:

> Actualization in a certain perfection may be such that it is known to exist and to be pleasurable; but, as long as it is not experientially realized, its manner is not conceived and its pleasure is not felt. For, that which one does not feel one neither desires not inclines toward. An example of this is the sexually impotent [man]. For he realizes full well that there is pleasure in sexual intercourse, but he does not desire it (...). The state of the blind with respect to beautiful pictures (...) is similar. (7) For this reason, the rational person ought not to imagine every pleasure is similar [to the pleasure the donkey has in its belly (...) and that the Lord of the worlds does not have within His authority and special property (...) something which is the ultimate in virtue, in nobility and goodness, which we exalt above naming it 'pleasure'. (...) For our state with respect is similar to the state of the deaf person who had never had any hearing. (IX.7.6-7)

For Ibn Sīnā it was clear that in the state of imperfection, one may only estimate what it means to be perfect. He therefore described what may distract

---

way to Heaven and that their intellect already lives there, that it came back to Allah. Cf. H. Corbin, *History of Islamic Philosophy* (New York: Kegan Paul International, 2006), pp. 22–38, 219–220.
25 Cf. Wisnovsky, *Avicenna's Metaphysics in Context*, pp. 114–115.
26 All quotations from Al-Ilāhiyyāt by M. E. Marmura.

one's intellectual faculty in discerning what are the perfect pleasures of the soul (IX.7.8-9):

> The perfection proper to the rational soul consists in its becoming an intellectual world, in which there is impressed the form of the whole; the order in the whole that is intellectually apprehended; and the good that emanates on the whole, beginning with the Principle of the whole [and] proceeding [further] (...). It thus becomes transformed into an intelligible world that parallels the existing world in its entirety, witnessing that which is absolute good, absolute beneficence, [and] true absolute beauty, becoming united with it. (IX.7.11)

It is also clear here, that this state of unification of the human rational soul with an "intellectual world" cannot be properly compared to anything (cf. IX.7.12-13). It may only be described by what is able to attain more intense and immutable knowledge, being individual intellectual faculty.

The above explains why Ibn Sīnā would see a way for human enhancement exactly in the faculty of intellect. He further described what is necessary to proceed: one may not be distracted by "appetites", "base things" or "sensuous, animal and irascible pleasures" (IX.7.15). One must be also aware that the state of perfected humans is truly inscrutable for those who did not attain it (IX.7.16-17) which is because human knowledge is based merely on a specific desire, and not a completely justified opinion. Ibn Sīnā stated that his view of the knowledge of the perfect state of a rational soul is just an "approximation" (IX.7.19) – he is not sure if having an idea of the principles of natural state or material world is a way toward knowledge of human perfection. What is important, is Ibn Sīnā's statement that "true happiness [ar. *as-saāda*] is not fully achieved except by rectifying the practical part of the soul" (IX.7.20), which is – "moral temperament" (IX.7.21) and "disposition (...) to preserve the rational soul in its natural constitution, together with the bestowal of the state of ascendancy and transcendence" (IX.7.22). In general, we may witness here the use of Aristotelian language to explain a plainly Neoplatonic position. In the last parts of this chapter, Ibn Sīnā describes the state of "imbecile" souls and their chances of attaining true happiness and state of perfection.

* * *

I will now discuss the points which may be relevant for contemporary discussions on human enhancement. Is it possible to apply some of Neoplatonic considerations in these debates, even if "science is likely to reshape our conceptions of justified morality"?[27] Can the Neoplatonic considerations of medieval Arabs coming from

---

27 J. Savulescu and I. Persson, "Moral Enhancement, Freedom and the God Machine," *The Monist*, Vol. 95, No. 3 (2012), pp. 399–421.

such a different environment in terms of the advancement of science and technology be of any value for us today? It is my theory that all enhancements which are designed to restrain the bad effects of our bodily conditions[28] would be welcomed by Neoplatonic philosophers, regardless of the religious sect they belonged to. As long as freedom, responsibility, and virtue are limited to the "kingdom" of intellect, they would see enhancing the capabilities of human bodies and brains as acceptable. Among the things which Neoplatonists would consider a problem was the one termed by Savulescu and Persson as a "strategic program to use knowledge from the science of morality to deliberately and effectively improve moral disposition",[29] and other general projects to "eliminate evil".[30] Generally speaking, a Neoplatonic thinker would not believe in the success of creating a better world but rather claim that a person should be responsible for their character, virtuous behavior, and "being better". We may refer here to Al-Fārābī's conception of the ideal state, where the conditions of enhancing human morality were listed. First of all, none of the means listed by the *Second Teacher* have a 'technical' character – human enhancement has a different character. The capacities we want to improve have to be used, exercised by each person on their own. According to Al-Fārābī, only the direction or target of these improvements are indicated by state leaders, but it is only us who personally put our souls – as well as our intellects – to work, rather than having them enhanced in a technical way by someone else. That is why moral dispositions of our souls and intellects may not be perfected through genetic modifications, medical treatments, or injections.

Moreover, for Al-Fārābī, a classical philosopher, the possibility (which he would deny) of enhancing morality and intellect would be problematic for two other reasons. First, "the better world projects" aim to create a society of equal chances and make all people commit fewer crimes, which for Al-Fārābī could only be partially possible. Even in a perfect state where all citizens possessed enhanced morality, inequality would still remain. As Al-Fārābī points out, every human person possess natural capacities, talents, and weaknesses, which could not be totally discarded neither by medical therapy nor genetic treatment. Of course, one may try, but one would need to take reality into account: pure equality is impossible to achieve.

---

28 Cf. R. Sitaram, A. Caria and N. Birbaumer, "Hemodynamic Brain-Computer Interfaces for Communication and Rehabilitation," *Neural Networks*, Vol. 22, No. 9 (2009), pp. 1320–1328.
29 Savulescu and Persson, "Moral Enhancement, Freedom and the God Machine," p. 13.
30 M. Walker, "Enhancing Genetic Virtue. A Project for Twenty-first Century Humanity?," *Politics and the Life Sciences*, Vol. 28, No. 2 (2009), p. 27.

The last problem would be connected with our understanding of what it means to have an enhanced moral character. Ibn Sīnā, several times (in the chapter of *Al-Ilāhiyyāt*) refers to the estimative character of the knowledge about a perfected soul (i.e., human person). It is based merely on a specific desire, which may be rationalized in a philosopher's armchair or scientific institution, but put into practice in the political arena will still distort the picture of the situation. We may not be sure if the effects of our actions towards a better world would be the same as the planned results. We would not be concerned in a philosophical or ethical way about the moral enhancement project, if there were nothing dubious about it. Ibn Sīnā would claim, that generally it is a great idea to pursue, but it is like treading on thin ice – problematic and based merely on an "approximation" (*Al-Ilāhiyyāt* IX.7.19).

In conclusion, it is worth adding that the works of early Arabic writers significantly affected the philosophy and science of Europe in the Middle Ages. About 1150, Dominicus Gundissalinus translated a couple of writings of the thinkers mentioned. The next were translated around 1190 by Gerardus of Cremona, which laid the foundations for many of the discussions of the time, mainly of a scientific character. Taking into account the popularity of Ibn-Sīnā, it should be admitted that the Arabic philosophers played a significant role in shaping not only science or metaphysics, but also the anthropology of the Middle Ages.

References

Primary sources

Aristotle. *Metaphysics*. Ed. W. D. Ross. Oxford: Clarendon Press, 1924.

Aristotle. "Theology." In: Plotini Opera – Tomus II: Enneades IV–V. Plotiniana Arabica ad codicum fidem anglice vertit G. Lewis. ed. H.-R. Schwyzer. Paris: Desclee De. Brouwer et Cie & L 'Edition Universelle, 1959. English translation, following the order of the text of Plotinus.

Al-Fārābī. *Ārā' Ahl al-Madīnah al-Fāḍilah* (*Al-Farabi on the Perfect* State). ed. R. Walzer. Oxford: Clarendon Press, 1975.

Al-Kindī. The Philosophical Works of Al-Kindī. Ed. and trans. P. Adamson and P. E. Pormann. Karachi: Oxford University Press, 2012.

Ibn-Sīnā.The Metaphysics of the Healing, a parallel English-Arabic text translated, introduced, and annotated by M. E. Marmura. Provo: Brigham Young University Press, 2005.

Plotinus. "Enneads." In: Plotini Opera, eds. P. Henry and H.-R. Schwyzer. Brussels, Paris: Museum Lessianum, Series Philosophica XXXV, 1951–1973.

Plotinus. *Enneads*. Trans. Stephen MacKenna and B. S. Page. London: Faber and Faber,1969.
Proclus. *in. Tim*. Ed. E. Diehl. Leipzig: Bibliotheca Teubneriana, 1903–1906.
Proclus. *Inst. Theol.* (*The Elements of Theology*). Ed. E. Dodds. Oxford: Clarendon Press, 1963.
Simplicius. *in Caelo* (*Commentary on Aristotle's on the Heavens*). Ed. J. L. Heiberg. Berlin: Reimer, 1894.

Studies

Adamson, P. *The Arabic Plotinus. A Study of the Theology of Aristotle and Related Texts*. London: Duckworth, 2002.
Adamson, P. "The Kindian Tradition. The Structure of Philosophy in Arabic Neoplatonism." In: *The Libraries of the Neoplatonists*, ed. C. d'Ancona. Leiden, Boston: Brill, 2007, pp. 351–371.
Adamson, P. *Studies on Plotinus and Al-Kindī*. Dorchester: Ashgate, 2014.
Adamson, P. and R. Taylor. *The Cambridge Companion to Arabic Philosophy*. Cambridge: Cambridge University Press, 2005.
Blumenthal, H. J. *Plotinus' Psychology: his Doctrines of the Embodied Soul*. The Hague: Martinus Nijhoff, 1971.
Cooper, J. *Pursuits of Wisdom. Six Ways of Life in Ancient Philosophy from Socrates to Plotinus*. Princeton: Princeton University Press, 2013.
Corbin, H. *History of Islamic Philosophy*. New York: Islamic Publications, 2006.
D'Ancona, C. "Greek into Arabic: Neoplatonism in Translation." In: *The Cambridge Companion to Arabic Philosophy*, eds. P. Adamson and R. Taylor. Cambridge: Cambridge University Press, 2005.
Fakhry, M. *History of Islamic Philosophy*. New York: Columbia University Press, 1983.
Gerson, L. P. "Foreword." In: *Neoplatonism*, ed. R. T. Wallis. London: Duckworth, 1995.
Gutas, D. "Geometry and the Rebirth of Philosophy in Arabic with al-Kindi." In: *Words, Texts and Concepts Cruising the Mediterranean Sea: Studies on the Sources, Contents and Influences of Islamic Civilization and Arabic Philosophy and Science*, eds. R. Arnzen, J. Thielmann. Leuven: Peeters, 2004, pp. 195–209.
Ivry, A. *Al-Kindi's Metaphysics*. Albany: State University of New York Press, 1974.

Leroux, G. "Human freedom in the thought of Plotinus." In: *The Cambridge Companion to Plotinus*, ed. L. P. Gerson. Cambridge: Cambridge University Press, 1996, pp. 292–301.

Netton, I. R. *Al-Fārābī and his School*. Surrey: Curzon, 1999.

O'Brien, E. *The Essential Plotinus*. New York: New American Library, 1964.

Rist, J. *Plotinus: The Road to Reality*. Cambridge: Cambridge University Press, 1967.

Runia, D. T. and M. Share. *Proclus: Commentary on Plato's Timaeus*. Vol II, B. 2: *Proclus on the Causes of the Cosmos and its Creation*. Trans. and notes by D. T. Runia and M. Share, Cambridge: Cambridge University Press, 2008.

Savulescu, J. and I. Persson. "Moral Enhancement, Freedom and the God Machine." *The Monist*. Vol. 95, No. 3, 2012, pp. 399–421.

Sitaram, R., A. Caria, N. Birbaumer. "Hemodynamic Brain-Computer Interfaces for Communication and Rehabilitation." *Neural Networks*, Vol. 22, No. 9, 2009, pp. 1320–1328.

Walker, M. "Enhancing Genetic Virtue. A Project for Twenty-first Century Humanity?" *Politics and the Life Sciences*, Vol. 28, No. 2, 2009, pp. 27–47.

Wallis, R. T. *Neoplatonism*, London: Duckworth, 1995.

Wisnovsky, R. *Avicenna's Metaphysics in Context*. New York: Cornell University Press, 2003.

Witt, R. E. "Iamblichus as a Forerunner of Julian." In: *De Jamblique a Proclus*, Vandoevres, Geneve: Fondation Hardt, 1974.

Jan Kiełbasa

The Perfection of Being and the Perfection of Action. Medieval Understanding of the Various Dimensions and Limits of Human Perfection

Abstract: The aim of this paper is to analyze the main issues connected with the medieval conception of man's perfection. It is argued that from the patristic period to the scholastics of the late Middle Ages, human perfection was considered, on the one hand, as one of the many relative and limited forms of perfection of creation, and, on the other hand, as a specific case with a special importance due to the iconic understanding of human nature as reflecting the nature of God – the one being infinitely and completely perfect. The research focuses on the conceptions of Saint Gregory of Nyssa and of Saint Thomas Aquinas and leads to the confirmation of a paradoxical – radically optimistic and, at the same time, realistic and sober – nature of the medieval doctrine of human perfection. From the normative point of view, as an ideal, ambitious project, or an ethical duty with eschatological consequences, man's perfection is the possibility of never-ending self-improvement and growth in goodness, directed towards an ultimate assimilation to the absolute divine perfection. However, as a matter of fact, human perfection must coexist with the limitations and deficiencies of human nature, and therefore it is always imperfect, i.e., partial and relative. According to the Church Fathers and masters of scholastic theology, there are many available forms of this specifically human partial perfection – in the structure and general functions of an individual human being, as well as in his concrete actions (for example, intellectual cognition, acts of the virtues, etc.); but the key dimension of this imperfect human perfection is love as the principle and purpose of human, and especially Christian, life. In conclusion, I seek to elucidate why the appeal to imitate Christ, to live an authentic Christian life based on charity, seems to be the typical answer given by medieval thinkers to the human aspiration of being perfect.

Keywords: Perfection, Perfecting factor, Completeness, Christian life, Charity

From its patristic origins to its mature, scholastic form, the medieval understanding of the concept of perfection was an important component of a multi-level construction, composed of the most general metaphysical considerations on the perfection of ontological structure and ontological orders, including theological (the absolute perfection of God), cosmological (the perfection of the created world, ranging from basic elements to celestial bodies and living organisms), and also anthropological considerations, particularly their ethical and ascetic aspects (human perfection in the order of existence and in the order of action – towards oneself and towards others, towards God). The subjects of considerations within this construction are, on the one hand, the unity of perfection centered in God, who is perfect in everyway (*universaliter perfectus, perfectissimus*)[1] and, on the other hand, the multiplicity and diversity of perfection ascribed to creatures, because the perfection of the universe required different degrees of perfection in things.[2] Human perfection itself is just one of the many specific forms of creatures' perfection, but has acquired a special importance due to the iconic understanding of human nature as reflecting the nature of God Himself. The present study is primarily dedicated to this particular embodiment of perfection.

In both Eastern and Western Christianity, the Church Fathers have a clear awareness of the difference between the deficiency-free, eternal, and absolute perfection of God, and the derivative and relative human perfection, which is postulated rather than already actualized. In their eyes, the latter is in fact only a task to be fulfilled, which, however, cannot be actually fulfilled given the human condition at the time. The one who found out that he had already attained perfection would, on the one hand, be subject to a dangerous delusion born of his pride and obscuring his actual imperfection, and on the other hand, he would lose motivation to continue self-improvement and fall into stagnation – a condition in which he would have nothing to strive for, nothing more to achieve, and nothing more to demand from himself. In the patristic teaching, however, human perfection should be the objective of a never-ending process of improvement, which cannot be exhausted in a state of some easy and soporific complacency, because the man who practices *prosoche*, i.e., is mindful, alert, and careful,[3] cannot be satisfied with any pretence of perfection or its substitute. The Church Fathers come to this conclusion all the easier because they perceive this process through the prism of the iconic nature of man, his being in God's image and likeness, meaning that man treats the universally perfect nature of God as an unattainable, though obliging God's guiding light. From this perspective, the human similarity to God (*homoiotes, similitudo*) should be understood dynamically, i.e., as a lifelong process of approaching this similarity, imitating the divine, perfect model, and assimilating oneself to it (*homoiosis, assimilatio*).

---

1 Cf. Thomas de Aquino, *Summa contra Gentiles* I, cap. 28.
2 Cf. Thomas de Aquino, *Summa theologiae* I, q. 89, a. 1, c: Perfectio universi exigebat, ut diversi gradus in rebus essent.
3 The postulate of *prosoche* – living an attentive and careful life to ensure that all aspects of one's own humanity are in the best condition – was formulated, among others, by early Christian apologists and the Church Fathers such as Clement of Alexandria, St. Basil the Great, and St. Gregory of Nyssa.

In the second half of the 4th century, St. Gregory of Nyssa, one of the Cappadocian Fathers, openly writes: "For he who has truly come to be in the image of God and who has in no way turned aside from the divine character bears in himself its distinguishing marks and shows in all things his conformity to the archetype."[4] In his writings, on the one hand, he stresses the incompleteness and inadequacy of each level of perfection attainable to man at any time in his life, thus showing the time limitlessness of the process of improvement and the actual inability of man to achieve absolute perfection; on the other hand, he postulates that man needs to make efforts to reach that unattainable ultimate perfection through continuous self-development, thus ascending to even higher levels of natural human potential and gradually gaining this partial human perfection. Neither this gradual improvement, nor the partial perfection should have limits, because "the Divine is by its very nature infinite."[5] Therefore, Gregory of Nyssa, who willingly compares the process of improvement to travelling, running, flying, climbing, or hiking up a hill,[6] states clearly that "the participant's desire itself necessarily has no stopping place but is stretched out with the limitless."[7] Therefore, while in the order of material things perfection is located within some well-defined limits, determined by some numerical measures, and treated as a beginning and an end of these things, in the ethical order it is otherwise, i.e., as a virtue representing human participation in goodness, understood in the original and proper manner – as the infinite goodness of God – it is not constrained by any limits.[8] Since perfection in the ethical sense – like its divine prototype – has no limits and no end, which for man would mean its complete and ultimate actualization, and the end of the process of improvement, it can be assumed, referring to Gregory of Nyssa, that man cannot, *de facto*, attain perfection.[9]

However, Gregory of Nyssa does not conclude that man, who is a finite being, should not strive for this unlimited and ultimately unattainable perfection. On the contrary, he invokes a call to Christians contained in one of the Gospels to encourage him all the more intensely to be perfect as his heavenly Father is (cf.

4　Cf. Gregorius Nyssenus, *De vita Moysi* II, 320 (PG 44, 429): Ho gar alethos kat'eikona theou gegonos, kai medamou paratrapeis tou theiou charakteros, eph'eautou ta gnorismata pherei kai symbainei dia panton tei homoiosei pros to archetypon.
5　Cf. Gregorius Nyssenus, *De vita Moysi*, I, 5 (PG 44, 300).
6　Cf. Gregorius Nyssenus, *De vita Moysi*, I, 5 (PG 44, 300); II, 106 (PG 44, 356–357); II, 152 (PG 44, 372); II, 225–227 (PG 44, 401); II, 242 (PG 44, 405).
7　Cf. Gregorius Nyssenus, *De vita Moysi*, I, 7 (PG 44, 301).
8　Cf. Gregorius Nyssenus, *De vita Moysi*, I, 5 (PG 44, 300); I, 7 (PG 44, 301).
9　Cf. Gregorius Nyssenus, *De vita Moysi*, I, 8 (PG 44, 301).

Matthew 5, 48: *Ginesthe teleioi hos kai ho Pater hymon ho ouranios teleios esti*). It seems that Gregory of Nyssa quite rationally explains this paradoxical call to strive for the unattainable goal to become like God, who can never be actually caught up with:

> Although on the whole my argument has shown that what is sought for is unattainable, one should not disregard the commandment of the Lord (...) For in the case of those things which are good by nature, even if men of understanding were not able to attain to everything, by attaining even a part they could yet gain a great deal. We should show great diligence not to fall away from the perfection which is attainable but to acquire as much as is possible: To that extent let us make progress within the realm of what we seek. The perfection of human nature consists perhaps in its very growth in goodness.[10]

A perfect life is thus possible on a human scale, when understood as a path to perfection (*hodos tei psychei pros teleiosin*) and continuous human growth toward that which is better (*pros to kreitton*).[11]

Gregory of Nyssa chooses the biblical figure of Moses, as a model of this kind of perfect life in the human sense, and the one who guides man toward perfection. Based on his own interpretation of Moses and his fate, the author assumes that this biblical hero can be considered a symbol and an example of such endless growth in goodness, because "at no time [Moses] stopped in his ascent, nor did he set a limit for himself in his upward course. Once having set foot on the ladder (...), he continually climbed to the step above and never ceased to rise higher."[12] Of course, Moses has a clearly defined target, i.e., he climbs towards God in hope of experiencing ("seeing") Him more closely and more directly, and to satisfy his desires aimed at God. And even if this experience is not fully achievable, and the desires can never be completely satisfied (which, according to Gregory, can be concluded from the story of Moses),[13] man will make such a significant progress on the path towards God – by purifying his own mind, effectively constraining his lower instincts and passions, as well as developing and expanding his virtues – that this process of improvement will not only allow him to live decently, but also to gradually imitate God. In other words, he will be considered the one who becomes God's friend by doing good and serving Him, yet not out of convenience, or out of fear of punishment, or expectation of a reward. According to Gregory of Nyssa, this is the purpose of living in accordance with the virtues

---

10 Cf. Gregorius Nyssenus, *De vita Moysi*, I, 9–10 (PG 44, 301).
11 Cf. Gregorius Nyssenus, *De vita Moysi*, II, 306 (PG 44, 425).
12 Cf. Gregorius Nyssenus, *De vita Moysi*, II, 227 (PG 44, 401).
13 Cf. Gregorius Nyssenus, *De vita Moysi*, II, 162–164 (PG 44, 376–377); II, 230 (PG 44, 401); II, 239 (PG 44, 404).

(*telos tou kat'areten biou*), and this is the perfection of human life (*he teleiotes tou biou*): to become pleasing to God and to become His friend (*to philon genesthai theoi*).[14] It can be argued that the model revealed in this early medieval understanding of perfection is specifically Christian, however, in some of its aspects it corresponds to the model developed in Platonic and Neo-Platonic traditions.

This understanding of human perfection indirectly indicates another dimension important for medieval thinkers, namely, its association with the project, or postulate, of a Christian life, or a life worthy of a Christian. This postulate is related to ideals, requirements, and duties derived directly or indirectly from biblical revelation and, in particular, from the teachings of Jesus Christ recorded in the Gospels and interpreted by successive generations of medieval philosophers and theologians. The very meaning of the term "Christian" binds everyone who shares his name with Christ as the founder of both the religion and the community to which one belongs, and seems to entail the duty to live "like Christ", according to his teaching and his deeds. Even at such a basic level, we can develop a justification for the religious and ethical postulate of imitating Christ (*mimesis tou Christou, imitatio Christi*), which was so popular throughout the Middle Ages – from the first apologists in the 2nd and the 3rd centuries to Thomas a Kempis who lived in the 15th century. On the same basis, we can also make inferences about the specifically Christian understanding of perfection in the Middle Ages: the perfection of man in that epoch is first and foremost the perfection of a Christian or the perfection of a Christian life. As early as in the fourth century, the above-mentioned Gregory of Nyssa claimed that the very name "Christian" – which is, in fact, the only really important name for a believer in Christ, regardless of their social status and functions[15] – shows a pattern of conduct, including basic ethical requirements: "We are sharers of the greatest and the most divine and the first of names, those honoured by the name of Christ being called Christians, it is necessary that there be seen in us also all of the connotations of this name, so that the title be not a misnomer in our case, but that our life be a testimony of it."[16] The key issue is thus an authentic Christian life assessed in terms of the consistency of one's thoughts, words, and deeds with the teaching and life of Christ. This is what confirms the right to name someone a Christian and testifies that Christian parameters of perfection are fulfilled: "The marks of the true Christian are all those we know in connection

---

14 Cf. Gregorius Nyssenus, *De vita Moysi*, II, 315 (PG 44, 428); II, 320 (PG 44, 428–429).
15 Cf. Gregorius Nyssenus, *De perfectione*, 2 (PG 46, 252).
16 Cf. Gregorius Nyssenus, *De perfectione*, 3 and 9 (PG 46, 252 and 256).

with Christ. (…) Thus, it is necessary for the Christian life to illustrate all the interpretative terms signifying Christ, some through imitation (*dia mimeseos*), others through worship (*dia proskyneseos*)."[17] The proposed Christian life, understood as an imitation of Christ, entails a constant endeavor to actualize those values that seem to adequately express the meanings implicitly contained in the name of the founder of Christianity; these include royal dignity (*basileia*), power, and wisdom, but also peace, humility, and patience. The greater the correspondence between these meanings and the Christian life, the more actual – existential rather than just declarative – the communion between them, and the more perfection there is in this life.[18] Therefore, Christian life can be recognized, for example, by the very fact that in his dealings, a Christian – who shares his name with Christ himself – reveals and follows within his measure the divine attributes of wisdom and power – the wisdom to choose what is better and the power to faithfully persist in this goodness and not to give in to evil.[19] Similarly, man reaches perfection when, following Christ, he establishes peace and becomes peace himself, as when he builds reconciliation and harmony between the various divided aspects of his human nature.[20]

In his other texts on what it means to be a Christian, and a perfect Christian in particular, Gregory of Nyssa also points to the close relationship between human perfection and imitating Christ and his divine attributes as the basic characteristic and the primary requirement of a Christian life.[21] In these texts, Gregory also boldly formulates the idea that Christianity is an imitation of the divine nature (*christianismos esti tes theias physeos mimesis*),[22] and that the name "Christian" contains a promise of an imitation of God.[23] It is worth noting on this occasion that the arguments of this saint and Doctor of the Church once again reveal a peculiar dialectic of the possible and the impossible, the achievable and the unachievable. On the one hand, treating the attempts to imitate God as an essential characteristic of Christianity and Christian life is not an usurpation or exaggeration (despite the obvious disparities between the two elements of this

---

17 Cf. Gregorius Nyssenus, *De perfectione*, 12 (PG 46, 256).
18 Cf. Gregorius Nyssenus, *De perfectione*, 85 (PG 46, 285).
19 Cf. Gregorius Nyssenus, *De perfectione*, 21–24 (PG 46, 260).
20 Cf. Gregorius Nyssenus, *De perfectione*, 25–26 (PG 46, 261).
21 Here I refer to *Deprofessione christiana* (also known as: *Quid nomen professione christianorum sibi velit*) and *De proposito secundum Deum* (in Migne's *Patrologia Graeca*: *De instituto Christiano*).
22 Cf. Gregorius Nyssenus, *De professione christiana*, 19 (PG 46, 244).
23 Cf. Gregorius Nyssenus, *De professione christiana*, 22 (PG 46, 245).

mimetic relationship), as it is grounded in the biblical vision of man created in the image and likeness of God; the basic intention of Christianity and the goal of perfect Christian life is to restore or reinstate this iconic relationship between God and man, obscured and distorted by the fall of man.[24] On the other hand, if the postulate of imitating God explicitly assumes that man should become similar to the divine original, this raises doubt whether man can assimilate himself into the original at all in his present condition, and therefore whether it is possible to effectively imitate God in the context of a Christian life.[25] Gregory proposes a twofold solution to this dilemma: a complete assimilation of man to God that makes him equal to the original and somehow equates human and divine natures, and an imitation of God understood as *synergeia* – entering into an interaction with God, taking good actions like those of God, and refraining from all evil deeds, words, and thoughts.[26] Gregory considers the former – the complete assimilation of man to God that somehow makes human nature equal to the divine nature – to be virtually impossible, but he also says that the evangelical call that man should imitate God and become like Him also in terms of perfection does not contain such a radical demand. The emphasis is rather on the second, weaker understanding of man's likeness to God, i.e., an imitation of the actions of God. This version of *imitatio Dei* is most attainable for man. Thus, man cannot attain complete perfection, which can only be actualized in God, but he can attain endless improvement and asymptotically approach the ideal of divine perfection, and even more so, he can and should imitate, within certain limits, the actions of God, although he cannot cross the border separating his nature as a creature from the nature of God – the Creator.

Gregory of Nyssa also approaches this duality of the human condition – the ability to imitate God and engage in never-ending improvement, and the imperfection of man that accompanies him at every stage of this process of improvement and does not allow him to catch up with the absolute perfection of God – from the other side, namely, the side of human mutability which at first glance seems to be man's burden and weakness, but after a careful look, it can also be a good opportunity and a chance for him. This is at least how this early Christian thinker from Cappadocia presents the issue in his ultimately optimistic interpretation:

---

24 Cf. Gregorius Nyssenus, *De professione christiana*, 20 (PG 46, 244)
25 Cf. Gregorius Nyssenus, *De professione christiana*, 25 (PG 46, 245)
26 Cf. Gregorius Nyssenus, *De professione christiana*, 26 (PG 46, 245).

But what if someone should say that the good is difficult to achieve, since only the Lord of creation is immutable, whereas human nature is unstable and subject to change, and ask how it is possible for the fixed and unchangeable to be achieved in the changeable nature? (...) For does man make a change only towards evil? Indeed, it would not be possible for him to be on the side of the good if he were by nature inclined only to a single one of the opposites. In fact, the fairest product of change (*to kalliston tes tropes ergon*) is the increase of goods (*en tois agathois auxesis*), the change to the better (*pros to kreitton alloiosis*) always changing what is nobly changed into something more divine. Therefore, I do not think that it is a fearful thing (I mean that our nature is changeable). The Logos shows that it would be a disadvantage for us not to be able to make a change for the better, as a kind of wing of flight to greater things. Therefore, let no one be grieved if he sees in his nature a penchant for change. Changing in everything for the better, let him exchange 'glory for glory,' becoming greater through daily increase, ever perfecting himself, and never arriving too quickly at the limit of perfection. For this is truly perfection: never to stop growing towards what is better and never placing any limit on perfection (*to medepote stenai pros to kreitton auxanomenon, mede tini perati periorisai ten teleioteta*).[27]

Therefore, according to Church Fathers, an elementary task and duty of one living a Christian life is to be conscious of the limits and imperfections of human nature and to know clearly the requirements and value of the virtue of humility and, at the same time, to strive infinitely toward a greater perfection (*aei teleioumenos*), to be always in moral progress (*kreitton ginomenos*) – intentionally and increasingly nearer to the perfection of God himself, as his image and imitator. This is the patristic legacy for the later generations of medieval philosophers.

The Late Middle Ages inherited from antiquity the basic understanding of perfection as a perfect state, a state of complete satisfaction and complete actualization. Medieval thinkers would often repeat after Aristotle that a thing is perfect if it has no deficiencies, i.e., if it lacks nothing.[28] In other words, the scholastic authors state that any entity is perfect if it has developed all of its natural possibilities and has attained its end so that nothing can be added to it or removed from it.[29] In this sense, perfection excludes any absence and any non-actualized potentiality, while confirming the completeness and the integrity of

---

27 Cf. Gregorius Nyssenus, *De perfectione*, pp. 86–89 (PG 46, 285).
28 Cf. Albertus Magnus, *Ethica*, lib. I, tract. 2, cap. 1: Perfectum est unumquodque, ut dicit Aristoteles, cum ei nihil deest. Similarly: Thomas de Aquino, *Summa theologiae* I, q. 5, a. 5, c: Perfectum autem dicitur cui nihil deest secundum modum suae perfectionis. See also: Bonaventura, *Itinerarium mentis in Deum*, cap. V, 6: Ideo perfectissimum: tali omnino nihil deficit, neque aliqua potest fieri additio. Primary source for all these medieval texts: Aristoteles, *Physica*, III, 6 (207a).
29 For twelfth-century scholastics see: Petrus Abaelardus, *Dialogus inter philosophum, iudaeum et christianum*: Perfectum quippe est, cui nihil addendum est. Similarly, for

being.[30] Complementarily, denial of perfection must mean some incompleteness of being, i.e., it is not completely what it should be. Any entity is imperfect if it lacks any of the things that belong to it according to its proper nature and should therefore be actualized in it.[31] Taking this characterization of the concept of perfection and its denial into account, it can be concluded that medieval authors understood perfection statically, i.e., only as some already achieved final state, the point of arrival, the closure of the entire "history" of being, the culmination and exhaustion of its innate development opportunities. From this point of view, imperfection might seem more promising and attractive when perceived as a condition in which an entity maintains its own dynamics, because as long as the teleological process leading to full actualization has not been completed, there are still gaps to be filled and opportunities to be used within this entity. However, the paradoxical conclusion that this dynamic imperfection has an alleged advantage over static perfection seems decidedly premature. Firstly, it ignores the fact that, for example, according to Thomas Aquinas, perfection has clearly an act-related nature[32] and thus it is not a permanent state of actualization, but a continuous actualization of individual entities and their innate potentialities. Secondly, as the final cause and an appetible object, perfection triggers the internal dynamics of the individual entities that seek and desire perfection.[33] Thirdly, with regard to the created world, perfection is sometimes described (e.g., by Thomas Aquinas) as a dynamic relationship between different entities or aspects of being, namely, between the perfecting factor *perfectivum*, a being that is subject to perfection or is in a position to be made perfect (*perfectibile*), and a being that has already been perfected (*perfectum*). In the case of man, this relationship occurs primarily in the mental and physical structure of his being: the soul is the actualizing factor that perfects the body, or even makes it *de facto* the body, i.e., something more

---

thirteenth-century scholastics: Albertus Magnus, *Ethica*, lib. I, tract. 1, cap. 1: Ultimum autem in unoquoque completum est et perfectum.
30 Cf. Thomas de Aquino, *Scriptum super Sententiis*, lib. I, d. 8, q. 3, a. 3 (expositio textus): Perfectio autem posse attendi tripliciter: vel secundum quod excluditur privatio vel non esse (…). Vel secundum quod excluditur potentialitas (…). Vel quantum ad integritatem ipsius esse.
31 Cf. Thomas de Aquino, *De Potentia*, q. 3, a. 1, ad 14: Imperfectio designat carentiam alicuius quod natum est haberi vel debet haberi.
32 Cf. Thomas de Aquino, *Summa theologiae*, q. 5, a. 1, c: Intantum est perfectum unumquodque, inquantum est actu. Complementarily: ibid., a. 3, c: Omnis actus perfectio quaedam est.
33 Cf. Thomas de Aquino, *Summa theologiae* I, q. 6, a. 1, c: Unumquodque autem appetit suam perfectionem.

than a random configuration of tangible material; the latter is in potency to be completely actualized and perfected by the soul, because it is not able to spontaneously expand and fulfill its potentialities by itself.[34] In turn, remaining outside its natural relationship with the body, the soul alone is neither *perfectivum*, i.e., it does not actualize and perfect the body, nor *perfectum*, i.e., it does not completely actualize the perfection of human nature. Although, according to Aquinas, it is capable of independent existence (*per se subsistit*) and action (*potest per se operari*), it is still only a part of human nature (*pars speciei humanae*), as man cannot be reduced to it both by definition and in his real existence.[35] Therefore, as a typically incomplete being, the soul can neither be perceived as mere perfection, nor does it have the ultimately perfect nature in its own, separate, extra-corporeal existence, as it is still more advantageous for it to remain within the body.[36] Therefore, the soul attains its perfection only when it takes the shape of the body and attains the perfection thereof.[37] When it enters into a relationship with the body as *perfectivum* – the actualizing and perfecting factor, for which the body is *perfectibile* – that which can be made perfect by the soul. This is how the first perfection of man as a complex entity is constituted – perfection in his very structure and the order of existence (*secundum esse*).

In addition to this example of the relational nature of perfection, which reveals itself already at the level of the constitution of a human being and of human nature itself, in his understanding of man, Thomas Aquinas assumes a specific relationality of perfection both in the order of cognition and in the order of desire: the object of cognition or desire turns out to be *perfectivum* both for the cognitive faculty and the appetitive faculty, which themselves appear to be *perfectibilia*. In other words, in this approach, an entity considered in the context

---

34 Cf. Thomas de Aquino, *Summa contra Gentiles* II, cap. 57: Corpus autem per animam fit actu ens de potentia existente (…) per animam fit vivens actu. Similarly: ibid., cap. 69: Corpus enim hominis non est idem actu praesente anima, et absente; sed anima facit ipsum actu esse.

35 Cf. Thomas de Aquino, *Summa theologiae* I, q. 75, a. 4, c: Homo non est anima tantum, sed est aliquid compositum ex anima et corpore.

36 Cf. Thomas de Aquino, *De unitate intellectus contra averroistas*, 112: Anima humana a corpore separata non habet ultimam perfectionem suae naturae, cum sit pars naturae humanae. Similarly: *Summa theologiae* I, q. 89, a. 1, c: Propter melius animae est ut corpori uniatur.

37 Cf. Thomas de Aquino, *De Potentia*, q. 3, a. 10, ad 11: Anima licet sit secundum se simplicior omni corpore, tamen est forma et perfectio corporis ex elementis compositi. Similarly: ibid., ad 12: Prima perfectio animae attenditur secundum esse suum naturale, quae quidem perfectio consistit in unione eius ad corpus.

of its cognoscibility, i.e., as truth, would be the factor that perfects the human mind, whereas an entity considered in the aspect of its desirability, i.e., as goodness, would be the factor that perfects the human will,[38] though in each case this process of perfecting would take place in a different way. Reason would be complemented and perfected by the true, i.e., adequate concept of the thing under study, recognized only in terms of its cognitive and abstracted form (*secundum rationem speciei*), while the will would be complemented and perfected by goodness identified with the desired thing, and the desire would include the thing in its real existence rather than in its form (*non solum secundum rationem speciei, sed etiam secundum esse quod habet in re*).[39] The perfection of man, or rather the perfection of his faculties – reason and will – and their activities, i.e., mental cognition and desire, would be somehow transferred or borrowed from what references these faculties and activities. It would therefore be inferred from the truth about things under study and the goodness ascribed to the things that are wanted, desired, and cherished. As *perfectiva*, truth and goodness alone would shape and perfect what is shapeable and perfectible – the *perfectibilia*, i.e., reason and will. It is only in this relationship and thanks to it that man would gain his second perfection – the perfection of an entity, which can be achieved through cognition and the desire to participate in truth and goodness, in their transcendental, broadest sense. The second dimension of human perfection is no longer based on his hylemorphic structure or the act of existence defined by form, but on the actions[40] man takes and the intentionality of his cognitive and volitional acts.

Finally, the human perfection that is gained in the learning of truth and in the desire for goodness (initially) or the enjoyment of this goodness (target) is referenced – like the perfection of other creatures – to its first and efficient cause, its master pattern, i.e., to God, the unique model of all perfection, in whom all perfections of his creatures pre-exist in a fuller and most appropriate way: *Omnes enim perfectiones in creaturis inventae in Deo praeexistunt secundum*

---

38 Cf. Thomas de Aquino, *De Veritate*, q. 21, a. 3, c: In utroque enim ostenditur habitudo perfectibilis ad perfectionem, quae est bonum vel verum; scilicet in appetitu boni et cognitione veri.
39 Cf. Thomas de Aquino, *De Veritate*, q. 21, a. 3, c: Tam verum quam bonum (…) habent rationem perfectivorum sive perfectionum. (…) verum sit perfectivum alicuius secundum rationem speciei, bonum autem non solum secundum rationem speciei, sed etiam secundum esse quod habet in re.
40 Cf. Thomas de Aquino, *De Malo*, q. 1, a. 4, c: Perfectio autem est duplex: scilicet prima quae est forma vel habitus, et secunda, quae est operatio.

altiorem modum (*Summa theologiae* I, q. 14, a. 11, c). Moreover, if some perfection co-exists with some imperfection (e.g., limitation, volatility, contingency) within a creature, it must be assumed *a priori* that the same perfection in God as its master pattern is free from any imperfections.[41] All perfections contained in God are at the same time – according to Thomas Aquinas – the one and only perfection of God, or the whole thinkable perfection centered in His essence and identical to this essence.[42] It is worth noting, however, that even in this seemingly completely unambiguous approach, the relationship between God and perfection is not free from controversy. On the one hand, God is said to be the most perfect being (*ens perfectissimum*), the perfection of all perfections (*perfectio omnium perfectionum*). On the other hand, Aquinas raises concerns as to whether the terms "perfection" or "perfect" can be used at all to describe God in the literal and strict sense, i.e., the one that assumes the developing and completing of that which was previously undeveloped or unfinished, or the actualizing of that which was previously only potential (if, according to Latin etymology, *esse perfectum* implicates always *esse factum*).[43] From this perspective, these terms could be used in a negative rather than a positive sense, i.e., only to express the absence of any deficiency and any potentiality in God.[44] The term "perfection" could also be used in reference to God provided that the meaning of perfection is extended to include not only that which has been made by someone or something and has been gradually reaching the state of actualization to become completely actualized overtime, but also that which has not been made by anything or anyone and has always been fully actualized without ever becoming or

---

41 Cf. Thomas de Aquino, *Scriptum super Sententiis*, I, d. 4, q. 1, a. 1, c: Cum omnis perfectio sit in Deo et nulla imperfectio, quidquid perfectionis invenitur in creatura, de Deo dici potest quantum ad hoc quod est perfectionis in ipsa, omni remota imperfectione.
42 Cf. Thomas de Aquino, *Summa theologiae* I, q. 6, a. 3, c: Solus Deus habet omnimodam perfectionem secundum suam essentiam.
43 Cf. Thomas de Aquino, *Summa contra Gentiles*, I, cap. 28: Perfectio Deo convenienter attribui non potest, si nominis significatio quantum ad sui originem attendatur; quod enim factum non est, nec perfectum posse dici videtur (...). Tunc recte perfectum esse dicitur quasi totaliter fatum, quando potentia totaliter est ad actum reduta, ut nihil de non esse retineat, sed habeat esse completum.
44 Cf. Thomas de Aquino, *De Veritate*, q. 2, a. 3, ad 13: Perfectionis nomen, si stricte accipiatur, in Deo poni non potest quia nihil est perfectum nisi quod est factum. Sed in Deo nomen perfectionis accipitur magis negative quam positive, ut dicatur perfectus, quia nihil deest ei ex omnibus; non quod sit in eo aliquid quod sit in potentia ad perfectionem, quod aliquo perficiatur quod sit actus eius.

attaining such a state.⁴⁵ Having made this stipulation, it can still be argued, by referring to Thomas Aquinas, that the absolute divine perfection is the efficient, formal, and final cause of the first and the second perfection of man, as well as his alleged ultimate perfection (*ultima perfectio*). This would involve his unification with purely mental entities, a kind of communication with them (*communicatio*). However, this ultimate perfection of man is clearly eschatological in nature, i.e., it is projected on another, non-empirical reality – some future life, available only to those redeemed; moreover, it is explicitly treated as a supernatural gift.⁴⁶

In his current condition, man gains perfection in the absolute sense (*simpliciter*) if he accepts and cultivates the virtue of love: in this strong sense, to be perfect means to be perfect in charity and to be perfectly alive with charity.⁴⁷ Naturally, man can also be perfect in a certain respect (*secundum quid*) when he develops his spiritual life.⁴⁸ Therefore, some may be perfect in knowledge (*secundum intellectus cognitionem aliqui dicuntur esse perfecti*), while others may be perfect in patience or any other virtue (*perfectu saliquis dici potest et secundum patientiam (...) et secundum quascumque alias virtutes*). There are therefore many available forms of perfection in the weak sense, which support various – cognitive and ethical – aspects of man's spiritual life. At first glance, surprisingly, Aquinas seems to approve of an even further-reaching relativization of the concept of perfection by accepting the term "perfect" even in relation to clearly negative disposition and behavior; for example, he does not dispute the fact that formulas such as "a perfect thief" are used in everyday language.⁴⁹ It seems, however, that this is just *modus loquendi*: after all, this kind of alleged perfection in evil in no way serves the spiritual life of man, as it does not raise his life to a higher level, either generally or in some particular aspect. Therefore, when Thomas Aquinas seriously considers the issue of different types of human

---

45 Cf. Thomas de Aquino, *Summa contra Gentiles* I, cap. 28: Per quandam igitur nominis extensionem perfectum dicitur non solum quod fiendo pervenit ad actum completum, sed id etiam quod est in actu completo absque omni factione.
46 Cf. Thomas de Aquino, *De Potentia*, q. 3, a. 10, ad 12: Ultima autem perfectio eius est in hoc quod communicat cum substantiis aliis intellectualibus; et illa perfectio dabitur ei in coelo.
47 Cf. Thomas de Aquino, *De perfectione spiritualis vitae*, cap. 1: Simpliciter igitur in spirituali vita perfectus est qui est in caritate perfectus.
48 Cf. Thomas de Aquino, *De perfectione spiritualis vitae*: Secundum quid autem perfectus dici potest, secundum quodcumque quod spirituali vitae adiungitur.
49 Cf. Thomas de Aquino, *De perfectione spiritualis vitae*: Etiam in malis aliquid dicitur esse perfectus, sicut dicitur aliquis perfectus fur aut latro.

perfection and distinguishes between *perfectio simpliciter* and *perfectio secundum quid*, he always does so in the context of the spiritual life of man and their impact on this life. It turns out that if the spiritual life is defined as Christian life, i.e., inspired by Christianity and worthy of a Christian, all its particular perfections are ultimately assigned to love, which seems to be the most important Christian value. Aquinas repeatedly emphasizes the absolute primacy of love as constitutive for the perfection of spiritual life, or simply Christian life.[50] According to him, the perfection of charity is therefore perfection in the absolute and strong sense, because this is primarily where Christian life is expressed and made real. He believes it to be the highest form of existence available to man in his present condition. Although other particular perfections, including those postulated in the Decalogue and the so-called evangelical counsels, participate in the development of man's spiritual life, they are still based on the perfection of love as their principle and purpose,[51] and turn out to be its instruments, means of expression, specific applications, and ways to reach or maintain it.[52]

Since the perfection of Christian life – which for Aquinas is the key dimension of human perfection – is based on charity, this perfection can be triple (*triplex perfectio*) due to differences in the strength, range, target, and the subjective foundation of love. The first of its kind – absolute perfection – involves the complete and unconditional internal actualization of goodness, which is inherently worthy of all love and is actually just as lovable (*tantum diligitur quantum diligibilis est*); thus understood, goodness as the object of love fills up the whole being, and no form of this goodness is ignored in this being. Such

---

50 Cf. Thomas de Aquino, *Summa theologiae* II II –ae, q. 184, a. 1, ad 2: Vita autem christiana specialiter in caritate consistit, per quam anima Deo coniugitur (…). Et ideo secundum caritatem simpliciter attenditur perfectio christianae vitae, sed secundum alias virtutes – secundum quid.

51 Cf. Thomas de Aquino, *Summa theologiae* II II –ae, q. 184: Perfectio caritatis est principium respectu perfectionis quae attenditur secundum alias virtutes. Similarly: Summa theologiae II II –ae, q. 184, ad 3: Patientia dicitur habere opus perfectum in ordine ad caritatem, inquantum scilicet ex abundantia caritatis provenit quod aliquis patienter toleret adversa.

52 Cf. Thomas de Aquino, *Summa theologiae* II II –ae, q. 184, a. 3, c: Secundario et instrumentaliter perfectio consistit in consiliis. Quae omnia, sicut et praecepta, ordinantur ad caritatem, sed aliter et aliter. Nam praecepta alia ordinantur ad removendum ea quae sunt caritati contraria, cum quibus scilicet caritas esse non potest, consilia autem ordinantur ad removendum impedimenta actus caritatis, quae tamen caritati non contrariantur. Similarly: ibid., ad 1: Consilia sunt quaedam instrumenta perveniendi ad perfectionem.

absolute perfection is not available either to man or to any other creature, as it belongs exclusively to God.[53] Man, however, may be entitled to the other two types of perfection, one of which is actualized under the current conditions of human life (*perfectio in hac vita* or – due to the transitory nature of the earthly human condition – *perfectio in via*), while the second type of perfection can be expected in some future life, or – according to Christian eschatology – "in the heavenly homeland" (*perfectio patriae* or *in patria*). In both cases, perfection means human love for God, who is goodness greater than man. While in one case – the possible perfection of some future life – such love would be always and exclusively actualized with all human love, in the second case – the perfection of our current condition – man's love is not always truly and completely focused on God, yet it is enough that it simply ignores or excludes everything that is contrary to such a focus.[54] Thus, this latter kind of perfection purifies human desires primarily of all that would significantly alter the focus of our love, turning it radically away from God and towards sin. Such purification accomplished in this life to prevent at least the most severe changes in the direction of love (or what is called mortal sin – *peccatum mortale* – in religious language) would be – according to Thomas Aquinas – a necessary condition for salvation (*est de necessitate salutis*). At the same time, it would at least partially remove minor obstacles in the human understanding of complete love for God (even if these obstacles, i.e., venial sins – *peccata venialia* – do not entirely exclude such a possibility). It is worth noting that the latter case would corroborate the thesis that perfection available to man in his current condition co-exists with some forms of imperfection associated with the limitations of our current life, including the limitations of power and the non-exclusivity of man's love for God.[55] From this point of view, medieval anthropology and, especially, the doctrine of human perfection, seem quite paradoxical – on the one hand, they are ambitious and radically optimistic

---

53 Cf. Thomas de Aquino, *Summa theologiae* II II -ae, q. 184, a. 2, c: Et talis perfectio non est possibilis alicui creaturae, sed competit soli Deo, in quo bonum inegraliter et essentialiter invenitur.
54 Cf. Thomas de Aquino, *Summa theologiae* II II -ae, q. 184, a. 2, c: Tertia autem perfectio est quae neque attenditur (…) secundum totalitatem ex parte diligentis quantum ad hoc quod semper actu fertur in Deum, sed quantum ad hoc quod excludantur ea quae repugnant motui dilectionis in Deum (…). Et talis perfectio potest in hac vita haberi.
55 Cf. Thomas de Aquino, *Summa theologiae* II II -ae, q. 184, ad. 2: illi qui sunt in hac vita perfecti, in multis dicuntur offendere secundum peccata venialia, quae consequuntur ex infirmitate praesentis vitae. Et quantum ad hoc etiam habent aliquid imperfectum per comparationem ad perfectionem patriae.

as an eschatological project, i.e., as an ideal, according to which man could develop his capacities infinitely and assimilate himself to his divine Creator and *ens perfectissimum*; but, on the other hand, they are rather sober and realistic as a description and evaluation of the actual state of man, clearly limiting his aspirations to be perfect. From the patristic period to the scholastics of the Late Middle Ages, the typical answer to these aspirations seems to be an appeal to imitate Christ and, at the same time, to accept – as specifically human – the perfection in proper imperfection.

References

Primary sources

Albertus Magnus. "Ethica." In: *B. Alberti Magni Opera omnia*, Vol. 7, ed. A. Borgnet, Parisiis: apud Ludovicum Vives bibliopolam editorem, 1891.

Bonaventura. "Itinerarium mentis in Deum." In: *Tria opuscula Seraphici Doctoris S. Bonaventurae* (editio secunda). Quaracchi: Ad Claras Aquas, 1896.

Gregorius Nyssenus. "De vita Moysi." In: *Patrologia Graeca*, Vol. 44, Parisiis, apud J.-P. Migne editorem, 1863.

Gregorius Nyssenus. "De perfectione." In: *Patrologia Graeca*, Vol. 46, Parisiis, apud J.-P. Migne editorem, 1863.

Gregorius Nyssenus. "De professione christiana." In: *Patrologia Graeca*, Vol. 46, Parisiis, apud J.-P. Migne editorem, 1863

Petrus Abaelardus. "Dialogus inter philosophum, iudaeum et christianum." In: *Patrologia Latina*, Vol. 178, Parisiis: apud Garnier Fratres, editores et J.-P. Migne successores, 1885

Thomas de Aquino. Pars Prima "Summae theologiae." In: Thomae Aquinatis *Opera omnia iussu Leonis XIII P. M. edita* (Editio Leonina), Vols. 4–5, Romae: Ex Typographia Polyglotta S. C. de Propaganda Fide, 1888–1889.

Thomas de Aquino. Secunda Secundae "Summae theologiae.". q. 123–189. In: Thomae Aquinatis *Opera omnia iussu Leonis XIII P. M. edita* (Editio Leonina), Vol. 10, Romae: Ex Typographia Polyglotta S. C. de Propaganda Fide, 1899.

Thomas de Aquino. "Summa contra Gentiles." Lib. 1–2. In: Thomae Aquinatis *Opera omnia iussu Leonis XIII P. M. edita* (Editio Leonina), Vol. 13, Romae: Typis Riccardi Garroni, 1918.

Thomas de Aquino. *Scriptum super libros Sententiarum magistri Petri Lombardi*, Vol. 1, ed. P. Mandonnet. Parisiis: P. Lethielleux, 1929.

Thomas de Aquino. "De Veritate." In: Thomae Aquinatis *Quaestiones disputatae*, Vol. 1, ed. R. Spiazzi. Taurini: Marietti, 1953.

Thomas de Aquino. "De Potentia." In: Thomae Aquinatis *Quaestiones disputatae*, Vol. 2, ed. P. M. Pession. Taurini, Romae: Marietti, 1965.

Thomas de Aquino. "De perfectione spiritualis vitae." In: Thomae Aquinatis *Opera omnia iussu Leonis XIII P. M. edita* (Editio Leonina), Vol. 41 B, Romae: Ex Typographia Polyglotta S. C. de Propaganda Fide, 1969.

Thomas de Aquino. "De unitate intellectus contra averroistas." In: Thomae Aquinatis *Opera omnia iussu Leonis XIII P.M. edita* (Editio Leonina), Vol. 43, Romae: Editori di San Tommaso, 1976.

Thomas de Aquino. "Quaestiones disputatae de malo." In: Thomae Aquinatis *Opera omnia iussu Leonis XIII P.M. edita* (Editio Leonina), Vol. 23, Roma: Comissio Leonina, Paris: J. Vrin, 1982.

Studies

Blanchette, Oliva. *The Perfection of the Universe According to Aquinas. A Teleological Cosmology*. Pennsylvania: The Pennsylvania State University Press, 1992.

Colish, Marcia L.*Studies in Scholasticism*. Aldershot: Ashgate, 2006.

Hamman, Adalbert G. *L'homme image de Dieu. Essai d'une anthropologie chrétienne dans l'Église des cinq premiers siècles*. Paris: Desclée, 1987.

Heine, Roland E. *Perfection in the Virtuous Life. A Study in the Relationship between Edification and Polemical Theology in Gregory of Nyssa "DevitaMoysi"*. Cambridge: Philadelphia Patristic Foundation, 1975.

Jaeger, Werner, *Two Rediscovered Works of Ancient Christian Literature: Gregory of Nyssa and Macarius*. Leiden: Brill, 1954.

Marty, François. *La perfection de l'homme selon saint Thomas d'Aquin: ses fondements ontologiques et leur vérification dans l'ordre actuel* (Analecta Gregoriana, Vol. 123) Roma: Presses de l'Université Grégorienne, 1962.

McInerny, Ralph. *Boethius and Aquinas*. Washington, D. C.: The Catholic University of America Press, 1990 (second edition: 2012).

McInerny, Ralph. *Ethica Thomistica: The Moral Philosophy of Thomas Aquinas* (revised edition). Washington, D. C.: The Catholic University Press, 1997.

Zachhuber, Johannes. *Human Nature in Gregory of Nyssa. Philosophical Background and Theological Significance* (Supplements to Vigiliae Christianae. Texts and Studies of Early Christian Life and Language, Vol. 46) Leiden, Boston, Köln: Brill, 2000.

Wojciech Załuski

The Idea of Human Perfection in Modern Philosophy

Abstract: The present paper provides a synthetic description of the history of the idea of human perfection in modern philosophy. The description is aimed at capturing the main currents in this history, which have been divided into the "anti-perfectionist" and the "pro-perfectionist". The former currents include: (1) the re-appearance of the pessimistic picture of human nature underlying skepticism with regard to the feasibility of the idea of human moral perfection; (2) the fall of the classical picture of the world leading to the rejection of the idea of human teleological/metaphysical perfection and to skepticism with regard to the possibility of the justification of the idea of human moral perfection; and (3) the discovery of the incompatibility and incommensurability of moral values leading to doubts regarding the coherence of the very idea of human moral perfection. The latter currents include: (1) the new ("Romantic") way of understanding human moral perfection based on the idea of radical freedom and (2) the appearance of the belief in the perfectibility of human beings, i.e., their unlimited capacity to perfect their moral, intellectual, and physical abilities. The article also outlines the complex relations between these currents.

Keywords: Human perfection, Human nature, Modernity, Perfectibility, Value pluralism

1 Introduction

The history of the idea of human perfection did not proceed in modern philosophy in one distinct direction; rather, one can discern in it several different and often mutually contradictory currents. These currents can be broadly and schematically divided into two kinds, which I shall call "anti-perfectionist", i.e., undermining in some way the idea of human perfection, and the "pro-perfectionist", i.e., endorsing this idea and transforming its sense (as compared with its earlier – ancient and medieval – senses). Before presenting these currents, I will clarify the two basic concepts of human perfection to be invoked in this article, viz. teleological/metaphysical perfection and moral perfection. The former refers to the kind of perfection which consists in a human being's reaching his/her "natural end" (*telos*), i.e., attaining the state in which his/her nature is fully developed, whereas the notion of human moral perfection refers to the state in which a human being fully realizes his/her moral duties. Now, the anti-perfectionist currents include: (1) the re-appearance of the pessimistic picture of human nature underlying skepticism regarding the feasibility of the idea of human moral perfection; (2) the fall of the classical picture of the world leading to the rejection of the idea of human teleological/metaphysical perfection and to the rise of skepticism towards the possibility of the justification of the idea of human moral perfection; (3) the discovery of the incompatibility and incommensurability of moral values giving rise to doubts regarding the coherence of the very idea of human moral perfection. The pro-perfectionist currents, in turn, include: (1) the new ("Romantic") way of understanding human moral perfection based on the idea of radical freedom; (2) the appearance of the belief in the perfectibility of human beings, i.e., their unlimited capacity to perfect their moral, intellectual, and physical abilities. In the following sections I shall successively discuss these currents, revealing their various, often complex, mutual relations.

2 Anti-perfectionist currents

2.1 The pessimistic view of human nature

The pessimistic view of human nature underlies skepticism regarding the possibility of man achieving moral perfection. It was endorsed by many modern thinkers: Martin Luther and John Calvin, for example, believed (following the Augustinian, anti-Pelagian, tradition) that human nature is deeply flawed – directed towards evil rather than good. Accordingly, in his earthly life man cannot achieve moral perfection; as Calvin wrote: "Even saints cannot perform one work, which, if judged on its merits, is not deserving of condemnation" (Calvin 2016, p. 340). Human beings can therefore be justified before God not by their works but *sola fide* (which itself is not their 'work' but a gift of divine grace). The pessimistic view was also assumed by Michel de Montaigne, who emphasized human intellectual frailty, and Blaise Pascal, who additionally stressed human moral corruption, manifesting itself in various forms of *amour propre*, viz. *libido sentiendi, libido sciendi, and libido dominandi*.[1] Although Pascal distanced himself from Montaigne, saying in *Entretien avec M. de Saci sur Épictète et Montaigne* that Montaigne saw only the misery of man, while in fact there is also greatness in man, stemming from his status of being God's creature, Pascal's view of human

---

1 Cf. the following quotation: "*Tout ce qui est au monde est concupiscence de la chair ou concupiscence des yeux ou orgueil de la vie. Libido sentiendi, libido sciendi, libido dominandi. Malheureuse la terre de malédiction que ces trois fleuves de feu embrasent plutôt qu'ils n'arrosen.*" B. Pascal, *Pensées, Les Provinciales* (Paris: Bokking International, 1995), Thought 458, p. 164.

nature was in fact equally or even more pessimistic than Montaigne's, precisely because of its insistence on man's moral corruption. On the ground of such views of human nature as those assumed by Reformation thinkers, Pascal, and many other representatives of modern philosophy (e.g., Hobbes or La Rochefoucauld) the very idea of human moral perfection, albeit not losing its coherence, proves to be an unattainable ideal, which human beings cannot even approximate to. It should be mentioned, however, that Montaigne would probably not be inclined to consider his view of man as pessimistic; he recognized human imperfections, but at the same time he seemed to have gladly accepted them, arguing that we should live as human beings, with all our imperfections, rather than strive to lead the life of morally perfect creatures (angels). In Montaigne's opinion, by attempting to achieve moral perfection we are more likely to fall to the level of beasts rather than achieve the status of angels. In this point, his view of human nature is different from the other views presented in this section: Calvin, Luther, or Pascal treated human moral perfection as a desirable but unattainable ideal, whereas Montaigne seems to have treated it as both undesirable and unattainable. The pessimistic picture of human nature took perhaps its most radical form in the thought of Joseph de Maistre, who more strongly than anyone else asserted the existence of destructive proclivities in human nature. According to de Maistre, man does not desire only happiness, peace, and harmony; equally strong, or even stronger are his desire of war, of sacrificing himself, of submitting himself to authority, of suffering. In his view humans are also irrational: they do not know what they want, they want what they do not want, and they do not want what they want.[2] As we shall see in point 3.2, this bleak view of human nature was not unanimously accepted by modern philosophers. In modern philosophy there were equally strong, optimistic tendencies in seeing man, leading to the endorsement (as feasible) of the idea of human moral perfection.

It bears stressing that the intermediary picture of human being, neither optimistic, nor pessimistic, was defended by Immanuel Kant. His conception of radical evil, presented in his *Religion within the Limits of Reason Alone*, sets him apart from the dominant in the 18th century, optimistic trend of thinking about man. Radical evil, as understood by Kant, consists in that human beings, even though they have the knowledge of moral law, follow the principle according to which they may deviate at times from this law. He asserted that radical evil manifests itself in three features exhibited by human beings, viz. *fragilitas*, i.e., their lack of

---

2 Cf., e.g. J. de Maistre, *Contre Rousseau. De l'état de nature* (Paris: Éditions mille et une nuits, 2008).

persistence in complying with the moral maxims they accepted, *impuritas*, which describes the fact that human motivation to do moral actions is often impure, combining morally proper and morally improper motives, and *vitiositas*, i.e., the human propensity to accept morally improper maxims.[3] Accordingly, Kant could say, in his *Idea for a Universal History with a Cosmopolitan Purpose*, that "nothing straight can be constructed from such warped wood as that which man is made of" (Kant 1999, p. 46). It must, however, be emphasized: Kant was not an anthropological pessimist either. In his view, human evil is radical *not* in the sense of being enormous or absolute (diabolic): Kant denied that man may ever behave like a devil, i.e., want evil for the sake of evil. Human evil is radical in the two, much more moderate senses, both of them deriving from the etymology of "radical", which is the Latin *radix* (root): Evil cannot be *rooted out* from human nature, and it pollutes *the root* of our moral maxims (i.e., we rarely comply with them for their own sake). Thus, even though the conception of radical evil sets Kant's anthropology apart from the optimistic 18th century philosophical anthropology, the former cannot be said to stand in strong opposition to the latter. Consequently, Kant could agree with the optimistic philosophers of the Enlightenment[4] that human history is a history of progress, i.e., that it evolves in the direction of the ever-greater ("perpetual") peace, freedom, and welfare (realized, in Kant's cosmopolitan vision, in the Federation of States). The difference between Kant and these philosophers lay in the fact that while Kant treated the idea of human progress as a "regulative idea", i.e., an idea which ought to govern our conduct, and thereby ought to be treated as realizable (since otherwise our moral efforts to attain it would be weaker), even though there is no certitude that it *will be* realized, the optimistic philosophers of the Enlightenment treated this idea as well grounded in empirical facts. It must be added, though, that in Kant's philosophy human progress is not a *purely* regulative idea; Kant also provided some empirical arguments to support it. He famously used the "invisible-hand-like" argument saying, in the style of Bernard Mandeville's claim, that "private vices beget public virtues", and that in order to bring about the development of human innate capacities, nature employs "unsocial sociability of men", i.e., "their tendency to come together in society, coupled, however, with a continual resistance which constantly threatens to break up this society" (Kant 1999, p. 44). Accordingly, "all the culture and art which adorn mankind and the finest social

---

3 Cf. I. Kant, *Political Writitngs*, trans. H. B. Nisbet (Cambridge: Cambridge University Press, 2008), pp. 15–48.
4 Their views are discussed at greater length in point 3.1.

order man creates are fruits of his unsociability" (Kant 1999, p. 46). Thus, in Kant's view, if it had not been for the various forms of unsociability, e.g., rivalry, competition, emulation, envious jealousy, or vanity, "all human talents would remain hidden forever in a dormant state, and men, as good-natured as the sheep they tended, would scarcely render their existence more valuable than that of their animals" (Kant 1999, p. 45).

It could be asked whether the assumption of a given – pessimistic, optimistic, or intermediary – view of human nature affects the *content* of the idea of human moral perfection, i.e., whether it leads to a more restrictive or more lax conception of moral duties. The proper answer to this question seems to be that there is no simple logical connection between a given conception of human nature and a given content of moral duties imposed on man. On the one hand, one could argue that the pessimistic view of human nature justifies a more restrictive conception of morality as being necessary to curb the immoral proclivities in human nature. On the other hand, however, one could argue the opposite, that since men are immoral, they are unlikely to observe strict moral rules imposed upon them and can only be expected to observe the requirements of a less restrictive nature. It seems that the first line of reasoning was more frequent. Luther and Calvin, for example, questioned the medieval distinction between strict duties (addressed to all) and counsels of perfection (more demanding duties addressed to those who desire to attain moral perfection), arguing that all men have the duty to be perfect (even though, as was already mentioned in this section, the Reformation thinkers asserted at the same time that the since the nature of human beings is deeply flawed, they are incapable of complying with these duties, and can only count on God's forgiveness to attain salvation).

2.2 The fall of the classical picture of the world

In ancient and medieval philosophy the central idea of human perfection was not moral but teleological/metaphysical, i.e., that human beings attain perfection, above all, by likening themselves to God (in this life, or more probably, in the afterlife), rather than leading a morally decent life. This likening could be attained, as was believed, for example by pursuing a purely contemplative life, enjoying the beatific vision of God, and reaching a kind of union with God. This view was strictly connected with the classical vision of the world based on two beliefs: the existence of transcendence, and the existence of "natural ends" ("final causes") ascribed to each species (including humans) as something that every representative of a given species must attain in order to fully realize its nature. This picture was undermined by the development of modern science, which dispensed both with the idea of "final causes" and with references to transcendence in its explanatory schemes. As a result, the very idea of "natural ends", and thereby of the human *telos*, lost its plausibility for those who treated science seriously. They could no longer regard the idea of teleological/metaphysical perfection as meaningful; instead, like Francis Bacon, they started to see in it a manifestation of man's anthropomorphic propensity (counted by Bacon among *idola tribus*) to use beyond its proper domain (i.e., the domain of human action) the category of ends. Accordingly, as was aptly noticed by John Passmore,

> beginning with the Renaissance, but with increasing confidence in the seventeenth century, men began to maintain that in their relationships to their fellow-men, rather than in their relationships to God, lay their hope of perfection. 'Perfection' was defined in moral rather than in metaphysical terms, and came gradually to be further particularized as 'doing the maximum of good' (Passmore 2000, p. 260)

Thus, in this "modern" view, a human being realizes his/her nature, not by trying to escape from the transient – earthly – reality to some "higher", transcendent realm, but by behaving morally in a social context.

At this point, three further remarks concerning the meaning of the fall of the classical picture of the world for the idea of human perfection may be presented.*Firstly*, the crisis of the idea of human teleological/metaphysical perfection was caused not only (though assuredly, primarily) by the fall of the classical picture of the world, but also by the emphasis placed by many modern thinkers (whose views were discussed in the previous section) on the moral corruption of human beings. Given man's ineradicable moral defects, it would be presumptuous to maintain, so the argument goes, that man can have some special spiritual experience of a union, or vision of God. The (modest) call of a man is not to aspire to such lofty, religious experiences, but to become engaged, to the best of man's (limited) moral capacities, in secular life, by realizing, with due diligence and in the spirit of faith, man's moral duties.

Secondly, the fall of the classical picture of the world had repercussions for the understanding of the idea of human moral perfection. The classical picture implied that moral values or norms are somehow built into the fabric of the universe, out of which they can be read or deduced. In other words, nature understood in a teleological manner provided patterns of moral behavior, and as such should be "imitated". However, since nature as understood on the grounds of modern science could no longer be conceived as a source of moral values, the division of the realm of facts and values took place. This division was not immediately realized by modern thinkers. Even philosophers of the French Enlightenment who were fascinated by the scientific achievements were inclined to look at nature in a moral manner, i.e., as providing patterns of morally proper behavior. The same, a *fortiori*, applies to such (sentimentalist) thinkers as Ashley Anthony Shaftesbury and Jean Jacques Rousseau who exhibited a far-reaching skepticism towards the completeness of the picture of the world provided by modern science. The division between both realms was perhaps most clearly realized by David Hume, who was arguably the first thinker to assert that "ought" cannot be derived from "is", and that, consequently, morality is not grounded in our reason purportedly discerning moral values in nature, but in our sentiments, which through the mechanism of projection morally "colorize" facts (being themselves morally neutral). As we can see, the fall of the classical picture of the world undermined not only the idea of human teleological/metaphysical perfection, but also made problematic the idea of human moral perfection by raising the question of its justifiability. The strongest blow to the idea of human moral perfection, however, was dealt by the thinkers who discovered something disquieting in the very realm of moral values; I shall discuss this discovery in the next section.

Thirdly, it may be worth, by way of digression, to point at an interesting connection between the fall of the classical picture of the world and the development of the idea God's perfection. In modern philosophy, the notion of perfection started to be treated for the first time in the history of European thought as part of God's definition, as an attribute distinguishing God from His creatures. In medieval – teleological – philosophy, which assumed that perfection is a quality of each individual entity realizing its "true nature" (*telos*) (each individual entity *with the exception of God*, who creates the things endowed with their *teloi* but Himself, as uncreated, has no *telos*), and that perfection is connected with finitude, "closure" (whereas God is infinite), God could not be defined by recourse to the concept of perfection; He had to be defined in some different way, e.g., as *ens quo maius cogitari nequit*, as an absolute, unconditional, ineffable, inconceivable entity, or as pure existence. It was not until Descartes that perfection started to be regarded as God's attribute (Spinoza and Leibniz followed in Descartes's footsteps in their definitions of God).[5] Therefore, in modern philosophy there occurred the passage from the teleological to a theological understanding of metaphysical perfection: the former implies that every entity (with the exception of God) can be perfect, the latter implies that only God can be perfect. One may speculate that this change in defining God was occasioned by the fall of the

---

5   Cf. W. Tatarkiewicz, *O doskonałości* (Warszawa: PWN, 1976), pp. 62–64.

classical picture of the world, which undermined the very notion of teleological perfection.

2.3 The discovery of the incompatibility and incommensurability of moral values

European philosophy, from Plato until modernity, was based, according to Isaiah Berlin (Berlin 2013a), on the three-fold assumption ("the Platonic ideal", *philosophia perennis*, or "the Ionian illusion", as Berlin called it), viz.: that all meaningful questions about facts and values can be answered, that these answers can be known by human beings, and that these answers are consistent with each other. The consequence of this assumption in the realm of moral discourse was the belief that moral values are consistent with each other, or, at worst, that even if they are not consistent (incompatible), they are commensurable, so that they can be ordered in a hierarchy, allowing one to easily solve their potential conflicts. Arguably, this view also underlies the idea of human moral perfection. It would be odd to maintain that there can be *more than one* idea of this type; it seems to lie in the very nature of this idea that it can be concretized in a unique way. Accordingly, if it turned out that moral values (the building-blocks of the idea of moral perfection) are not only multiple but may also be incompatible *and* incommensurable, the very idea of human moral perfection would prove to be incoherent. According to Berlin (2013e, 2013f), values are indeed incompatible and incommensurable, and this discovery was made in modern philosophy by Niccolo Machiavelli and (independently) by Giambattista Vico and Johann Gottfried von Herder, although the consequences of this discovery for the moral discourse, and perhaps also the discovery itself, were not fully realized either by Machiavelli, or Vico and Herder. Machiavelli distinguished between pagan morality and Christian morality, and even though his natural sympathy was on the side of the former, he believed, according to Berlin, that they are equally objective but inconsistent and incommensurable. Pagan morality embraces such values as pride, magnanimity, strength, courage, energy, persistence in the face of adversity, justice, success in public life, fame, whereas Christian values are mercy, self-sacrifice, love of God, forgiveness, humility, contempt for earthly pleasures, treating afterlife as the goal of human existence. The discovery of the fact that moral values may be objective but at the same time incompatible and incommensurable (like, e.g., equality and freedom, mercy and justice, tolerance and ardor, knowledge and happiness, safety and freedom, devotion to science, art, and devotion to family) had, in the long run, serious consequences for the idea of human moral perfection: it cast doubt on the belief in its coherence.[6] For instance, the idea of human moral perfection can be constructed on the basis of pagan or Christian morality, and the result will be two essentially different, incompatible and incommensurable ideas, the choice between which will always be arbitrary by virtue of the fact that there exist no higher-order moral criteria allowing one to rationally decide which of these two ideas stands morally higher. Berlin maintained that Machiavelli, Vico and Herder were not relativists, since they recognized the objectivity of moral values, but at the same time asserted that these values are multiple, incompatible and incommensurable: they can therefore be regarded as moral pluralists *avant la lettre*.[7] Interestingly, the full consequences of the discovery of the incompatibility and incommensurability of moral values were not fully realized until the middle of the 20th century, when Isaiah Berlin, shocked by this discovery, highlighted its crucial importance for moral and political philosophy. He showed, among other things, that the optimistic view according to which conflicts and tragedy can be eliminated from human life if only human beings become morally improved is untenable, since conflicts and tragedy may be caused not only by the wrong behavior of their victims, but also by the features of the very realm of moral values (their being often incompatible and incommensurable). Berlin also pointed out that all attempts at constructing an ideal world (an utopia)[8] are not only doomed to failure in practice, but are also entirely arbitrary in their axiological dimension, since they ignore the fact that values are incompatible and incommensurable, and as such can underlie multiple, and equally justifiable,

---

6   Of course, those who reject the view that values are incompatible and incommensurable will not be prone to use the term 'the discovery of the fact'; I have used this term because I share Berlin's view.
7   According to Isaiah Berlin, "Alleged Relativism in Eighteenth-Century European Thought," in: I. Berlin, *The Crooked Timber of Humanity: Chapters in the History of Ideas* (Princeton, NJ: Princeton University Press, 2013), pp. 73–94., the crucial difference between Vico and Herder on the one hand, and Montesquieu and Hume on the other, is that while the former were moral pluralists, the latter were moral monists, believing that even if there is not one basic moral value to which all other values can be reduced, all moral values are compatible. Furthermore, Montesquieu and Hume believed, in contrast to Vico and Herder, that various people, nations, civilizations accept in fact the same general values, and differ only in the choice of the means for their realization.
8   The projects of such a world – various utopias – were very popular in modern philosophy; suffice it to mention Thomas More's *Utopia*, Tomasso Campanella's *The City of the Sun*, or Francis Bacon's *New Atlantis*.

projects of a perfect society.⁹ Finally, Berlin argued that the idea of the progress of human species is problematic, since it implies that there exists an objective hierarchy of moral values that can serve as a benchmark for determining the direction of the progress.

3 Pro-perfectionist currents

3.1 The new way of understanding human moral perfection

According to Berlin (2013d, 2013g), two ideas of modern philosophy were especially important. The first one (discussed in section 2.3) was the discovery of the inconsistency and incommensurability of moral values, threatening the very consistency of the idea of human moral perfection. The second one was the claim raised by the Romantic thinkers (especially Friedrich Schiller, Friedrich Schlegel, and Johann Gottlieb Fichte), that value sare not objective but constitute an expression of human will. This – in fact metaethical – "Romantic" claim was motivated by two different developments in modern philosophy: the aforementioned appearance of the awareness of the multiplicity of moral values (the most forceful expressions of this awareness being the philosophies of Giambattista Vico and Johann Gottfried von Herder), which led to skepticism regarding their objectivity, and the emphasis on the role of will in human activity.¹⁰ The combination of these developments gave rise to the belief that moral values cannot be discovered (by intuition, reasoning, or studying holy books), because they have no independent existence: they are created by human beings and shaped by their will. This Romantic view of human values can be viewed as underlying a new idea of human moral perfection, according to which a human being is perfect not by virtue of his/her realizing some pre-established moral values (such as justice or tolerance), because such values do not exist objectively, but by virtue of the qualities of his/her choice of the values he/she proposes to realize and his/her further attitude to the chosen values. If the agent's choice is truly authentic, sincere, passionate, concordant with his/her deepest feelings and convictions, and if the chosen values are thereafter closely observed by him/her to the point of a readiness to die for them, then the agent is morally praiseworthy. What is

---

9 Cf. I. Berlin, "The Decline of Utopian Ideas in the West," in: I. Berlin, *The Crooked Timber of Humanity: Chapters in the History of Ideas* (Princeton, NJ: Princeton University Press, 2013), pp. 21–50.
10 Kant also emphasized the role of human will, but will for him was still understood in a rationalistic manner – as reason in action. Will became independent on reason in the thought of the Romantic thinkers.

revolutionary in this view, according to Berlin, is that it implies that human perfection consists of fidelity to the chosen values irrespective of what these values are (the only condition is that they must be chosen authentically). The question whether the chosen value is true or false is regarded as irrelevant, or rather mistaken, since the values, as created by human beings, cannot be assessed in these epistemic categories: they do not exist independently and objectively, and thereby cannot be a criterion for assessing our choice of values. Furthermore, the fuller expression of our "inner world" they constitute, the higher evaluation they deserve, since their individual character guarantees that they are really our values, and not a conformist choice. One could, of course, argue that Romantic thinkers in fact implicitly assumed some values as objective (e.g., fidelity, authenticity), but in response to this objection they could point out that these values are of a different (higher-order) category than the "traditional" values: their domain is the sphere of the choice of values rather than the sphere of action. It is worth noticing that Romantic thinkers rejected not only the idea of an objective world of moral values, but also the idea, common both to the classical and the scientific (modern) picture of the world, that reality has a rational and durable structure which can be discovered by the human mind. Romantic thinkers believed that the structure is only an appearance: the reality, in its core, is an irrational chaos, on which humans can impose a structure according to their will. This idea of the "chaotic reality" is in fact closely related to the "expressivist" view of values: if there is no "nature", to which our choice should conform, then we are radically free in choosing our values, according to which we can thereafter shape formless and value-free reality.

Berlin maintained that the "Romantic" idea of human moral perfection as a fidelity to freely chosen moral values was absent throughout most of modern philosophy, and appeared only at the end of the era of Enlightenment. This view does not seem fully convincing. Arguably, the pre-romantic idea of human perfection can be found in the thought of the Renaissance philosophers, such as Coluccio Salutati, Giannozzo Manetti, and especially Pico della Mirandola, who in his Oratio de hominis dignitate, extolled human beings' capacity for freely choosing their own nature as their most distinctive feature, arguing that this freedom in fact exhausts the concept of human nature, and thereby, at least implicitly, denying the classical view that human beings have a pre-established nature, telos which they are supposed to realize. Thus, the cult of freedom was a characteristic feature of Italian Renaissance philosophy, anticipating in this regard the Romantic movement. Berlin was right in so far that the Renaissance thinkers emphasized only one of the two elements of the Romantic idea of moral perfection, viz. freedom in choosing moral values, but not fidelity to the chosen values.

3.2 The perfectibility of man

As mentioned in section 2.1, many modern thinkers set forth a pessimistic view of human nature, and thereby were skeptical towards the feasibility of the ideal of human moral perfection. It should be noticed, however that, arguably, this picture is not most typical for modern philosophy in general and, assuredly, it is not most typical for the philosophy of Enlightenment. The philosophers of the 18th century were, in general, prone to support an optimistic view of human nature, according to which human beings are either born good, innocent (their natural feelings are sublime and noble, they desire peace, welfare, freedom, justice, happiness equality[11]), or are born morally neutral (their mind is a moral *tabula rasa*, with only one natural disposition, viz. to pursue what brings them pleasure and to avoid what gives them pain) and thereafter can be shaped in the desired direction by education.[12] According to this view (in either version), all human vices, e.g., dishonesty, greediness, or envy, are caused by bad social institutions, and thereby can be eradicated if these institutions are improved. In addition

---

11 This view was defended, for example, by the Cambridge Platonists (e.g., Ralph Cudworth and Richard Clumberland) and many philosophers of the French Enlightenment.
12 The most famous adherent of the latter view was John Locke, who in his treatise *Some Thoughts Concerning Education* has, as John Passmore put it, "opened up, in principle, the possibility of perfecting men by the application of readily intelligible, humanly controllable mechanisms. All that is required is that there should be an educator, or a social group, able and willing to teach the child what to purse and what to avoid" (J. Passmore, *The Perfectibility of Man* (Indianapolis: Liberty Fund, 2000), pp. 249–250). Since education is a process of habit-formation, proper education involves developing in the child morally proper habits, i.e., virtues. According to Locke, in the process of habit formation, the educator should use various forms of pleasure and pain (not only physical, but also psychological, such as shame) to direct a child's behavior in a desirable direction. Locke believed the object of formation by education are not only our emotions, but also our power of willing, because as John Passmore put it presenting Locke's view, "willing is one of our habits, not, as Pelagius had thought, the work of an innate faculty which can cut across our habits" (Passmore, *The Perfectibility of Man*, p. 255). A similar account of the role education in the formation of human character, according to which man is all that he has learnt, and thereby innate differences (if there are any) between individuals play no role in the development of their character, was held, e.g., by David Hartley. In the 20th century this view became developed by behavioral psychologists.

to this view, most 18th century anthropological optimists assumed the theory called "ethical intellectualism", according to which immoral acts are the effects of mistaken judgments: we act wrongly not because we are moved by irrepressible passions or corrupted will but because we erroneously judge the wrong course of action to be a good one. Ethical intellectualism underlies the conviction shared by most 18th century philosophers that moral vices can be dispelled by reason, i.e., that moral knowledge is a sufficient condition for liberating us from immorality. The above views constitute a basis for the thesis, characteristic for the 18th century philosophers, about *the perfectibility of man*, i.e., about human beings' capacity to develop their moral, intellectual and physical abilities to a virtually infinite degree. In William Godwin's famous definition of the term "perfectible", presented in his treatise from 1796, it "expresses the faculty of being continually made better and receiving perpetual improvement. The term 'perfectible', thus explained, not only does not imply the capacity of being brought to perfection, but stands in direct opposition to it. If we could arrive at perfection, there would be an end to our improvement" (quoted after Passmore 2000, p. 241).

The most eloquent defense of the thesis about the perfectibility of men and a scrutiny of its implications was provided by Antoine Nicolas Condorcet (1957), who sketched "a historical picture of the progress of the human mind", arguing that, given the developments in science and in moral thought, one has strong grounds to believe that human beings will be perfected infinitely in the direction of the state in which "truth, virtue, and happiness" will be triumphant.[13] As he explained, if people are free from prejudices, have more or less equal material resources, equal access to education, equal legal rights, and the level of their welfare is high due to the progress in practical sciences, then they will have no reason to commit evil – antisocial – acts. As he put it, "Nature has determined no limit to our hopes and possibilities" (Condorcet 1957, p. 213). Condorcet's firm belief was that the progress in morality is inextricably linked to the progress in education: they reinforce each other. In his view, the progress in morality consists not only in developing a human being's moral sensitivity, but also in improving legal rules. Condorcet, somewhat naively, believed that the improvement of legal rules will be made possible by the probabilistic calculus and combinatorics, enabling

---

13 Similar views were defended by, e.g., Anne-Robert-Jacques Turgot, Richard Price, and Joseph Priestley, whom Condorcet called the "the first and most perfect apostles" of the doctrine about the unlimited possibility of perfecting human species (cf. A.N. Condorcet, *Szkic obrazu postępu ducha ludzkiego poprzez dzieje* (*Esquisse d'un tableau historique des progrès de l'esprit humain*), trans. Ewa Hartleb (Warszawa: PWN, 1957), p. 174).

the exact predictions of the effects of applying various rules and thereby their evaluation. He also believed that the application of these mathematical tools will enable making a selection of the optimal system of legal rules from a virtually infinite number of systems respecting "general principles of equality and natural rights" (Condorcet 1957, p. 232).[14] Condorcet asserted as well that there will be a change of rules concerning procreation: should humans become aware of the potential danger of overpopulation and understand that children should be conceived only if they have a chance of pursuing a happy life, they will be able to refrain (if necessary) from procreation – from "filling the earth with useless and unhappy human beings is irresponsible" (Condorcet 1957, p. 232). Interestingly, Condorcet also suggested, in anticipation of the contemporary adherents of cognitive enhancement, that science may enable human beings to exceed the biologically determined border of their moral, intellectual, and physical capacities. In his philosophy, therefore, the concept of human perfection has not only a moral sense but also acquires, arguably for the first time in the history of European thought, the intellectual and physical sense.

The optimistic view of human nature did not, however, lead to the belief in the beneficial effects of the development of civilization. For instance, Rousseau, in his *Discours sur l'origine et les fondements de l'inégalité parmi les hommes*, claimed that the progress of human civilization does not amount to *real* progress of human beings. It unduly complicates human feelings, in particular generating and exacerbating man's *amour propre*, which is one of the main sources of unhappiness of man (*amour propre* is, according to Rousseau, an unnatural form of self-love, which is an effect of the socialization process and whose satisfaction depends on a comparison with others: an agent derives his/her self-love if he/she has a feeling of superiority over other people). However, Rousseau did not postulate a return to the blissful state of a "noble savage" – the state of happy solitude and freedom, which he deemed impossible., Instead he rather, postulated cherishing our noble – natural, spontaneous – sentiments, for instance, *amour de soi* – a natural, good form of self-love, whose satisfaction does not depend on comparison with others, and pity – the innate aversion to see one's fellow-men suffer. Such sentiments would flourish, as he believed, in the state constructed in accordance with his version of *contrat social*. In such a state there

---

14 According to Condorcet, the optimal system is the one which "provides the best protection of the general principles of equality and natural rights, enables exercising them in a wide extent, best guarantees tranquility and welfare of individuals, and power, peace, and wealth of the nations" (Condorcet, *Szkic obrazu postępu ducha ludzkiego poprzez dzieje*, p. 232).

will be no conflicts between private and public interest, since human beings will discern the irrationality of pursuing their egoistic, asocial goals. It is worth stressing that Rousseau also used the term "perfectibility", but gave it a different sense than Condorcet and Goodwin. For Rousseau, perfectibility is just a desire and capacity, characteristic for human beings, to improve their natural condition, their position in the world; it is not, like for Condorcet and Goodwin, the ability to perfect oneself to a virtually infinite degree. Perfectibility is therefore an ambivalent feature of humans: it is a source of most of the misfortunes they experienced in history, but in Rousseau's view it is not unequivocally bad. Its positive side is that it may enable human beings to reach the state of freedom and equality, which will have one important advantage over the state of nature: while in the latter they were basically amoral, neither virtuous nor vicious, exhibiting only natural goodness, in the *contrat social* state they will exhibit a higher level of goodness, viz. moral goodness.

4 Final remarks

From among the currents discussed above, the ones that deserve a special emphasis due to their utterly original character and the profound influence on the further development of philosophy are the fall of the classical picture of the world, the discovery of the incompatibility and incommensurability of values, the "Romantic" account of values, and the idea of the perfectibility of man. It is worth noticing that the first three had wider consequences than the last one (which was strictly connected with the idea of human perfection): they had an impact not only on the development of the idea of human perfection, but also played an important role in discussions concerning many other moral problems, e.g., concerning the status of values, the relation between facts and values, and the relations between values. To give one example, the influence of the Romantic view of values on the further developments of philosophy and culture was enormous, and can be noticed, for example, in Kierkegaard's and Nietzsche's thought, in existentialism, and in various emotivist metaethical theories. The Romantic view that the best thing human beings can do is to remain faithful, at any price, to their freely chosen ideals, whatever these ideals are, had also immense influence on the course of history. However, as noted by Berlin, after the tragic events of the first half of the 20th century, which arguably were at least partly caused by the distorted understanding of the Romantic view of values, this view lost much of its attractiveness and popularity, and the traditional conception of universal moral values, existing independently of human choices, became dominant again.

References

Berlin, I. "The Pursuit of the Ideal." In: *The Crooked Timber of Humanity: Chapters in the History of Ideas*. I. Berlin. Princeton, New Jersey: Princeton University Press, 2013a, pp. 1–20.

Berlin, I. "The Decline of Utopian Ideas in the West." In: *The Crooked Timber of Humanity: Chapters in the History of Ideas*. I. Berlin. Princeton, New Jersey: Princeton University Press, 2013b, pp. 21–50.

Berlin, I. "Alleged Relativism in Eighteenth-Century European Thought." In: *The Crooked Timber of Humanity: Chapters in the History of Ideas*. I. Berlin. Princeton, New Jersey: Princeton University Press, 2013c, pp. 73–94.

Berlin, I. "The Apotheosis of the Romantic Will: The Revolt against the Myth of an Ideal World." In: *The Crooked Timber of Humanity: Chapters in the History of Ideas*. I. Berlin. Princeton, New Jersey: Princeton University Press, 2013d, pp. 219–252.

Berlin, I. "The Counter-Enlightenment" In: *Against the Current: Essays in the History of Ideas*. I. Berlin. Princeton, New Jersey: Princeton University Press, 2013e, pp. 1–32.

Berlin, I. "The Originality of Machiavelli." In: *Against the Current: Essays in the History of Ideas*. I. Berlin, Princeton, New Jersey: Princeton University Press, 2013f, pp. 33–100.

Berlin, I. *The Roots of Romanticism*. Princeton, New Jersey: Princeton University Press, 2013g.

Calvin, J. *Institutes of the Christian Religion*. Transl. H. Beveridge. Ontario, Canada: Devoted Publishing, 2016.

Condorcet, A. N. *Szkic obrazu postępu ducha ludzkiego poprzez dzieje* (*Esquisse d'un tableau historique des progrès de l'esprit humain*).Transl. E. Hartleb. Warsaw: PWN, 1957.

Kant, I. "Idea for a Universal History with a Cosmopolitan Purpose." In: *Political Writings*. I. Kant. Transl. H. B. Nisbet. Cambridge: Cambridge University Press, 1999.

Kant, I. *Religion within the Limits of Reason Alone*. Transl. T. M. Greene and H. H. Hudson, New York: HarperOne, 2008.

Locke, J. *Some Thoughts Concerning Education*. New York: Dover Publications, 2007.

de Maistre, J. *Contre Rousseau. De l'état de nature*. Paris: Éditions mille et une nuits, 2008.

Pascal, B. "Pensées." In: *Pensées, Les Provinciales*. B. Pascal. Paris: Bookking International, 1995, pp. 15–332.

Pascal, B. *Entretien avec M. de Saci sur Épictète et Montaigne*. Cambridge: Cambridge University Press, 2013.

Passmore, J. *The Perfectibility of Man*. Indianapolis: Liberty Fund, 2000.

Rousseau, J. J. *Discours sur l'origine et les fondements de l'inégalité parmi les hommes*. Paris: Gallimard, 1969.

Tatarkiewicz, W. *O doskonałości*. Warszawa: PWN, 1976.

Part II Enhancement Methods: The Promise and Limitations of Enhancing Ourselves

Robert Audi

Moral Philosophy as a Framework for Approaching Human Enhancement[1]

Abstract: Among the greatest challenges to ethics is how to guide scientific research. There are, for instance, questions of how human and animal subjects should be treated, questions concerning the impact of research or its results on the environment, and the prior question of what research program should be undertaken in the first place. In this domain, there is perhaps no issue more worrisome than whether scientific researchers should try to achieve human enhancement and, if so, in what ways. This paper will outline a framework for approaching this question and will explore some instructive hypothetical cases. I begin with a sketch of some major orienting options among ethical theories, namely, virtue ethics, Kantian ethics, utilitarianism, and common-sense intuitionism. Reflection reveals both differences among the four kinds of ethical theory just noted, and some degree of complementarity. I consider some important differences and, in this light, explore the possibility of a unified view that captures the best elements in each. I argue for a unified pluralistic theory that may help us in biomedical decisions (including enhancement decisions) – namely, a Kantian intuitionism. The rough idea is that in making moral or morally significant decisions – which include most decisions in bioethics – we should do what accords with the Rossian duties, if there is no conflict among them and, if there is, resolve that conflict with the help of Kant's Humanity Formula interpreted in light of a conception of what it is to treat persons as means and as ends. No historically influential position in ethics is by itself adequate to guide biomedical decisions, and the best response to this is to formulate a view that captures the best elements in each. Doing this reduces dependence on appeals to practical wisdom and moral intuition, but it is naive to think that these appeals can be entirely eliminated. The result is a moral theory that facilitates everyday decisions by starting with basic obligations but also provides a theoretical framework, using notions Kant rightly made central, for dealing with difficult decisions in which such obligations conflict. With this framework in view, I discuss the idea of human enhancement, in particular one of its most controversial aspects, namely, moral enhancement, which is a significant area of bioethical interest.

Keywords: Virtue ethics, Kantian ethics, Utilitarianism, Intuitionism, Genetic engineering

---

1 For valuable discussion of many of the topics in this paper I thank Marta Soniewicka and Adriana Warmbier.

1 Four major kinds of ethical theory

In broad outline, both moral philosophy and ethics, as studied outside philosophy, have centered on four kinds of ethical theory, with three – virtue ethics, Kantianism, and utilitarianism – apparently dominant in orienting moral analysis.[2] Often the term "approach" is more apt for cases in which decision-makers have an ethical orientation grounded in one or another kind of theory but do not self-consciously hold a theory as such. Here I review these kinds of theory, note some of their strengths and weaknesses, and proceed to sketch a decision framework that draws on the best elements in each. With that before us, I will consider some issues concerning genetic enhancement.

1.1 Virtue ethics

In both Western and Eastern philosophy, virtue ethics has long appealed to many thinkers. Plato, Aristotle, and Confucius developed moral views of this sort. Aristotle, whose influence in bioethics since the 1970s has been major, described (for instance) just acts as the kind that a just person would perform; he denied that a just person is to be defined as one who performs just acts. He apparently took moral traits of character to be ethically more basic than moral acts. He said, for instance, regarding the types of acts that are right, "Actions (…) are called just and temperate when they are such as the just and the temperate person would do" (Nicomachean Ethics 1105b 7–8).[3] He took virtues, such as justice and temperance, rather than acts, as ethically central. This idea apparently underlies a number of his claims, for instance that "virtue makes us aim at the right target, and practical wisdom makes us use the right means" (NE 1144a).

For a virtue ethics, agents and their traits, as opposed to rules of action, are morally basic. We are to understand what it is to behave justly through studying the nature and tendencies of the just person. We do not, for instance, first construct a notion of just deeds as those that, say, treat people equally, and then define

---

2   This section and the next draw heavily on, though they also refine and somewhat revise, points made in my *Moral Value and Human Diversity* (New York and Oxford: Oxford University Press, 2007), esp. chapter 1. These sections are also shortened versions of the first three sections - similarly related to *Moral Value and Human Diversity* - of my related paper *Ethical Theory and Moral Intuitions in Biomedical Decision-Making*, in *The Ethics of Reproductive Genetics – Between Utility, Principles, and Virtues*. ed. Marta Soniewicka (Springer, 2018).
3   Aristotle, *Nicomachean Ethics*, trans. and ed. R. Crisp (Cambridge: Cambridge University Press, 2000).

a just person as one who characteristically does deeds of this sort. The person of practical wisdom is the chief role model in ethics; such people exemplify all of the moral virtues and also tend to be good advisors in ethical decisions.

Aristotle understood the virtues in the context of his theory of the good for human beings. For him, happiness ("flourishing" in some translations) is central. Happiness is our final unifying end: we may seek other things for their own sake, but only when "through them" we can achieve happiness. Happiness is not, however, a passive state; it requires a life in which "actions and activities ... that involve reason" (which is our distinctive characteristic) are central; the "human good," then, proves to be "activity of the soul (roughly, mind) in accord with virtue" (NE 1098b 14–17).

Virtue here is best understood as "excellence", and there is no doubt that Aristotle is a source of conceptions of excellence that bear on what constitute reasonable targets of enhancement. In general, if we conceive enhancement of our species in Aristotelian terms, it should conduce to our excellence. What constitutes excellence, then, is an important guiding standard for determining permissible and worthwhile efforts toward enhancement. Here one may focus on the intellectual, moral, aesthetic, religious, or indeed athletic domain – the last being the one in which excellence is perhaps clearest, at least if we are guided by the sports already pursued and imagine retaining their current rules. We know, for instance, what, in tennis, would constitute a higher level of skill than we now see, and sports are typically designed with criteria for degrees of skill implicit in their rules.

Human life as a whole is much more complicated than an athletic contest. There is far more controversy about what constitutes excellence in most of the other domains of life, and many thinkers considering the question of human enhancement would argue that the moral or the religious, point of view, or some combination of the two perspectives, should have priority in determining what constitutes excellence. In practice, one might expect both scientists and, especially, those who fund research directed toward enhancement, to seek consensus on what constitutes enhancement. Consensus, however, even when it rises to unanimity, is fallible. This is a point unlikely to be ignored by virtue ethicists, and I suggest that we need a sound ethical approach that we may hope will earn consensus, even though it should not depend on doing so.

If, however, in thinking about human excellence, we take traits as ethically (or otherwise) more basic than acts, how can we determine what counts as, for instance, being generous or honorable? Virtue ethics has resources for answering this, including the appeal to practical wisdom as applied to the context of decision. A person of practical wisdom is a paradigm of one having virtue, and Aristotle calls virtue a state that decides, consisting of a mean, the mean relative to us, which is defined by reason. It is a mean between two vices, one of excess and one of deficiency (see NE 1097a). Consider beneficence: if, relative to my resources, I am selfish and ignore others' needs, this is a deficiency; if I give so much at once that I am prevented from significantly helping others later, I am excessive. Good ethical decisions, on this view, may be seen in the light of such comparisons.

The problem just posed is not that the ideal of excellence is inappropriate as a central one for enhancement decisions. It is that the ideal by itself, even taken in a single domain of human life, is insufficiently determinate in implications for practical decision and needs supplementation by standards of a different if complementary kind. The main contrasting approaches, not surprisingly, are *rule theories* – so called because they make rules of action central and demand specific kinds of deeds. Three categories of rule ethics should be briefly considered. We can then seek to frame a theory that draws strength both from them and from virtue ethics.

1.2 Kantian ethics

When virtue ethics is contrasted with rule ethics, it is both natural and common to consider Kantian ethics (which will also be discussed, in a different way, in Section 3). Central to it is Kant's Categorical Imperative. It says that we must always act in such a way that we can rationally will the principle we are acting on to be a universal law: "Act as if the maxim of your action [roughly the principle underlying it] were to become through your will a universal law of nature."[4] Thus, I should not make a lying promise to keep medical information confidential if I could not rationally universalize the underlying principle, say that when I can gain advantage only by making a lying promise to keep a confidence, I will make such a promise. I can see why the Imperative disallows this by noting that we count on sincere promises from others and cannot rationally endorse the universality of a deceitful practice that would victimize us.

Kant also gave a less abstract formulation of the Categorical Imperative, the Humanity Formula (Sec. 429): "Act in such a way that you always treat humanity, whether in your own person or in the person of any other, never merely as a means, but always at the same time as an end." Negatively, the Imperative prohibits *using* people, for example by breaking promises given to them, for one's

---

4   See I. Kant, *Groundwork for the Metaphysics of Morals*, ed. and trans. A. W. Wood (New Haven: Yale University Press, 2002), sec. 422.

convenience. Positively, treating people as ends clearly requires caring about their good *for their sake* rather than using them instrumentally.[5]

This formulation, more obviously than the other quoted here, applies to oneself as well as others. It requires a kind of respect for persons, and this includes self-respect – an attitude that bears on how the voluntary sale of organs is to be viewed.[6] If we take Kant's two formulations together (and he considered them equivalent), then apparently we must not only treat persons as ends but – as the rational universalizability of our principles would suggest – treat them *equally*, so that everyone matters, and matters equally.

1.3 Utilitarianism

A very different kind of rule theory is suggested by the question: what *good* are rules unless they contribute to our well-being – unless (above all) following them enhances human happiness and reduces human suffering? This kind of concern leads to *utilitarianism*, the position of Jeremy Bentham and John Stuart Mill. In Mill's words:

> The creed which accepts as the foundation of morals "utility" (…) holds that actions are right in proportion as they tend to promote happiness, wrong as they tend to produce the reverse of happiness. By happiness is intended pleasure, and the absence of pain.[7]

If one act produces more happiness than another, it is preferable, other things equal. If the first also produces suffering, other things are not equal. We have to weigh good consequences of our projected acts against any bad ones and subtract the negative value from the positive. In making enhancement decisions, then, we would ideally at the same time produce pleasure *and* reduce suffering.

The ethical aim for action is to find options second to none in total value understood in terms of happiness.[8] It is noteworthy, however, that on this view

---

5   The notions of treating persons as ends, and of treating them merely as means, can be clarified even independently of Kant's ethical writings. For an indication of how, and references to literature on Kantian ethics, see chapter 3 of my *The Good in the Right: A Theory of Intuition and Intrinsic Value* (Princeton: Princeton University Press, 2004).
6   I have discussed a number of issues concerning this matter in "The Ethics of Organ Transplantation," *Utilitas*, Vol. 8, No. 2 (1996), pp. 141–158.
7   J. S. Mill, *Utilitarianism*, ed. O. Piest (New York: Macmillan, 1957), p. 10.
8   Mill's quoted formulation is less clear than mine in the preceding text; that represents a major kind of utilitarianism – though not the only kind found in Mill – as a sort of *ethics by cost-benefit analysis*. What makes this ethics rather than a kind of economics is that it makes goodness, not profit, the standard of conduct.

no specific act-type, such as lying, has a moral nature in itself: moral properties are never basic; they are all derivative from considerations of utility, in this case hedonic considerations. This does not prevent utilitarians from condemnation of lying in general. The point is that it is derivatively, not intrinsically, wrong.

The derivative character of moral properties for utilitarianism is often misunderstood. This misunderstanding is abetted by another: utilitarianism is commonly formulated as the position that for an act to be morally right is for it to produce "the greatest good for the greatest number".[9] This misrepresents the view. Utilitarians are concerned, above all, to maximize the good. Some ways to produce it, say by providing education for all children, are undoubtedly better than others because (other things equal) they favorably affect more people; but the idea that doing (or producing) good for more rather than fewer people is not a *basic* concern, nor is it appropriate to defining the position. For instance, if providing public libraries only in highly educated communities would produce more good (say, in stimulating innovations and productivity) than providing them equally to a whole population, the former would be preferred.[10]

1.4 Common-sense intuitionism

Suppose one agrees with virtue theorists that there are as many different dimensions of morality as there are moral virtues, and with rule theorists in holding that morality demands that we have and act on *principles*. This may lead to the kind of common-sense ethical theory set out by W. D. Ross.[11] His approach – a pluralistic multiple-rule view – is to categorize our basic duties (moral obligations). He did this by considering the kinds of grounds on which moral obligations rest; for instance, making a promise to help you do an experiment is a ground of an obligation to do it, injuring someone in rushing to a train

---

9 Joseph DesJardins, e.g., in his *An Introduction to Business Ethics* says, "Utilitarianism is typically identified with the policy of 'maximizing the overall good' or, in a slightly different version, of producing 'the greatest good for the greatest number'". See J. DesJardins, *An Introduction to Business Ethics* (New York: McGraw-Hill Education, 2005), p. 30. He does not discuss the difference (which is far from slight), and he discusses utilitarianism in relation to both characterizations.
10 This assumes that the narrow distribution of libraries would not create a degree of resentment that would cause suffering so great as to outweigh the benefits of favoring the educated. Utilitarians always seek to consider the total effect of a possible action; the point here is that inequality of distribution is not *automatically* or in itself to be avoided. The overall good is the sole standard of conduct.
11 W. D. Ross, *The Right and the Good* (Oxford: Oxford University Press, 1930), pp. 29–30.

is a ground of an obligation to make reparations, and seeing someone bleeding by the wayside, as the Good Samaritan did, provides a ground – even if not necessarily a predominating one – of an obligation to help. For Ross, the basic obligations are to (1) keep promises, (2) act justly, (3) do good deeds towards others, (4) express gratitude for services rendered, (5) avoid injuring others, (6) make reparations for wrong-doing, (7) avoid lying, and (more positively) (8) improve oneself. He took it to be intuitively clear and indeed self-evident that we have these obligations: you can see this by engaging in sufficiently clear and deep reflection - a kind of intuitive thinking - on the moral concepts in question. Hence the name "intuitionism" for the position that morality is to be conceived in terms of the principles expressing these commonly recognized obligations.[12]

Ross knew that *prima facie* obligations can conflict. Consider the Good Samaritan. He went to great lengths to help a wounded stranger (*Luke* 10,30–35). Suppose he had promised to help his daughter harvest her olives and was unable to do this given the delay caused by ministering to the stranger. Ross thought that where two or more duties (his term for obligations) conflict, we often need practical wisdom (wisdom in human affairs) to determine which duty is *final*, that is, which duty is, all things considered, the one we ought to fulfill, as opposed to our "*prima facie* duty", our duty relative to any one kind of moral ground in the situation, here a wounded stranger's needs and a promise to one's daughter. Our final duty is what we ought to do "in the end", and it will be the same as our *prima facie* duty *if* no other such duty of equal weight conflicts with that. If I promise to write a paper for you and have no conflicting duty, writing it is what I ought to do.

2 Differences and commonalities among the "pure" forms of the theories

Reflection reveals both differences among the four kinds of moral theory just described and some degree of complementarity. Let us first consider some differences and, in that light, explore the possibility of a unified view that captures the best of each kind of theory.

To see some differences among these basic kinds of ethical view, consider a case in which your grandfather (who has outlived your parents) puts you in charge of directing his medical treatment if he becomes incompetent. You have promised to let him die with dignity if he is suffering, unable to communicate,

---

12 For a detailed account of Ross's intuitionism and a defense of a view that incorporates major elements of it, see my *The Good in the Right*, esp. chapters 2 and 3.

and clearly terminally ill. His lung disease prevents normal breathing, and putting him on a respirator is suggested. He suffers when conscious, cannot communicate or even understand what is said to him, and is being fed through tubes. Many facts that such a case presents cannot be filled in here, but we can see some differences between the approaches. Take common-sense intuitionism first, since it views our promissory obligations as a morally basic kind. Unless we find some conflicting obligation of equal weight, we must do as we promised and decline to allow a respirator. Imagine, however, that other grandchildren have asked to come to him one last time and need a day to make the trip. Here one might have an obligation of beneficence – to do something good for them – that would favor a respirator if he would otherwise die too quickly. Suppose one could confine its use for this short-term purpose. Allowing its use might then be consistent with the original promise.

A virtue ethics could (though it need not) lead one to a similar decision. The virtue of fidelity is the one most relevant here. Fidelity to one's word is central, but the virtue is broader and encompasses loyalty to others. There is a virtue of beneficence as well, and this would incline one much as the Rossian duty of beneficence would. One's central focus, however, would be on what kind of person to be in the situation; this conception is to lead one to the right deed. The procedure is not to consider types of action and bring rules to bear on them. It is crucial to see that as different as these approaches are, they may, like different ways of building a bridge, take one to the same destination.

The Kantian and utilitarian accounts both differ strikingly from the intuitionist and virtue ethical views. They are each, what might be called, "master principle theories" of right action, whereas the latter are highly pluralistic.[13] For the virtue approach, there is a plurality of virtues central for ethical thinking; for intuitionism there is a plurality of rules. This is not to imply that the decision you should make must differ depending on which of the master principle views is your guide. On the kind of utilitarian view sketched, our focus must be on the good to be done by making one decision rather than another. We might now focus on how much suffering the patient will endure in the extra day on the respirator, and we might compare that with the suffering of the grandchildren

---

13 This contrast is not sharp (and deserves analysis not possible here). Even supposing Kant's formulations of the Categorical Imperative are all equivalent, he appeals (in the intrinsic end formulation of it) to a plurality of moral considerations, e.g., an obligation to avoid treating people merely as means and an obligation (not entailed by that) to treat them as ends. For Mill, there is also at least the plurality that comes from taking value to have both negative and positive dimensions.

if they cannot get to him before he dies. We might also think about the effects of the example we set if we delay (or if we do not). Being seen as breaking a promise can have very bad consequences. For utilitarianism, however, even the pressure for hospital space and the costs of the extra medical care will be significant considerations. None of these things need be *irrelevant* on the other views; but for utilitarianism, facts are relevant on the basis of their bearing on the consequences of our options for the happiness of all affected, not on the basis of their bearing on whether we are keeping a promise, being virtuous, or following a rule that is universalizable in the way Kant intended. This makes a great difference in approach, and even if one often reaches the same moral destination, one may not do so in just any causally effective way. We could be influenced by the monetary costs much more than on the other views, perhaps thinking of how much good could be done with the savings. For intuitionism, by contrast, the obligation of beneficence – which is the overarching obligation for utilitarianism – is only one important moral consideration here; the promise also has moral weight, and even a duty of gratitude toward the grandfather may add to the grounds for adhering to the original promise.

In the kind of broad terminology required by a single essay, we might identify main sources of dissatisfaction with the four kinds of theory as follows – and here I choose the terms of description with bioethical and research policy decisions foremost in mind. Virtue ethics is plausibly considered insufficiently determinate in its implications for many practical decisions. Kantian ethics is caught between, on one interpretation, excessive formality and serious vagueness and, given the easiest remedy for these, rigorism. Utilitarianism seems often in danger of compromising justice in the quest to maximize non-moral value. Intuitionism appears to lack a procedure for resolving conflicts of obligation, which Ross and other proponents grant are a challenge for the theory.

3 Can the merits of the leading theories be combined?

We should begin by noting that some of the moral virtues stressed by Aristotle and other virtue ethicists are reflected in Kantian categorical imperatives with a small "c", for instance, the imperatives to avoid lying and to do good deeds – imperatives of veracity and beneficence, which each represent moral virtues. Less noticed are the "secondary rules" stressed by Mill at the end of chapter two in *Utilitarianism*. As to intuitionism, surely the intuitive common-sense standards articulated by Ross are common coin among the major theories and many major religions. These are: 1. Justice: including the positive obligation to prevent and rectify injustice, and the negative obligation to avoid commission of injustice;

2. Non-injury: roughly, the obligation to avoid harming others; 3. Fidelity: the obligation to keep one's promises; 4. Veracity: avoidance of lying – this obligation, like that of fidelity (under which Ross subsumed it) is a kind of fidelity to our word; 5. Reparation: the obligation to make amends for wrong-doing; 6. Beneficence: the obligation to contribute to virtue, knowledge, or pleasure in others; 7. Self-improvement: the obligation to better oneself; and 8. Gratitude: the obligation to respond in an appropriately appreciative way to those who do good deeds toward us.

One further comment is needed here.[14] Justice should be taken to entail not only treating people in accord with their merit (as Ross put it), but also equally in some (doubtless related) proportionate sense.[15] Equal treatment is something to which people are sensitive even early in life. Take small children of the same age who are given different privileges in playing with toys. They will tend to compare the toys each may play with and become upset if theirs is visibly less elaborate. Resentment of preferentially unequal treatment may well trace to just such cases.

There are, on my view, other Rossian elements – roughly, principles of the same *a priori* status and also apparently basic in guiding intuition and inference: 9. Liberty: the obligation to preserve and enhance it; and 10. Respectfulness: the indefinitely many obligations of *manner* to treat persons in respectful ways. Obligations of manner concern the *way* we do what is obligatory as opposed to *what* we must do. Clearly one can do the right thing for the right reason but in a morally unacceptable *way* – say, telling a patient that painful chemotherapy is the only hope for remission, yet in a way that projects annoyance and no perceptible concern for the suffering to occur.[16]

Is there any prospect of a unified pluralistic theory that aids us in enhancement policy and other biomedical decisions? I have argued in previous work that there is: a Kantian intuitionism.[17] The rough idea is that in making moral or morally significant decisions – which includes most decisions in bioethics – we should do what accords with the Rossian duties if there is no conflict among them and, if

---

14 Detailed discussion of Ross's principles, together with a rationale for my two added ones, is provided in chapter 5 of my *The Good in the Right*.
15 For a fine-grained and illuminating theory of equality, see L. Temkin's *Inequality* (Oxford: Oxford University Press, 1997).
16 Detailed discussion of what constitutes a manner of action and how it is morally important is provided in my *Means, Ends, and Persons: The Meaning and Psychological Dimensions of Kant's Humanity Formula* (New York and Oxford: Oxford University Press, 2016).
17 See Audi, *The Good in the Right*, esp. chapter 2.

there is, resolve that conflict with the help of the Humanity Formula interpreted in the light of a conception of what it is to treat persons merely as means and, by contrast, as ends. For instance, sometimes there is no moral conflict regarding the obligations simply to tell the truth, to treat patients equally, and to abide by contractual promises. I have provided a detailed account of the Humanity Formula in descriptive terms (overlapping Kant's but also quite different), so that its application can be guided without presupposing moral judgments.

What, then, is the overarching standard of obligation? It is to fulfill one's obligation under the intuitive categories just presented, with any conflicts between or among them to be resolved at least in part by determining whether the decision treats all concerned as ends in themselves and, especially, avoids treating anyone merely as a means. One might like to eliminate "at least in part", but here I think we must grant that in certain cases practical wisdom is needed to supplement, perhaps through moral imagination, our reflections in accordance with principles. I take such wisdom to yield moral intuitions that are not idiosyncratic and can be defended, and here I would argue that the universalizability formulation of the Categorical Imperative is useful. Practical decision will have a basis in the facts of the case, and these – when adequately considered – will often make it obvious what to do given the primacy of one of the intuitive principles or the overall preferability of one option. Kantian intuitionism, taken together with sufficiently described facts of a case, also enables us to frame a generalization for like cases. If that generalization is not rationally universalizable within the framework of values implicit in Kantian intuitionism – which, like the ten Rossian principles, views persons as ends in the sense demanding respect – then it is morally unacceptable.[18]

---

18 In earlier work, I formulated a principle based on the idea that there are at least three conceptually independent factors that a good ethical theory should take into account: happiness – roughly, welfare, conceived in terms of pleasure, pain, and suffering; justice, conceived largely as requiring equal treatment of persons; and freedom. These are all reflected on the Rossian list of basic obligations. On this approach – call it *pluralist universalism* – the broadest moral principle requires optimizing happiness so far as possible without either producing injustice or curtailing freedom (including one's own); and this principle is to be *internalized* – roughly, automatically presupposed and normally also a strong motivator – in a way that yields moral virtue. As a priority rule for achieving a balance among the three values I offered this: considerations of justice and freedom take priority over considerations of happiness; justice and freedom (presumably) do not conflict because justice requires the highest level of freedom possible within the limits of peaceful coexistence, and this is as much freedom as any reasonable ideal of liberty demands. The Kantian intuitionism proposed in this paper is richer

4 The notion of enhancement

Enhancement of living beings (in the sense most relevant for science policy) can be conceived as their structural alteration, say by genetic modifications, in a way that increases the probability of their realizing excellences of a determinate kind. These might be characterized following Aristotle or in other ways.[19] As our discussion suggests, enhancement may be global – applying to our general capacity – or specific, concentrated on one or more particular merits. Moral philosophy has standardly taken human persons to be its primary concern, and, if we are guided by it, then human excellences will be what guides discussion. But one might also take a transhumanistic view on which there is no need to be limited by projections concerning the human species. I will consider this view at certain points, but my main concern is with the ethics of enhancement concerning human beings broadly conceived.

It must be kept in mind, however, that what begins as enhancement of human beings can yield organisms that are not human beings. This is not to say that we are sacred in a sense implying that we cannot be transcended; but we have rights that might be abridged or even trampled by superior beings, and superiority of the kind in question does not imply more extensive or weightier moral rights. One question here is, of course, how *gradual* such enhancement might be. If sufficiently so, no one generation would differ dramatically from the next, even if, in time, there is a transhuman species.

The second point is this. It is certainly arguable that a reasonable limit on enhancement aspirations is (at present) the highest level now clearly morally acceptable for people raising children. If we may aspire to have children (in the normal way) who achieve a certain high level, it would seem that this level, as opposed to some transhumanistic level, might be a justifiable aim of

---

and more powerful than the pluralist universalism just stated; but that principle may be useful for those wanting a simpler formulation, provided the Kantian intuitionist framework is understood as crucial for its adequate application.

19 Here it is instructive to compare Allen Buchanan's characterization: "a biomedical enhancement is a deliberate intervention, applying biomedical science, which aims to improve an existing capacity that most or all human beings typically have, or to create a new capacity, by acting directly on the body or brain." See his *Beyond Humanity?: The Ethics of Biomedical Enhancement* (Oxford; Oxford University Press, 2011), p. 23. This does not entail any *success* in the aim. His later, more generic characterization (closer to mine) does entail some success: "to enhance human beings is to expand their capabilities – to enable them to do what normal human beings have hitherto been unable to do," p. 39.

enhancement efforts. We have already had geniuses, and (one might hope) more have been good than bad. This provides *prima facie* reason to think that properly controlled enhancement efforts going only that far might be permissible. The same kind of argument, however, might be reiterated when the first enhanced generation is at hand, with the result that, by degrees, we reach a transhuman species. Clearly, then, ethics calls for a kind of guarantee that *any* species-wide enhancement (and, for that matter, even individual programs of child-rearing) avoids creating a situation in which the superior beings wrong those who have not been enhanced.[20]

In the literature on human enhancement, a number of writers (some to be cited below) are concerned with moral enhancement. One might wonder whether a reason for this is that it might seem paradoxical for anyone to object from the moral point of view to competent attempts to improve us morally. It is not paradoxical, as I hope to show, but clearly some enhancement efforts might be aimed at non-moral improvement, and that should be considered here as well.

One matter should be set aside for purposes of this discussion: the permissibility of therapeutic and preventive enhancements. If the term "enhancement" seems to be stretched here, bear in mind that we can speak of enhancing resistance to disease and of enhancing fortitude – where fortitude, though a virtue, is a matter of bearing up under pressure or pain and not of exercising a "positive" excellence. Provided the risks are suitably low – a condition not easily achieved in biomedicine, to be sure – it does seem morally permissible to alter our genetic make-up so that we resist such diseases as cancer or so that we are easily cured of diseases to which we are subject. In practice, however, such genetic change is unlikely to occur without discovery of means to make unrelated alterations; nor is the notion of disease either wholly clear or entirely neutral in a moral sense. This is evident when one considers what constitutes "mental illness". Thus, even though therapeutic and preventive "enhancement" is in some cases likely permissible, I would resist endorsing any blanket moral approval of it. Each case must be carefully scrutinized. One person's eccentricity might be another's

---

20 Buchanan notes that even if creation of transhumans yielded beings of higher *moral status* that would not nullify the *moral standing* possessed by all human beings; see, e.g., Buchanan, *Beyond Humanity?*, pp. 209–211. Still, there might be differences in what he calls "moral considerability", and these are a basis on which "postpersons" might be argued to merit preferential treatment of a kind may be reasonably considered invidious. See, e.g., Buchanan, *Beyond Humanity?*, pp. 220–225. On this score, it is important whether our ethical orientation is determined by utilitarianism, as I have held that it should not be.

mental illness. There are some uncontroversial cases of health and disease; but the concepts themselves, understood in the comprehensive way required for enhancement discussions, are normative and require ethical scrutiny.

In addition to the important distinction between therapeutic and non-therapeutic interventions, and between enhancements of a single organism, as opposed to genetic enhancement of a category or even whole species, there is a distinction between genetic changes that can be prevented from taking effect in behavior and those than cannot be thus suppressed. Buchanan notes that "Scientists already know how to block the expression of genes they insert into laboratory animals"[21] and there is no reason to doubt that some negative elements in genetic enhancements might be similarly suppressible.

On the side of positive enhancement, it is fruitful to consider moral enhancement as the least objectionable case. What is it, and what is the case for undertaking it? One writer, arguing for "The Genetic Virtue Project" grants that "there is much disagreement as to what exactly fairness amounts to" but remains optimistic about genetic enhancement of "the virtue of caring [which] refers to behavior that seeks to promote the good of others".[22] The writer surely means something like a tendency to promote that good – caring *about*, as opposed to caring *for*, which is possible for someone acting selfishly and only for profit – is not behavior on any plausible interpretation. In any case, the paper provides no account of the good of others sufficiently plausible to guide genetic alteration. Nor is there any attention to the point that even a highly virtuous person needs intellectual skills and adequate information to succeed in caring behavior.

In a related paper supporting the idea that we might seek by genetic means to achieve moral enhancement, Thomas Douglas tells us that "A person morally enhances herself if she alters herself in a way that may be reasonably expected to result in her having morally better motives".[23] This writer not only does not indicate what constitutes a morally better motive but also specifically avoids commitment to a view on that.[24] I find this puzzling. Some notion of the morally

---

21 Buchanan, *Beyond Humanity?*, p. 173. He adds that "administration of a drug" may suffice (p. 173). To be sure, drugs may have bad side-effects and might be needed repeatedly or in the next generation, but the distinction in question is still ethically significant.
22 See M. Walker, "Enhancing Genetic Virtue: A Project for Twenty-first Century Humanity," *Politics and the Life Sciences* Vol. 2, No. 28 (2009), p. 30.
23 T. Douglas, "Moral Enhancement," *Journal of Applied Philosophy*, Vol. 25, No. 3 (2008), p. 229.
24 Douglas, "Moral Enhancement", p. 229.

good is needed in order to make morally permissible decisions on whether a technique is worth undertaking and, once it is instituted, whether it has worked at all. We are told that "a person's having morally better motives will tend to be to the advantage of others".[25] This is plausible enough for the conception of goodness that goes with the moral theory I have sketched earlier, but policy decisions governing major scientific research programs depend on agreement on goals, and I believe that may be too much to hope for regarding the battery of motives in question.

The specter of human manipulation raises the question whether genetic alteration of human beings, especially in making us better, is a threat to our dignity. Many religious people are likely to think so, on the ground that human dignity depends on our having a human nature deigned by God. On my view, the notion of human dignity is explicable in non-theological terms (which does not exclude the possibility that we possess dignity because of divine action); and the property of having dignity is also non-relational and perhaps better called a *status*. The notion is, however, insufficiently clear to be as good a basis for decisions on enhancement policy as the framework proposed in this paper.[26]

There is also an even deeper issue regarding the attempt at moral enhancement through biomedical techniques. Given how motives conflict – as where obligations themselves do – judgment is needed to determine what counts as appropriate action in such cases. Surely, enhancing motives alone will not necessarily improve moral conduct. It is also possible that in strengthening one kind of motive one weakens another kind, or produces an imbalance that reduces behavioral efficiency or impairs moral judgment, or both. Indeed, motives, as distinct from traits of character including virtues, have intentional objects. What is the content of the motives to be strengthened by genetic alteration? To do more for others? For one's family? For those to whom one has obligations? All of the above? And can we induce such specific desires in people without manipulating them, especially if we produce in them motives that dominate in their behavior over conflicting motives such as those favoring one's own projects over service to others? These are important matters, and they need resolution not to my knowledge forthcoming in the relevant literature.

---

25 Douglas, "Moral Enhancement", p. 230.
26 For a valuable exploration of the notion of dignity and its bearing on enhancement, see N. Bostrom, "Dignity and Enhancement," in: *Human Dignity and Bioethics: Essays Commissioned by the President's Council on Bioethics* (Washington, D.C. 2008), pp. 173–207. This paper reinforces my view that the explication of dignity does not depend on (though it may of course be clarified by) theology.

5 The enhancement case from a broadly utilitarian perspective

As one might expect from what has already been said about utilitarianism in a strongly naturalistic form (as with classical versions), some of the strongest support for moral and other kinds of enhancement comes from this kind of ethical view. In part because utilitarianism lends itself so readily to a naturalistic understanding, and in part because it promises quantification and objective measurement as a scientific basis for ethical decisions, it tends to be favored among scientists, or at least social scientists. This makes it especially appropriate to consider the view of Savulescu and Persson, who address enhancement mainly from a utilitarian perspective.[27] They maintain, e.g., that "to be morally enhanced is to have those dispositions which make it more likely that you will arrive at the correct judgment of what it is right to do and more likely to act on that judgment".[28] They wisely leave this characterization neutral toward major ethical theories, but the only theory they consider is utilitarianism, which they characterize as the view that "the right action is that action which maximizes utility". They go on to work mainly with preference utilitarianism, on which right action "satisfies maximally the preferences of everyone affected by the action, where the preferences of everyone count equally".[29] They also grant that "utilitarianism is very demanding as a moral theory – even if someone else gains [only] slightly greater preference satisfaction than you do, you should act so as to satisfy their preferences rather than your own".[30]

One thing I miss in this scheme is any account of how one can genetically modify people to be better utilitarians while building the right tendency to hold back from the approach to using oneself merely as a means to maximizing utility – the overall obligation one has under the theory (it is often not realized how unimportant we are as individuals given the vast number of people – and presumably of other sentient beings – whose well-being counts as much as ours, and the injunction to maximize overall human well-being). There are also serious problems regarding preference satisfaction. Do we, for instance, take a preference on the part of one person as automatically more important than a weaker preference in someone else? This is the impression created by

---

27 For critical discussion of their view, quite different from mine, see R. Powell and A. Buchanan, *The Evolution of Human Enhancement* (forthcoming).
28 See J. Savulescu and I. Persson, "Moral Enhancement, Freedom and the God Machine," *The Monist*, Vol. 95, No. 3, p. 403.
29 Savulescu and Persson, "Moral Enhancement, Freedom and the God Machine," p. 404.
30 Savulescu and Persson, "Moral Enhancement, Freedom and the God Machine," p. 404.

the quotation on demandingness, but it has the morally counterintuitive consequence that these with stronger desires are to be given greater importance in moral decisions than those whose desires are moderate in strength. We also need a way of discounting artificial strengthening of desires in anticipation of a utilitarian calculation. Surely, we cannot enhance our moral importance simply by intensifying our preferences.

The neutrality that Savulescu and Persson seek on specific moral positions seems to disappear when they move on, a mere paragraph later, to say that "it is a prerequisite of moral action that one should sacrifice/constrain one's own self-interest for some moral code for the benefit of others".[31] What counts as benefitting others? If it is just a matter of their preference satisfaction, then we have the problem of how to discount irrational preferences. Hedonistic utilitarianism has a way of dealing with this, though (as suggested above) it is unsatisfactory. I might add here something uncontroversial for at least certain kinds of case: that one could *rationally* have preferences for the pursuit of intellectual excellence even if so doing costs more pain than it promises pleasure, in which case the hedonistic criterion for rational preference would yield implausible results.

To their credit, Savulescu and Persson exhibit a sense of the variety of changes needed to make people more moral. As "necessary for moral behaviour" they cite "willingness to co-operate with other people and impulse control".[32] They grant, however, that "both these traits could be used for nefarious purposes to increase immoral behavior, making someone, for example, a more effective criminal".[33] One also misses in this context any criterion of what is in our "interest", something crucial both for understanding beneficence and even for understanding self-interest in the prudential sense in which it is not simply a matter of what satisfies basic desires or basic preferences.

It is also to their credit that they bring out that there is no *necessary* undermining of liberty entailed simply by genetic enhancement regardless of its nature, and it is plausible to hold, as they do, that enhancement by itself does not entail the inability to act for "the same reasons as those of us who are most moral today do".[34] They also see that the case of Ulysses and the Sirens shows that not all desires should be satisfied: "we see in this case how it is necessary to frustrate some of a person's desires, even his strongest desires, if we are to respect

---

31 Savulescu and Persson, "Moral Enhancement, Freedom and the God Machine,"p. 404.
32 Savulescu and Persson, "Moral Enhancement, Freedom and the God Machine," p. 405.
33 Savulescu and Persson, "Moral Enhancement, Freedom and the God Machine," p. 405.
34 Savulescu and Persson, "Moral Enhancement, Freedom and the God Machine," p. 406.

his autonomy".³⁵ The problem they do not adequately deal with is how to make ethical determinations of what desires are to be satisfied and what kind of desires should be restrained, or even eliminated.

In this last case, however, one wonders how autonomy is to be rationalized under any version of naturalistic utilitarianism. Why should autonomy be important at all other than because respecting it is favorable to overall preference satisfaction or overall increase in pleasure? Neither seems to be the point in the classical example of Ulysses, and in any event the appeal to autonomy as an independent value is not appropriate to utilitarianism. The impression one may well have from the article as a whole is that it makes a better case for therapeutic or preventive enhancement than for the positive kind of enhancement it seeks (in a significantly qualified way) to support, and that the strength of the case depends on appeals to such notions as benefit, interest, and autonomy understood in ways that do not derive from the utilitarian theory which is the favored example of a guiding perspective. This interpretation leaves open how good the case is for endorsing genetic enhancement even for moral traits. I do not see a strong case for this, in part because the desired traits seem difficult to "operationalize" sufficiently for genetic "targeting". Another concern is that we should hope to enhance moral conduct itself and not just good traits that will lead to it only on the basis of adequate information and judgments of a kind that, given incommensurable obligations and values, cannot be codified. Even beneficence can overbalance the obligations one has to specially related people – or so it would seem once a plurality of non-hierarchical obligations is recognized.

As it happens, Persson and Savulescu published a paper four years earlier, which gives a quite different impression. In "The Perils of Cognitive Enhancement and the Urgent Imperative to Enhance the Moral Character of Humanity"³⁶ they stress the dangers of cognitive enhancement, though they "do not deny that cognitive enhancement is indispensable for moral enhancement".³⁷ Given this caution, which seems altogether reasonable, it is perhaps surprising that they seem as favorable to moral enhancement as they do. In my view, cognitive enhancement – except by normal educational means – is not indispensable for some degree of moral improvement in human beings – a kind of improvement that

---

35 Savulescu and Persson, "Moral Enhancement, Freedom and the God Machine," p. 410.
36 I. Persson and J. Savulescu, "The Perils of Cognitive Enhancement and the Urgent Imperative to Enhance the Moral Character of Humanity," *Journal of Applied Philosophy*, Vol. 25. No. 3 (2008), pp. 162–177.
37 Persson and Savulescu, "The Perils of Cognitive Enhancement and the Urgent Imperative to Enhance the Moral Character of Humanity," p. 173.

does not (at least not clearly) require genetic alterations in the human species. The need for such improvement, however, is a point of agreement among us. What I doubt here is that we have good reason to think that genetic enhancement is likely to conduce to the needed improvements.

6 A wider perspective on the ethics of enhancement

So far, our discussion has been conceptual and, in broad terms, normative. Moreover, I have written as if there might be some hope of stopping experiments aimed at enhancement and, in some cases, yielding irreversible changes in at least experimental subjects, but perhaps also in many other persons. History has shown, however, that biomedical technology has a momentum of its own, and we would do well to think in terms of *when* major enhancement efforts have major effects, not *whether* there will be such efforts.

In one domain I have not discussed, given my concern mainly with genetic enhancement, performance-enhancing drugs have already come into widespread uses. I refer to sports.[38] Here it might seem that the issue is mainly parentalism ("paternalism" in the usual terminology). If athletes are willing to risk harm for better performance, is that their own business? Arguably, it is, *provided* that all athletes, or at least all who compete with one another, follow the same rules. This is crucial: the issue is nothing less than how sports is conceived, say as a contest on a "level playing field" where all seek to excel by talent and disciplined training, or as a spectacle in which the level of performance is primary, no matter what its basis. The major issue is not one of whether to allow enhancing drugs, but of how to decide what counts as sports and as fairness therein. This is a cultural decision, though to be sure one with broad ethical implications.

Sport is not the only domain of life in which enhancement issues have international importance. If, as seems likely, scientific work, in and beyond the realm of sports, will be devoted to enhancement whether or not such enhancement is ethically desirable, and with or without the approval of governments, we must ask what mechanisms can be put in place to minimize possible harms and direct the main enhancement work toward the well-being of all concerned. Here I concur with Allen Buchanan on the urgent need for institutions that govern efforts toward enhancement. He emphasizes, moreover, the need for at least one internationally constituted and internationally effective institution, and here I would

---

38 For a wide-ranging discussion of enhancement particularly pertinent to the case of doping in sports, see T. H. Murray, "Enhancement," in: B. Steinbock, ed., *The Oxford Handbook of Bioethics*, (Oxford: Oxford University Press, 2009), pp. 491–515.

emphasize the need – especially given the tendency for scientists, often with the concurrence of the public, to regulate themselves – for moral philosophers to play a significant role in policy decisions. His sketch of a projected "Global Institute for Justice in Innovation" suggests how an organization could serve this function and could also bring the results of innovations in enhancement to a wider population than would benefit from them if they were limited to individual nations or, as has happened with medication and medical services, to those with extensive funds. To be sure, coordinated national institutions concerned with innovations can do some of what an international body can do, but the globalized world we live in surely makes desirable the establishment of an international organization of the kind Buchanan describes.[39]

The powerful nationalism in many countries and the often-conflicting national interests of almost any pair of nations one choses indicate a moral principle that conscientious researchers should adhere to. The principle I suggest concerns institutional support for scientific research, including biomedical efforts toward enhancement. In rough formulation, the principle is that the less the likelihood of institutional support (1) to assure the safety of the projected work and of its living subjects, (2) to support just distribution of any positive results, and (3) to provide compensation for damage done by their work to property, (non-human) animals, or, especially, persons, the stronger the case for abstaining from that work. Condition (1) includes protecting human and animal subjects from pain, disability, and death; condition (2) includes avoidance of using positive results for the benefit just of a privileged class of people; and condition (3) includes both financial and medical resources devoted to subjects who suffer from experiments gone bad, or persons who suffer from collateral harms. This principle is readily defensible on the basis of the Kantian intuitionism proposed above, but it can also be defended from other ethical perspectives. Subsidiary and supplemental principles are needed as well. Scientific work should not only be directed toward what is morally permissible, but also done in a context in which there are procedures and resources for compensating those who are harmed, as seems inevitable, by unforeseen effects of permissible scientific efforts. If harms or even morally unacceptable distribution of benefits are likely, we can at least minimize wrongdoing by establishing procedures and resources for rectification.

---

39 See Buchanan, *Beyond Humanity?*, chapter 8.

7 Concluding remarks

Human enhancement by genetic means may be a biomedical aim that, for some researchers, is irresistible and may lead to some alterations in human beings that are irreversible, at least for a time. The latter possibility makes genetic enhancement efforts a potentially dangerous intervention in "nature". Even the kinds of enhancement, whether genetic or not, that are uncontroversially aimed at prevention of disease, disability, or reduction of suffering may go wrong. There is also a serious question whether an adequately clear distinction can be made between enhancements of negative kinds and positive enhancements, such as increases in intelligence or other capacities. Even from the point of view of classical utilitarianism, the generally admitted unclarity about how to enhance the ratio of the good to the bad indicates a need for hesitation regarding enhancement. But that point of view is in any case morally unsound if virtue ethics, Kantian ethics, or anything close to common-sense intuitionism is our guide. My own moral theory, based on integrating the intuitive moral principles formulated by W. D. Ross – clarified and supplemented by my own reflections – with Kant's Humanity Formula, as I understand it, provides better guidance. The theory certainly takes the moral status of persons as a moral anchor not to be dislodged. This leaves us free to consider genetic engineering of restricted kinds, but it induces caution about changing our traits, or even increasing our intelligence, before we can guarantee their constructive use within the moral standards needed to license enhancement efforts in the first place. It is possible that scientific interventions in human life may lead to our abandoning morality, as we know it. This could be a result of our being changed into something termed "transhuman". But the possibility of abandoning moral principles does not falsify them. The moral truths central in ethics, as I have represented it, do not seem contingent. They are in any case the basis on which we should decide what constitutes enhancement and what risks are morally permissible in attempts to achieve it. On this basis, I do not see an adequate case at present for genetic enhancements of the sweeping kind that may be considered permissible if we are guided by classical utilitarianism. Even therapeutic enhancements of any permanent kind must be undertaken with great care and guided by a different and wider ethical perspective. Human persons are certainly improvable, but if their moral status and basic equality in human rights are to be preserved, then many kinds of enhancement that may be biomedically feasible and apparently beneficial to humanity should not be undertaken.

References

Aristotle. *Nicomachean Ethics.* Trans. and ed. R. Crisp. Cambridge: Cambridge University Press, 2000.

Audi, R. "The Ethics of Organ Transplantation." *Utilitas*, Vol. 8, No. 2, 1996, pp. 141–158.

Audi, R. *The Good in the Right: A Theory of Intuition and Intrinsic Value.* Princeton: Princeton University Press, 2004.

Audi, R. *Moral Value and Human Diversity.* New York and Oxford: Oxford University Press, 2007.

Audi, R. *Means, Ends, and Persons: The Meaning and Psychological Dimensions of Kant's Humanity Formula.* New York and Oxford: Oxford University Press, 2016.

Audi, R. "Ethical Theory and Moral Intuitions in Biomedical Decision-Making." In: *The Ethics of Reproductive Genetics – Between Utility, Principles, and Virtues*, ed. Marta Soniewicka. Springer, 2018, pp. 3–21.

Bostrom, N. "Dignity and Enhancement." In: *Human Dignity and Bioethics: Essays Commissioned by the President's Council on Bioethics.* Washington, D. C.: 2008, pp. 173–207.

Buchanan, A. *Beyond Humanity?: The Ethics of Biomedical Enhancement.* Oxford: Oxford University Press, 2011.

DesJardins, J. *An Introduction to Business Ethics.* New York: McGraw-Hill Education, 2005.

Douglas, T. "Moral Enhancement." *Journal of Applied Philosophy*, Vol. 25, No. 3, 2008, pp. 228–245.

Kant, I. *Groundwork for the Metaphysics of Morals.* Trans. and ed. A. W. Wood. New Haven: Yale University Press, 2002.

Mill, J. S. *Utilitarianism.* ed. O. Piest. New York: Macmillan, 1957.

Murray, T. H. "Enhancement." In: *The Oxford Handbook of Bioethics*, ed. B. Steinbock. Oxford: Oxford University Press, 2009.

Persson, I. and J. Savulescu. "The Perils of Cognitive Enhancement and the Urgent Imperative to Enhance the Moral Character of Humanity." *Journal of Applied Philosophy* Vol. 25, No. 3, 2008, pp. 162–177.

Powell, R. and A. Buchanan. *The Evolution of Human Enhancement* (forthcoming).

Ross, W. D. *The Right and the Good.* Oxford: Oxford University Press, 1930.

Sandel, M. *The Case Against Perfection. Ethics in the Age of Genetic Engineering.* Cambridge, Massachusetts and London: Harvard University Press, 2007.

Savulescu, J. and I. Persson. "Moral Enhancement, Freedom and the God Machine." *The Monist*, Vol. 95, No. 3, 2012, pp. 399–421.

Temkin, L. *Inequality*. Oxford: Oxford University Press, 1997.

Walker, M. "Enhancing Genetic Virtue: A Project for Twenty-first Century Humanity." *Politics and the Life Sciences*, Vol. 2, No. 28, 2009, pp. 27–47.

Thomas Douglas
Enhancing Moral Conformity and Enhancing Moral Worth[1]

Abstract: It is plausible that we have moral reasons to become better at conforming to our moral reasons. However, it is not always clear what means to greater moral conformity we should adopt. John Harris has recently argued that we have reason to adopt traditional, deliberative means in preference to means that alter our affective or conative states directly – that is, without engaging our deliberative faculties. One of Harris' concerns about direct means is that they would produce only a superficial kind of moral improvement. Though they might increase our moral conformity, there is some deeper kind of moral improvement that they would fail to produce, or would produce to a lesser degree than more traditional means. I consider whether this concern might be justified by appeal to the concept of moral worth. I assess three attempts to show that, even where they were equally effective at increasing one's moral conformity, direct interventions would be less conducive to moral worth than typical deliberative alternatives. Each of these attempts is inspired by Kant's views on moral worth. Each, I argue, fails.

Keywords: Neuroenhancement, Moral enhancement, Moral improvement, Moral worth, Kant

Morality gives us reasons to do, and not to do, certain things. It may also sometimes give us reasons to do certain things from certain motives, but let us focus, for the moment, on moral reasons to act that are insensitive to one's motives for acting.

Let us say that an agent *conforms to morality* or *morally conforms* to the extent that her conduct coincides with these moral reasons. An agent fully conforms to morality on a given occasion when she performs an act that is at least as well supported by moral reasons as any alternative act, and she fully conforms to morality over a period of time when she performs a series of acts that is at least as well supported by moral reasons as any alternative series.[2]

Most of us would, given a moment's reflection, have little difficulty identifying various ways in which we regularly fail to fully conform to morality. Perhaps we

---

1 This article was previously published as "Enhancing Moral Conformity and Enhancing Moral Worth,"*Neuroethics*, Vol. 7 (2014), pp. 75–91.
2 In each case, "act" should be understood to include inaction.

are insufficiently attentive friends. Perhaps we labor under subconscious sexual and racial biases that lead to subtly discriminatory behavior. Or perhaps we do too little to prevent or correct global problems like environmental destruction and developing-world poverty. We may not think of these moral failures, taken in isolation, as particularly grievous, but we should acknowledge that they can aggregate with devastating effect. Arguably, our failures of moral conformity are, taken together, a driving force behind climate change and global poverty. It is also increasingly recognized that many of history's greatest atrocities – ranging from the First World War to the Final Solution to the Cultural Revolution – were made possible by the ordinary moral failures of ordinary people [1-3].

It is plausible that we have reasons to correct our moral failures, bringing it about that we better conform to morality.[3] However, this is not to say that we ought to pursue greater moral conformity by any means available. There may be some means to increased moral conformity that we have conclusive moral reasons to avoid, and even among means that are not absolutely ruled out in this way, some means might be better supported by moral reasons than others. There is, in my view, much interesting work to be done in assessing the morality of different possible means to greater moral conformity.[4]

In a recent series of articles, John Harris has begun to do this work [4–6]. Harris has argued that we have reason to adopt traditional, deliberative means of increasing moral conformity in preference to certain more novel means that have recently been discussed by a number of other authors [7–13].[5] He does not precisely delineate the classes of intervention that he favors and disfavors. However, he does identify some members of each class. For example, he explicitly places within the favored category attempts to increase one's moral conformity though development of a "sophisticated understanding of cause and effect" and through "self-education, wide reading and engagement with the world" [4:104].[6] On the other hand, he raises concerns about a class of interventions that has been defended by Thomas Douglas. Douglas argues that it would sometimes

---

3   "We" refers here to all moral agents who do not already fully conform to morality. This category plausibly includes all mentally competent adult persons that have ever existed.
4   In this paper I consider only interventions that aim to increase one's own moral conformity. I remain silent on interventions that aim to increase the moral conformity of *others*.
5   Harris does not specify the nature of these reasons, but I take them to be *pro tanto* moral reasons.
6   See also J. Harris, "What it's like to be good," *Cambridge Quarterly of Healthcare Ethics*, Vol. 21, No. 3 (2012).

be permissible for individuals to directly influence their emotions – for example, through the use of neurally active drugs – in ways that can be expected to leave them with morally better motives or conduct [9].[7] But Harris objects to interventions that are "targeted on the emotions" in this way [6:1]. He allows that the voluntary use of such interventions to increase one's moral conformity might sometimes be morally permissible or even desirable. But he argues that we nevertheless have reason to *prefer* more traditional, deliberative means to increased moral conformity.[8]

In this article, I assess one concern that might be offered in support of this view. I call this the Superficiality Concern.

1 The superficiality concern

Harris does not unambiguously state the Superficiality Concern, but it can be distilled from a number of asides that he offers while setting out other concerns.[9] The most revealing passages can be found in a discussion of Douglas' definition of a class of interventions – emotional moral enhancements – as interventions that (i) "will expectably leave an individual with more moral (*viz.*, morally better) motives or behavior than she would otherwise have had" and (ii) operate via the direct modulation of emotions [10:3]. The distinctive feature of *emotional moral enhancements*, according to Douglas, is that "once the enhancement has been initiated, there is no further need for cognition: emotions are modified

---

7 See, for similar arguments, H. S. Faust, "Should we select for genetic moral enhancement? A thought experiment using the MoralKinder (MK+) Haplotype," *Theoretical Medicine and Bioethics* Vol. 29, No. 6 (2008), pp. 397–416, and D. DeGrazia, "Moral Enhancement, Freedom, and What We (Should) Value in Moral Behavior," *Journal of Medical Ethics*, Vol. 40, 2014, pp. 361–368.
8 Relatedly, R. Sparrow, *Better Living Through Chemistry? A Reply to Savulescu and Persson on 'Moral Enhancement' Journal of Applied Philosophy*, Vol. 31, Issue 1, 2014, pp. 23–32 and "(Im)moral Technology? Thought Experiments and the Future of 'Mind Control'," in: *The Future of Bioethics: International Dialogues*, ed. Akira Akabayashi. Oxford: Oxford University Press, 2014, pp. 113–119. argues that we should prefer political means of improving moral conformity to pharmaceutical or neurotechnological ones. As we shall see, Sparrow's concerns regarding pharmaceutical and neurotechnological means substantially overlap with Harris' concerns regarding the direct modulation of emotions.
9 These are the concerns that the kinds of interventions defended by Douglas would restrict freedom and that attempts at such enhancements would frequently misfire, bringing about morally worse, not better, motives and conduct.

directly".[10] Harris objects that "[t]his so-called distinctive feature (...) shows that this concept cannot be moral enhancement properly so called at all". An intervention that operates in this way "is hardly an enhancement, and certainly not one that has much to do with morality" [5:4]. Indeed, he maintains that "the notion of moral behaviour has been attenuated to a vanishing point" once one claims that such behavior could be produced by directly altering emotions [5:6]; "tinkering with the emotions is not a form of moral enhancement at all. It is more like the threat of punishment: it may make immoral behaviour less likely, but it does not enhance morality" [6:3-4].

One way of reading these passages would see them as an outright denial of the possibility of morally improving motivation or conduct by directly manipulating emotions. However, surely Harris would accept that direct manipulation of emotions could result in at least one kind of moral improvement: it could increase the moral *conformity* of one's conduct.

Note first that the enhancement of moral conformity through directly modulating emotions is nomologically possible – that is to say, it does not violate any laws of nature. Emotions are mental states, mental states are normally taken to be either constitutively or causally dependent on brain states,[11] and brain states are in principle susceptible to direct modulation. Thus, it is nomologically possible for direct interventions to alter the emotions. And it is surely also nomologically possible for the alteration of one's emotions to affect the moral conformity of one's conduct.

Perhaps Harris' thought was not that it is nomologically impossible to increase moral conformity through direct emotion-modulation, but simply that this is

---

10 This passage is cited in Harris, "What it's like to be good?," p. 4 and attributed to T. Douglas, "Moral enhancement via direct emotion modulation: A reply to John Harris," *Bioethics*, Vol. 27, No. 3 (2011). However, the published version of the latter paper does not include this passage. Presumably, it comes from an earlier version of Douglas' article.

11 There are variants of mind-body dualism which deny that mental states are causally or constitutively dependent on brain states. For example, G. W. F. von Leibniz, "New system, and explanation of the new system" in *Philosophical writings*, ed. G. H. R. Parkinson, trans. Mary Morris (London: Dent, 1973), famously subscribed to mind-body parallelism, a version of dualism which takes mental states and bodily states to be causally independent of one another. However, such views are now philosophically obsolete. Contemporary dualists are typically either epiphenomenalists or interactionists, and both of these views allow that all mental phenomena have physical causes.

unlikely to become technologically feasible. However, this suggestion also seems dubious. Consider this case:[12]

> Andrew is a doctor working in multi-racial area. He was brought up in a racist environment and emotional responses introduced during his childhood still have a biasing influence on his conduct. For example, they incline him to take more care in treating White patients than Black patients. Andrew is aware of this aspect of his psychology and suspects it to be morally problematic. Hoping to mitigate his bias, he embarks on new programme developed by neuroscientists. He first observes stimuli that elicit racial aversion (such as photos of mixed race couples and civil rights protests) while undergoing high-resolution brain scanning to determine which neural connections mediate the aversion. Those connections are then selectively attenuated via regular sessions of transcranial electrical brain modulation. This programme significantly weakens his disposition to racial aversion and does indeed lead him to treat his Black and White patients more equally.

It is somewhat plausible that an intervention of the kind described here would increase Andrew's moral conformity, and it is not fantastic to suppose that such an intervention might be developed in the future. After all, transcranial electrical brain modulation can already be used to alter rather specific mental abilities such as numerical competence [17] and the ability to deceive others [18].

It seems difficult to deny the possibility of enhancing moral conformity through direct emotional modulation, and this is so whether possibility is understood as nomological possibility or as likely technical feasibility.[13] However, there is a more plausible way of understanding Harris' concern. We could instead take Harris' claim that the direct modulation of emotions could not produce "moral

---

12 The case is modified from Douglas, "Moral enhancement via direct emotion modulation", p. 2.
13 It might, however, plausibly be argued that no means of directly modulating emotions that are likely to be developed in the foreseeable future would produce *reliable* moral conformity. Some authors hold that (i) reliable moral conformity can only be achieved through the exercise of moral judgment, and (ii) that moral judgment cannot be codified as a simple decision-making procedure See, for example, R. Hursthouse, "Virtue theory and abortion," *Philosophy & Public Affairs*, Vol. 20, No. 3, pp. 230-231. It might follow that the only reliable way to improve moral conformity is to refine one's noncodifiable moral judgment, and it might seem unlikely that any direct emotion-modulating intervention developed in the foreseeable future could do this (I address this worry in section 7 below). However, even those who take this line would surely accept that these interventions could increase moral conformity in a semi-reliable way, for example, because they induce motivational states that happen to roughly mirror those that would have been produced by the exercise of mature moral judgment.

enhancement properly so-called" to be the claim that, although such modulation could increase moral conformity, there is some deeper variety of moral improvement that it would not produce, or, at least, that it would produce to a lesser degree than the more traditional ways of improving moral conformity that Harris favors.[14] Thus, it would result in a kind of moral improvement that is, in one respect at least, more superficial than that produced by these more traditional means.

It will be helpful to introduce some terminology here. Call an intervention undergone by some agent a *conformity enhancement* if and only if (i) one of the agent's aims, in undergoing the intervention, is to increase her moral conformity during some extended future time period, and (ii) the intervention succeeds in realizing that aim. The Superficiality Concern, as I will understand it, maintains that, though all conformity enhancements by definition increase moral conformity, some kinds of conformity enhancement – those that employ certain *direct* means – fail to produce, or fail to produce to the same degree, a deeper kind of moral improvement that is typically produced by traditional, deliberative conformity enhancements.

Similar claims have been made by other authors concerned by certain means of enhancing moral conformity. For example, Fabrice Jotterand [20:8] argues that neurotechnological interventions intended to increase moral conformity are "unlikely to morally enhance people in the true meaning of the word". Similarly, Robert Sparrow [14], suggests that

> "while there is indeed evidence that certain pharmaceutical and neuro-scientific interventions can alter dispositions and behaviour in ways that we may be inclined to morally evaluate positively, this falls well short of constituting 'moral bioenhancement' in any interesting sense (...). [T]he prospect of making people 'more moral' through pharmaceutical or surgical interventions is slim indeed."

He argues further that whereas commentators in this area have often supposed that "altering behaviour – to prevent someone acting immorally or to ensure that they do the right thing in some particular circumstances – is 'moral enhancement'", this is too quick:

---

14 Harris is most naturally interpreted as claiming that there is some deeper variety of moral improvement that the direct modulation of emotions could not produce *at all*, however, I here attribute to him only the weaker view that such interventions would not produce this deeper variety of moral improvement *to as great a degree* as the more traditional, deliberative interventions that he favors. As we shall see, even this weaker claim is difficult enough to sustain.

the use of the sedative gas can prevent someone completing an assault and we would hardly think that this was a case of moral enhancement. At the very least, moral bioenhancement must improve people's motivations.

However, (...) even altering motivation as well as behaviour seems to fall significantly short of enhancing individuals' morality. We are (...) all familiar with drugs that can alter how we feel (...) [A]nyone who has had a few glasses of beer knows that drugs can make us feel love where we would otherwise feel apathy or brave where we would normally be scared. In some circumstances, these chemically influenced emotions may even motivate us to do the right thing. Yet, again, it stretches credulity to call this "moral enhancement" (...). A stiff shot of whiskey might allow us to summon up the "courage" required to act morally in some particular instance but it will not succeed in making us "more moral".

The Superficiality Concern, as I have outlined it so far, is thus not unique to Harris. However, in discussing that Concern in what follows, I will guided primarily by Harris' discussion of the concern, drawing on other authors only insofar as their worries overlap with his. Accordingly, unless otherwise specified, "the Superficiality Concern" refers to Harris' variant of the concern.

2 The scope of the superficiality concern

More on the content of the Superficiality Concern will follow, but first, it will be useful to say something about which conformity enhancements fall within the scope of the Concern, and which fall without it.

It seems clear that Harris would raise the concern in relation to the intervention undergone by Andrew in the case set out above; this case is only slightly modified from one offered by Douglas as an example of an emotional moral enhancement, and Harris does nothing to exclude that case from the scope of his concerns. I take it that Harris is also committed to raising his concerns (including the Superficiality Concern) in relation to this case:

> *Bryony* is a student from a wealthy family. She suspects she ought to do more to help the global poor. She does occasionally do *something* to help, for example, giving small amounts to support famine relief when approached by charities, but most of the time, the world's most unfortunate are far from her thoughts, and when they do cross her mind, she has trouble drumming up the sort of sympathy that might motivate greater sacrifices on her part. In an attempt to remedy this, she sets up her television so that it regularly displays disturbing and graphic images of the effects of poverty, though for such brief periods that she does not consciously recognise them. Nevertheless, through subliminal effects, the images do increase her feelings of sympathy, and these feelings stimulate her to make a large donation to Oxfam.

Unlike most of the putatively problematic conformity enhancements discussed by Harris, Bryony's intervention does not employ biomedical technologies. However, it does manipulate emotions directly, where directness is understood, as by Harris and his interlocutors, as implying that once the intervention is set in motion, it requires no further engagement of deliberative faculties.[15] This suggests that it would fall within the scope of Harris' concerns.

On the other hand, as we have seen, Harris raises no concerns regarding – and indeed endorses – interventions that increase moral conformity through "self-education". I take it, then, that he would have no problem with the intervention described in this case:

> Like Bryony, *Chloe* is a student who suspects she ought to do more to help the global poor, but has trouble drumming up much sympathy for them. In an attempt to remedy this, she goes to her local library and borrows a number of books containing first-hand accounts of life in poverty. Reading and reflecting on this literature augments her feelings of sympathy, and these feelings stimulate her to make a large donation to Oxfam.[16]

This seems an uncontroversial example of self-education.

What is less clear is where Harris would place conformity enhancements that act directly on mental states, but not on emotions. Such, interventions might instead directly alter desires, intentions, or beliefs. These interventions would presumably not qualify as self-education, since that plausibly implies

---

15 Harris does, at one point, suggest that environmental manipulations might be immune to the concerns he raises where the decision to undergo such an intervention is itself motivated by deliberation or, as he puts it here, is the product of a "self-conscious strategy" (J. Harris, "Ethics is for bad guys!' Putting the 'moral' into moral enhancement," *Bioethics*. Vol. 27, No. 3 (2013), p. 2). However, it is difficult to see how he can consistently take this view, since he raises his concerns regarding biomedical interventions to manipulate emotions even where the decision to undergo the intervention is the result of deliberation. Harris also explicitly excludes from the scope of his concerns a case that is rather like the case of Bryony but which involves conscious processing of images of poverty. However, he excludes this case on the grounds that viewing and reflecting on the images counts as a kind of conscious deliberation (Harris, "Ethics is for bad guys", pp. 2–3). It is difficult to see how he could say the same about the case of Bryony, where the images exert their effect subliminally.

16 Some might object to Chloe's intervention on the grounds that she appears to be driven by a desire to become more moral, a motive that some have found problematic. See, for discussion, K. Sorensen, "The paradox of moral worth," *Journal of Philosophy*, Vol. 101, No. 9 (2004), pp. 465–483. However, Harris alludes to no such concern, so I see no reason to attribute it to him.

the alteration of mental states *though deliberation*. But nor are they explicitly mentioned by Harris as among the interventions which raise the Superficiality Concern.

There is, however, some textual support for interpreting the Superficiality Concern to be broader than a concern about only the direct manipulation of emotions. For example, in discussing his concerns about the direct modulation of emotions, Harris frequently adverts to the thought that these "bypass" moral reasoning, or moral reflection, or the exercise of moral judgment [5:2,4-5; 22:E183]. Presumably the thought is that these interventions are used to bring about mental transformations of a sort that could otherwise be achieved through these forms of moral deliberation. But the kinds of mental transformations typically induced by moral deliberation include not just changes in emotional states, but also, at the very least, changes in conative states, such as desires and intentions.

In addition, Harris elsewhere characterizes his concern as attaching to interventions that operate "directly on the mainsprings of action" [5:2]. This suggests that his concern is with the direct modulation of any motivating mental states, and again, these include conative states as well as affective ones.

In what follows, I will assume that Harris would raise the Superficiality Concern in relation to all conformity enhancements that operate by directly altering affective or conative states.[17] And I take it that an intervention directly modulates an affective or conative state just in case, once the intervention has been initiated, it alters that state without requiring the exercise of deliberative faculties. I will refer to conformity enhancements that meet these conditions as *brute* conformity enhancements and will take the interventions undergone by Andrew and Bryony above to be examples of such enhancements.

Brute conformity enhancements can be contrasted with what I will call deliberative conformity enhancements: conformity enhancements that consist in moral deliberation. These conformity enhancements might involve moral reasoning, introspective reflection on one's moral failures, or calm moral discussion with others. I take it that Chloe's intervention is a deliberative conformity enhancement. Harris' Superficiality Concern, as I will understand it, is that there is some important variety of moral improvement that brute conformity

---

17 I remain agnostic on whether normative judgments are or comprise conative or affective states, and I allow that, if they do not, then they do not fall within the scope of Harris' concern.

enhancements fail to produce, or produce only to a lesser degree than typical deliberative conformity enhancements.

It is not clear whether Harris would extend this concern to all brute conformity enhancements that might *in principle* be developed in the future – because they are nomologically possible – or only to those that might plausibly be developed within some restricted time frame. Clearly, his claim would be more difficult to sustain if it were interpreted in the former way. I thus opt for the latter interpretation in order to present his concern in the most plausible light. I henceforth take Harris to be raising the Superficiality Concern regarding all brute conformity enhancements of a sort that might plausibly be developed within the medium term future – the next 50 years or so. Unless otherwise specified, "brute conformity enhancements" refers only to these enhancements.

3 The superficiality concern and moral worth

In alleging that brute conformity enhancements are superficial, one is alleging that these enhancements fail to induce (as much of) some deeper kind of moral improvement. But which deeper kind of moral improvement, exactly? Possible candidates for deeper moral improvements might include increases in the moral virtue, moral responsibility, moral understanding, moral knowledge and perhaps even moral status of the agent, as well as increases in the moral virtue and moral worth of the agent's conduct. In what follows I focus on the last of these; I flesh out the Superficiality Concern as a concern that brute conformity enhancements are less conducive to morally worthy conduct than typical deliberative enhancements. (I believe, however, that much of what I will say could be re-framed in terms of moral virtue or moral responsibilitywithout substantially affecting the arguments.)

The distinction between conforming to morality and acting in a way that has moral worth has been a commonplace since Kant. To say that an action has "moral worth" is, in standard philosophical usage, to say that it reflects well, morally, on the agent – that the agent merits moral praise for having done that act.[18] It is possible for two acts to accord equally well with moral reasons, yet for one to have greater moral worth than the other. Nomy Arpaly gives the example of two people who donate to Oxfam. The two donate the same amount of money,

---

18 Note that, although this is a standard way of understanding moral worth, it is not the only way in which it has been understood in recent philosophical literature. For an example of an alternative understanding, see, for instance, Robert N. Johnson, "Kant's conception of merit," *Pacific Philosophical Quarterly* Vol. 77 (1996): pp. 313–337.

but one donates to improve the state of the world, while the other does so merely because her accountant advises it [24:69]. It is plausible that the actions of the two donors conform equally well to morality, but that the first agent's action has greater moral worth.

This sort of case opens the door to the possibility that different conformity enhancements might be equally effective in increasing moral conformity yet have different effects on moral worth. This in turn raises the possibility of fleshing out the Superficiality Concern as follows:

> (*The Moral Worth Claim*) For all brute conformity enhancements likely to be developed in the medium-term future, whenever an agent has a choice between pursuing that conformity enhancement or achieving the same increment in moral conformity via a typical deliberative conformity enhancement, adopting the brute conformity enhancement will result in less morally worthy conduct.

To avoid repeatedly stating this rather cumbersome claim, I will sometimes paraphrase it as follows: brute conformity enhancements are less conducive to moral worth – or confer less moral worth on the agent's subsequent conduct – than typical deliberative conformity enhancements.

Harris does not himself explicitly flesh out his Superficiality Concern in the way that the Moral Worth Claim fleshes it out. However, this way of spelling out the Concern does sit well with some of what he says in support of it. For example, Harris notes, in defence of his Superficiality Concern, that

> [o]ne can accidentally discover something of scientific importance, but one cannot be scientific, one cannot do science, accidentally. Doing science is a deliberative and disciplined process. It involves, for example, doing things like formulating and testing a hypothesis and looking for disconfirmatory evidence as well as for confirmatory evidence (...) Being moral is like being scientific [5:6].

Harris suggests here that a brute intervention could not help one to "be moral" in part because it could at best lead one to do something of moral importance *accidentally*. He does not say exactly what he means by "being moral", but his worry would make perfect sense if he were equating "being moral" with "acting in a morally worthy way", for it has often been thought that moral worth requires non-accidental moral conformity. While one can accidentally conform to morality, if one does, one's conduct will lack moral worth.[19]

The Moral Worth Claim also sits well with one of Harris' other concerns about brute conformity enhancements. Harris is concerned that brute conformity enhancements, or at least certain among them, might diminish or restrict the

---

19 I discuss this view further in section 7 below.

freedom to do wrong, and perhaps one reason why the freedom to do wrong is valuable is that it enables *rightful* action to have moral worth. If we were not free to do wrong, then arguably nothing we did would have moral worth.

Robert Sparrow can be interpreted as appealing more directly to the concept of moral worth to support his variants of the Superficiality Concern. He claims, for example, that acting morally

> requires that agents should respond in the right way to counterfactuals: if we praise someone for helping another person who is in need, our assessment that their action is morally admirable rests upon the thought that they should not have been motivated to help them in the same way if the other person were not in need. It is difficult to see how any pharmaceutical could cause us to have the appropriate beliefs about what moral action would consist in, not only in the current circumstances that we face but also in others that are both relevantly similar and dissimilar. It would be a good drug, indeed, that made us feel love only for what is worthy of love and brave only in the service of a just cause.[14]

Sparrow's use of the terms "praise" and "morally admirable" in this passage strongly suggests that, by acting morally, he means acting in a morally worthy way.

It may be reasonable, then, to read Harris as implicitly endorsing the Moral Worth Claim, and Sparrow as explicitly endorsing it, or at least something close to it. Moreover, even if these authors do not endorse the Claim, it strikes me as among the more plausible ways of spelling out Harris' Superficiality Concern, and indeed of supporting his view that we have reason to adopt deliberative conformity enhancements in preference to brute ones. Thus, it seems worth considering whether the Moral Worth Claim is correct. In what follows, I will assess three attempts to show that it is.

Throughout, I will simply grant that the Moral Worth Claim would indeed, if correct, support the view that Harris' ultimately wishes to defend; I take this to be the view that we would have reason to adopt a typical deliberative conformity enhancement in preference to any alternative brute conformity enhancement that might plausibly be developed in the medium-term future. In fact, it is not obvious that the Moral Worth Claim does support this view, for it is not obvious that we have any reason to promote moral worth in our own future conduct. Morally worthy conduct is conduct that merits praise; it is not clear that it also merits promotion.[20] However, some have argued that we do have reasons to

---

20 For claims that moral worth does not merit promotion, see, for example, work by Richard Henson, "What Kant Might Have Said: Moral Worth and the Overdetermination of Dutiful Action," *Philosophical Review*, Vol. 88, No. 1 (1979), pp. 39–54, and Allen Wood, *Kant's Ethical Thought* (Cambridge: Cambridge University Press, 1999), p. 30; "Moral

promote moral worth [e.g., 28:15-17; 29], and for the sake of argument, I shall assume that they are correct.

This assumption places some constraints on what will qualify as an adequate defence of the Moral Worth Claim. Defences will need to be consistent with this claim bearing positively on Harris' view about the preferability of deliberative conformity enhancements. For example, a defence should not establish the Moral Worth Claim at the price of conceding that we have no reason at all to promote moral worth in our future conduct. Moreover, since our motive for assessing the Moral Worth Claim derives from its putative support for Harris' view that we should prefer deliberative conformity enhancements, it is natural to require also that any adequate defence of the Moral Worth Claim would allow this claim to play *an interesting role* in supporting Harris' view. For example, a defence should not establish the Moral Worth Claim while also establishing that the Claim provides only exceptionally weak support for Harris' view. I will expect attempts to defend the Moral Worth Claim to be consistent with this Claim bearing both positively and interestingly on Harris' view that we should prefer typical deliberative conformity enhancements to brute alternatives.

4 Acting from the right motives

Why might the conduct produced by a brute conformity enhancement have less moral worth than the conduct produced by a deliberative conformity enhancement even where the conduct conforms equally well to morality? The obvious place to begin the search for an answer to this question is with Kant's idea that to have moral worth, an act must be done for the right reason or from the right motive. There is little agreement about what sorts of motives are the right ones, but

---

Worth, Moral Merit, and Acting from Duty" in: *The Free Development of Each: Studies on Freedom, Right, and Ethics in Classical German Philosophy*. Oxford: Oxford Unviersity Press, 2014. Allen Wood ("Moral Worth, Moral Merit, and Acting from Duty") puts the point particularly forcefully in the latter work, claiming that "if a moral agent is dedicated to a meritorious end (…) – for instance, relieving the suffering of many people – then she will naturally care much more about this end than she does whether some of her actions taken toward it have moral worth because they are done from duty. Indeed, Kant's theory does not justify the agent's concern with this at all, unless the case is one where she will fail to act in conformity with duty unless she acts from duty, and then it is dutifulness itself, not action from duty or moral worth, that matters to the agent. Discussions of the beginning of the Groundwork (…) often treat moral worth as something agents have reason to want their actions to have. But (…) I think this is a thoroughly misguided thought."

on one prevalent view, Kant's own view, the action must be done from the motive of duty. Kant held that "if any action is to be morally good, it is not enough that it should *conform* to the moral law – it must also be done *for the sake of the moral law*" [30:4:390].[21]

The view that morally worthy actions must be done from the motive of duty – henceforth the "Kantian view" – has often been taken to support the view that moral worth requires deliberation. It is perhaps not obvious that acting from the motive of duty must involve moral reasoning or any other deliberative process. However, on one standard Kantian position, it must; an agent acting from the motive of duty deliberates about what morality requires or recommends. For example, on Barbara Herman's early interpretation of Kant, "[f]or an action to have moral worth, moral considerations must determine how the agent *conceives* of his action (he *understands* his action to be what morality requires), and this conception of his action must then determine what he does" [31:375, my italics]. We should not ascribe moral worth to the conduct of a "man of sympathetic temper (...) whose helpful actions (...) are motivated by his natural response to the plight of others". Why should we not ascribe moral worth in this case? Because this man "acts because he is, literally, moved by others' distress. There need be no moral component in his *conception* of what he does" [31:376-7, my italics]. Note the central role given here to the deliberative concepts of "conceiving" and "understanding". On Herman's view, it will plainly not be enough, for one to act from the motive of duty, that one acts on impulses or inclinations that reliably track duty. One must *think* about one's duty.

This might be thought to explain why brute conformity enhancements are less conducive to moral worth than typical deliberative ones. After all, the conduct produced by brute conformity enhancements was arrived at in part through non-deliberative means. For example, the decision by Bryony, the apathetic student, to make a large donation to Oxfam was arrived at in part through subconscious processes caused by her subliminal imagery program.

There is a difficulty with this explanation, however. Some brute conformity enhancements – including ones that might plausibly become technologically feasible in the medium-term future – might operate precisely by facilitating the sort of deliberation that the Kantian, as we understand her, takes to be necessary for moral worth. Our earlier example of Andrew, the biased doctor, might, depending on how the details are filled out, be just such a case. The brain

---

21 Volume and page numbers are for the Prussian Academy edition of Kant's collected works. Italics in the original.

modulation program that attenuates Andrew's racial aversion may help to promote moral conformity precisely because it removes one impediment to the sort of sound moral deliberation that the Kantian values.[22] Even in the case of Bryony, the apathetic student who embarks on a program of subliminal imagery, it seems possible that the brute conformity enhancement operates by promoting sound moral reasoning. Perhaps by increasing her feelings of sympathy for strangers, Bryony's subliminal imagery program stimulates her to engage in Kantian-style moral reasoning about how to respond to global poverty. If Andrew's and Bryony's interventions operate as I have just suggested they might, there seems no reason to deny, on the basis of the Kantian view, that their subsequent actions have moral worth. True, brute processes played a role in bringing about these actions. But the proximate aetiology of the agent's action was entirely deliberative in each case. This, plausibly, is all that is necessary to act from the motive of duty. After all, that motive is standardly (though not, as we shall see, universally) taken to be a *proximate* cause of action.[23]

Nevertheless, Harris would, I take it, object to these interventions. Harris appears to regard the Superficiality Concern as most serious in cases where an intervention directly alters the agent's affective or conative states, *and* those alterations in turn directly affect the agent's conduct, without the need for deliberation [5:2-5; 6:4]. Thus, the sorts of cases I have just discussed would perhaps not attract his *most serious* censure. In these cases, changes to affective and conative states affect conduct only indirectly, by facilitating good deliberation. However, these cases nevertheless fall within the scope of his Superficiality Concern. As we have seen, Harris presents his Superficiality Concern as attaching to all conformity enhancements that directly alter emotions, and he can be naturally interpreted as raising it also in regard to those that directly alter conative states. He does not restrict his Superficiality Concern to the subset of these interventions in which alterations to affective and conative states themselves directly influence

---

22 Indeed, Douglas' initial proposal was that moral enhancements might operate by mitigating emotions that serve has barriers to good motivation on any plausible account of good motivation (including a broadly Kantian one).

23 In more recent work, Barbara Herman, "Making room for character," in: *Aristotle, Kant, and the Stoics: Rethinking happiness and duty*, ed. Stephen P. Engstrom, and Jennifer Whiting (Cambridge: Cambridge University Press, 1996), has considered the possibility that the "motive of duty" might be understood not as a proximate cause of an action, but as something that is dispersed among various other causes of action, both proximate and distal. I consider the relevance of distal motives to moral worth in the next section.

action.[24] Moreover, if he did restrict the scope of the Concern in this way, he would render it dialectically uninteresting, since none of the authors targeted by Harris have defended conformity enhancements that directly modulate *action*.[25]

5 Bypassing deliberation

Even if we accept that (i) morally worthy actions must be done from the motive of duty and (ii) acting from the motive of duty requires deliberation, it seems possible that brute conformity enhancements could be highly conducive to moral worth: they could bring it about that one acts from the (necessarily deliberative) motive of duty. Thus, though (i) and (ii) would plausibly support a restricted variant of the Moral Worth Claim – one that applies only to brute conformity enhancements that directly influence action – they do not support that claim in its original, more general form. The defender of that claim will need to establish that, even where brute conformity enhancements influence action by enabling sound moral deliberation, they fail to be as conducive to moral worth as typical deliberative conformity enhancements.

Although the Kantian view outlined above, as standardly interpreted, does not directly support this position, some ideas that have been thought to underpin that view may support it. One of these is the idea that the causal history or aetiology of an action matters in determining its moral worth. In respect of its proximate aetiology, the conduct produced by brute conformity enhancements might well be beyond reproach, meeting the Kantian requirement that morally worthy actions be done from the (deliberative) motive of duty. However, perhaps there is a problem further back in the aetiology of the conduct. Indeed, it has been argued that Kant himself should be understood as being more focused on distal motivation than my discussion in the previous section implies [27]. On this interpretation, the distal aetiology of an action can influence its moral worth.

---

24 Harris does exclude, from the scope of his concerns, conformity enhancements that operate via the biomedical enhancement of cognitive capacity (Harris, "What it's like to be good," p. 9; Harris, "'Ethics is for bad guys!,'" p. 4). This might lead one to suppose that he would have no objection to conformity enhancements that directly manipulate mental states *and thereby facilitate good deliberation.*
25 Indeed, Douglas' initial proposal was that "moral enhancements" might work by attenuating emotions that serve as barriers to good motivation on any plausible account of good motivation, including a broadly Kantian one which takes sound moral reasoning to be the only good motive (T. Douglas, "Moral enhancement," *Journal of Applied Philosophy*, Vol. 25, No. 3 (2008), pp. 228–245).

What might be wrong with the distal aetiology of the conduct produced by brute conformity enhancements? One suggestion would be that the conduct does not *originate* in the deliberation of the agent. This would, I think, be a somewhat promising way of objecting to the imposition of brute conformity enhancements on *others*. Where *A* imposes a brute conformity enhancement on *B*, *B*'s subsequent conduct might be thought to originate not in *B*'s deliberation, but in *A*'s, and this might be thought to detract from its moral worth. However, as the basis for a general worry about brute conformity enhancements, the suggestion seems unpromising. Though it is not at all clear how we should understand the origin of an item of conduct, on any plausible characterization, it seems that the conduct induced by brute conformity enhancements *could* originate in the deliberation of the agent. This is because the decision to engage in the brute enhancement may itself be arrived at through deliberation.

A more promising suggestion would be that the problem with conduct induced by brute conformity enhancements is that its aetiology was not deliberative all the way down. That is to say, some steps in the aetiology of that conduct that could in principle have been accomplished through deliberation are bypassed – they are taken through non-deliberative means.[26] Andrew, the biased doctor, attenuates his racial aversion, and mitigates his biased conduct, through a program of electrical brain modulation. He might, perhaps, have achieved the same attenuation of racial aversion through deliberation, for example, by reflecting on his racial aversions, and perhaps by reading about their likely effects. But Andrew did not take these deliberative steps – he bypassed them. He does leave himself with some deliberative work to do. Following the electrical brain modulation program, he must still deliberate about, for example, how to treat a patient on a particular occasion. Thus, he has not *entirely* bypassed deliberative processes. However, in attenuating his racial aversion via the use of pharmaceuticals, he has used brute means to make some progress towards moral conformity in the sense that he has strengthened his disposition to morally conform. This is progress that he could in principle have made deliberatively, for example, through introspective reflection or reflective engagement with literature.

---

26 Of course, it is plausible that no conduct is deliberative "all the way down" in the strong sense of being motivated wholly through deliberation, and not at all through nondeliberative channels. The relevant point here is that the actions that result from brute conformity enhancements might be thought less deliberative than they could have been.

162   Thomas Douglas

6 Avoiding effort

Why should bypassing deliberation limit the moral worth of one's subsequent conduct? One answer is suggested by reflecting on the following pair of cases:

Compared to his peers, *David* conducts himself in a way that accords well with the moral reasons that apply to him. Indeed, he finds it easy to morally conform since he was brought up in a nurturing family where responsibility and moral sensitivity were encouraged and his role models seldom exhibited or endorsed objectionable moral attitudes. He also lives in a society that has internalised few problematic norms and encourages moral reflection and open moral discussion. It is not that he *automatically* does what morality requires; he frequently has to deliberate about what to do. But his deliberation is seldom biased or disrupted by powerful impulses or misguided social pressures, and sound deliberation is facilitated by the ease with which he is able to imagine the consequences of his actions and empathise with those he affects.

Unlike David, *Felix* was raised in a dysfunctional family where violence was openly encouraged, bigoted attitudes were routinely expressed and endorsed, and moral sensitivity was viewed as a sign of weakness. He also lives in a society that has embraced an objectionable moral code, so social pressures militate strongly in favour of moral nonconformity. Nevertheless, Felix frequently engages in moral deliberation and, despite the distorting influence of deeply engrained emotions and the consistently negative influence of those around him, he is able to conform well to morality – as well, in fact, as David.

Some would, I think, intuit that, at least in one respect, Felix's conduct has greater moral worth than David's. One natural way of accounting for this intuitive response would be to hold that expending moral effort – effort to morally conform – confers moral worth on the resulting actions. Felix's actions have (in one respect) greater moral worth than David's because he expended, on average, greater moral effort in bringing about that conduct.

Others, I suspect, would reject the intuitive response that I have just noted, but in any case, the view that expending moral effort confers moral worth on one's actions has enough philosophical support that it ought to be taken seriously.[27] Moreover, it might seem that, if that view is correct, it will lend support to the Moral Worth Claim. Deliberation is typically an effortful process, but perhaps

---

27 See, for example, K. Sorensen, "Effort and moral worth," *Ethical Theory and Moral Practice* Vol. 13, No.1 (2010), pp. 89–109. Kant himself also offers some support for this view in *The Metaphysics of Morals*, writing that "[t]he greater the natural obstacles (of sensibility) (...) so much the more merit is to be accounted for a good deed" (I. Kant, *Metaphysics of Morals*, trans. M. Gregor (New York: Cambridge University Press, 1991), Vol. 6, p. 228. Volume and page numbers are for the Prussian Academy edition of Kant's collected works).

directly altering one's affective or conative states is not. It might be thought that when an individual could achieve a given increment in moral conformity though either undergoing a brute conformity enhancement or engaging in deliberation, the brute conformity enhancement will invariably involve exerting less effort.[28] If this is so, then we might have good grounds to suppose that brute conformity enhancements are less conducive to moral worth than typical deliberative conformity enhancements.[29] Undergoing a brute conformity enhancement will make one more like David and less like Felix.

One problem faced by this line or argument is that, intuitively, actions can possess a very high degree of moral worth even if they are relatively effortless.[30] David's moral conformity required less moral effort than Felix's, and perhaps this makes David's conduct somewhat less worthy, at least in one respect. Nevertheless, it is plausible that David's actions frequently possess a very high degree of moral worth. Even those who defend the view that moral effort confers moral worth allow that relatively effortless actions can also be highly morally worthy, because moral effort is not the *only* ground of moral worth. Indeed, on one view, high levels of moral worth are attained at both ends of the spectrum of moral effort, namely, in (i) cases where heroic levels of moral effort are exerted, *and* (ii) cases in which very little moral effort is exerted, because little is needed: the agent's motives are so well-aligned with her moral reasons [33]. If this view is correct, then actions produced through brute conformity enhancements might possess a high degree of moral worth even if the brute conformity enhancement significantly diminishes the amount of effort necessary to perform those actions.

It might be objected at this point that, though actions produced by brute conformity enhancements could be highly morally worthy, they are nevertheless *less* morally worthy than comparable actions produced through effortful deliberative moral remedies. For example, it might be held that, though there are routes to

---

28 It is not at all obvious that brute conformity enhancements would always involve less effort than alternative deliberative ones. One might imagine, for example, that some individuals would have to exert rather great moral effort in order to undergo a brute conformity enhancement because, say, they feel repulsed by the thought of directly influencing their conative of affective states.
29 Though this would not follow trivially, for it could be that moral effort expended is only one factor that confers moral worth. Thus, even if brute conformity enhancements invariably produce action that is less morally worthy in respect of the effort exerted in bringing it about, it might be more morally worthy in other respects that more than compensate for this.
30 I thank an anonymous reviewer for pressing me to discuss this point.

moral worth besides the exertion of moral effort, *ceteris paribus*, more moral effort results in greater moral worth. Thus, suppose *Ervin* started out with a psychology rather like Felix's, so found it difficult to conform to morality, but then underwent a conformity enhancement, which left him with a psychology like David's. Ervin's later actions, like David's, might possess a high degree of moral worth. Yet it might be thought that they would have been even worthier had he achieved the same transformation (to a David-like psychological set-up) through effortful, deliberative means rather than effortless, brute ones.

I am not convinced that this response can succeed. An Aristotelian might, for example, maintain that, even if exerting effort can confer some degree of moral worth on one's actions, the morally *worthiest* actions are those whose aetiology features no significant moral effort. I cannot defend that view here. But it is, I think, a reasonable one. Moreover, if this view is correct, then, even if adopting a brute conformity enhancement in preference to a deliberative alternative avoids the exertion of moral effort, it may have no negative influence on moral worth. However, I will not pursue this thought. Instead, I turn to what is, I think, an even more serious problem for the effort-based argument for the Moral Worth Claim.

The problems that I have been discussing stem from the fact that high degrees of moral effort are clearly not *necessary* for an action to have high moral worth, and may not even be necessary for the action to have maximal moral worth. But there is another problem: exerting moral effort does not always confer moral worth, that is to say, it is not *sufficient* for it.

Thus, recall Chloe, who achieved an increment in moral conformity though reading and reflecting on first-hand reports of life in poverty. Suppose that Chloe could also have achieved this increment through pure introspective reflection – that is, *without* the aid of literature. These options, both deliberative, would have been equally effective in increasing moral conformity, but suppose that the purely introspective route to increased conformity would have required greater effort. If this additional effort would have been gratuitous – that is to say, if, leaving aside considerations of moral worth, Chloe had no more reason to adopt the more effortful route than the less effortful one – then it seems very doubtful whether exerting that additional effort would have conferred any moral worth on her subsequent conduct. It would surely be surprising if moral worth could be bought through the exertion of effort that there is no moral reason to exert.

Consideration of this case undermines the simple view that moral effort confers moral worth since it suggests that gratuitous moral effort does not. However, it also suggests a more plausible view – namely, the view that *nongratuitous* moral effort confers moral worth. (Note that this modified view might also seem able to account for the intuitive responses that I speculated some might have to the *David* and *Felix* cases, for we might well suspect that Felix has exerted more nongratuitous effort to morally conform than has David.)

Importantly, however, it seems doubtful whether this modified view will be helpful to anyone who wishes to appeal to the Moral Worth Claim in order to defend Harris' view – namely, the view we have reason to prefer deliberative conformity enhancements to brute alternatives.

To see the problem, suppose that an agent can bring about a given increment in moral conformity either through a (more effortful) deliberative route or a (less effortful) brute intervention. Suppose initially that an agent has more reason, leaving aside considerations of moral worth, to adopt the deliberative route than the brute alternative. In that case, the additional effort entailed by the deliberative route will be nongratuitous and will thus contribute to the moral worth of the agent's subsequent conduct.

But now suppose instead that, leaving aside considerations of moral worth, the agent does not have more reason to adopt the deliberative route. In this second variant of the case, if the agent opts for the more effortful deliberative route, she will simply be exerting gratuitous effort, and this will confer no moral worth on her subsequent conduct.

Thus we see that adopting a more effortful deliberative conformity enhancement in preference to a less effortful brute alternative confers greater moral worth on one's subsequent conduct only if one already has most reason to prefer that option. Insofar as exerting nongratuitous effort is what matters for moral worth, considerations of moral worth will at most add a supplementary reason to prefer the deliberative route. They will never give an agent most reason to prefer the deliberative route when she would not already have had most reason to do so.

This effectively relegates the Moral Worth Claim to an accessory role in justifying Harris' view that we have reason to adopt deliberative conformity enhancements in preference to brute ones. An appeal to effort may support the Moral Worth Claim *if* we assume that there would be more reason to engage in a typical deliberative conformity enhancement than any brute alternative. This assumption guarantees that the additional effort associated with the deliberative route is nongratuitous. However, if we assume this, then there is no need to appeal to the Moral Worth Claim in order to justify Harris view; we already have good grounds to accept it. I take it, then, that a defender of Harris' view would not want to defend the Moral Worth Claim in this way. Doing so would deprive the claim of most of its interest as a basis for preferring deliberative conformity enhancements.

7 Unreliable moral conformity

An alternative defence of the Moral Worth Claim would appeal to the thought that brute conformity enhancements would produce less reliable moral conformity than typical deliberative conformity enhancements.

According to the Kantian view described earlier, an action must be done from the motive of duty if it is to have moral worth. One of the thoughts that has often been taken to support this view is the thought that actions which accidentally conform to morality lack moral worth. As Barbara Herman puts it, the action's moral conformity must be "the nonaccidental effect of the agent's concern" – "we need to know that it was no accident that the agent acted as duty required" [31:366,368].

In Kant's famous example, a shopkeeper charges his customers fair prices – in conformity with morality – but does so solely in order to maximize his own profit. Kant maintains that the shopkeeper's actions lack moral worth. Herman explains why as follows:

> the moral fault with the profit motive is that it is unreliable. When it leads to dutiful actions, it does so for circumstantial reasons (...) This example suggests the need for a motive that will guarantee that the right action will be done [31:363].

Turning to Kant's "sympathetic man", whose natural inclinations lead him to help those in distress, Herman again cites the accidental nature of the man's moral conformity in explaining why his conduct lacks moral worth:

> He acts because he is, literally, moved by others' distress. There need be no moral component in his conception of what he does. Therefore, nothing in what motivates him would prevent his acting in a morally impermissible way if that were helpful to others, and it is to be regarded as a bit of good luck that he happens to have the inclination to act as morality requires [31:377].

This idea – that the moral conformity of an action cannot be an accident if the action is to have moral worth – has been found plausible by many, including some who reject the Kantian view that morally worthy actions must be done from the motive of duty [e.g., 35:206]. There are different ways in which we might make sense of the idea of nonaccidental or reliable moral conformity, but on one standard account, for an agent's moral conformity to be reliable, it must be the case that the agent would also have morally conformed in possible worlds apart from the actual one.[31] Determining precisely which possible worlds bear on the reliability of an agent's moral conformity is a complicated matter, and

---

31 On a variant of this view, what matters is whether the agent would have morally conformed *for the right reasons* in other possible worlds. I do not explicitly discuss

I cannot address it here. I simply assume that it is in principle possible to provide an account of reliable conformity in terms of counterfactuals, and also that it is in principle possible, by enumerating the number of relevant counterfactual scenarios in which the agent would have conformed, to use this account to generate a measure of the reliability of an agent's moral conformity.

These thoughts on the *reliability* of moral conformity are relevant to our assessment of the Moral Worth Objection to brute conformity enhancements, for it might plausibly be thought that brute conformity enhancements would invariably fail to produce such reliable moral conformity, understood in counterfactual terms, as can be achieved through deliberation. Deliberative conformity enhancements frequently work by enhancing an agent's moral knowledge, moral understanding or moral judgment – henceforth, collectively, her moral-epistemic resources. Through deliberation, the agent comes to know that she has certain moral reasons, acquires an understanding of why she has certain moral reasons, or becomes better at assessing and weighing moral reasons. These moral-epistemic resources are all-purpose tools that help her to morally conform in many circumstances. If an agent knows what moral reasons there are and understands why, and if she is good at assessing and weighing these moral reasons, then, provided she is also somewhat disposed to act in accordance with her moral judgments, she will be well-placed to do what she has most moral reason to do in almost any circumstance. Admittedly, moral-epistemic resources do not translate into moral conformity in all circumstances. One can know what moral reasons there are, or even correctly judge what morality requires of one in a particular case, yet fail to morally conform, for example, due to weakness of will. Still, an agent who possesses substantial moral-epistemic resources will generally be disposed to morally conform across a wide range of possible circumstances.[32]

---

this view in what follows. However, this view could be substituted for the one I do discuss without substantially affecting my arguments.

32 I am suggesting here that possession of the moral-epistemic resources might contribute to reliable moral conformity and thereby be relevant to moral worth. Some would argue that the moral-epistemic resources are relevant to moral worth as well or instead in a more direct way. For example, Alison Hills has recently argued that acting on the basis of moral understanding contributes directly to the moral worth of one's action (A. Hills, "Moral testimony and moral epistemology," *Ethics*, Vol. 120, No. 1 (2009), pp. 94–127). I will not explicitly address the view that moral-epistemic resources are directly relevant to moral worth, however, what I say below regarding the view that they are *indirectly* relevant applies equally to the direct view.

On the other hand, it might be thought that *brute* conformity enhancements would not normally operate by enhancing the agent's moral-epistemic resources. Rather, they would typically work by removing some relatively straightforward affective or conative obstacle to moral conformity. The most obvious examples of such obstacles might include a tendency toward impulsive violence, strongly xenophobic sentiments or a disinclination to feel sympathy for strangers. But note that these are not *universal* barriers to moral conformity. They obstruct moral conformity only in certain circumstances. Consider the tendency towards impulsive violence. While this may often be a barrier to moral conformity, there are circumstances in which it might instead be conducive to such conformity; for example, when one is fighting a just war, or perhaps when one is confronted with one person assaulting another on the street. Consider alternatively a tendency to be indifferent to the suffering of others. Though this might often be a barrier to moral conformity, there are circumstances in which it would be conducive to it: these may include those circumstances in which one is an emergency medic surrounded by severe pain and suffering or a judge charged with impartially weighing the claims of plaintiff and defendant. In these settings, a degree of indifference to the suffering of strangers may lead to greater moral conformity.

Though brute interventions, which attenuate the tendency toward impulsive violence or lessen indifference to the suffering of strangers might *in fact* increase moral conformity, the moral conformity they produce will be highly contingent on what circumstances obtain. This is one important respect in which any moral conformity produced by such interventions is *unreliable* – perhaps more unreliable, typically, than that produced by deliberative interventions.

There is also another. Whether tendencies towards impulsive violence and indifference to the suffering of strangers impede moral conformity depends on the degree to which those tendencies are present. For example, though a strong tendency towards impulsive violence is unlikely to be conducive to moral conformity, a milder tendency of the same kind may well be conducive to it, for example, because it helps to prevent excessively submissive conduct. Similarly, though thoroughgoing indifference to the suffering of strangers may well impede moral conformity, *some* tendency to ignore the suffering of strangers is presumably conducive to moral conformity; an individual overwhelmed by sympathetic responses to the suffering of others is unlikely to conform well to morality.

These thoughts suggest that there is a further respect in which brute conformity enhancements may produce less reliable moral conformity than deliberative ones. The moral conformity produced by these interventions may be more contingent on the degree to which they alter the targeted psychological trait. Consider the earlier case of Bryony. We supposed that her sympathy-enhancing intervention increased her moral conformity because it enabled her to better conform to her moral reasons to provide humanitarian aid. However, it is easy to imagine that, had her sympathies become a little stronger, she might instead have conformed less well. Perhaps she would then have been overwhelmed, and thus paralyzed, by those sympathies. By contrast, when an individual increases her moral conformity by augmenting her moral-epistemic resources, her moral conformitywill generally be more robust across different magnitudes of change. For example, it will normally be the case that her moral conformity would also have increased had she augmented her moral-epistemic resources to a slightly greater or slightly lesser degree.

There are, then, at least two *prima facie* reasons to suppose that brute conformity enhancements will produce less reliable moral conformity than typical deliberative alternatives: it may be that the moral conformity induced by brute conformity enhancements is both more contingent on the circumstances and more sensitive to the magnitude of the transformation that the enhancement induces.[33] If the moral worth of an act that conforms to morality is a function of the reliability of that conformity, these considerations will help to explain why brute conformity enhancements are less conducive to moral worth than their deliberative counterparts.

In response to this attempt to justify the Moral Worth Claim, an Aristotelian thought concerning the connection between moral conformity and moral knowledge might be inserted into the discussion. On one Aristotelian account of moral education, moral knowledge is acquired in part *through* conforming to morality. In the moral sphere, we "learn by doing".[34] If this account is correct, we should expect that brute conformity enhancements, like paradigmatic deliberative ones, will typically produce moral knowledge, and thus, other things being equal, augment our moral-epistemic resources. Brute conformity enhancements by definition enhance moral conformity, and moral conformity is itself conducive to the acquisition of moral knowledge. Thus, it might seem, the problems about unreliability have been overstated. Moral knowledge reliably produces

---

33 Note that the point here is not that those who have undergone brute conformity enhancements fail to act on reliable *motives*. The problem is rather that in someone who acts on a reliable motive as a result of having undergone some brute intervention, the presence of the reliable motive will itself be problematically contingent on circumstances.

34 See, for the classic presentation of this interpretation, M. F. Burnyeat, "Aristotle on learning to be good," in: *Essays on Aristotle's Ethics*, ed. A. O. Rorty, (Berkeley, CA: University of California Press, 1980), pp. 69–92.

moral conformity, and brute conformity enhancements tend to produce moral knowledge.

This reply engages in a problematic form of bootstrapping, however. We were interested in assessing the reliability of the moral conformity produced by brute conformity enhancements. It will not help, in addressing this issue, to maintain that *insofar as these interventions do increase moral conformity* they will also produce moral knowledge, which is a reliable promoter of moral conformity. Nothing has been done to resolve the initial unreliability: the unreliability with which the brute psychological change induced by the intervention produces moral conformity.

There is, however, an alternative, more persuasive response to the present attempt to justify the Moral Worth Claim. The response begins with the thought that brute conformity enhancements could operate by lowering barriers to moral knowledge, moral understanding or moral judgment. For example, they could improve moral conformity by attenuating some emotion or desire that acts as a barrier to clear thinking or vivid imagination, both of which plausibly facilitate the acquisition of moral knowledge, understanding and judgment. Of course, even brute enhancements of this sort would, in an important sense, be unreliable. Though they might in fact augment one's moral-epistemic resources, they would not reliably do so. For example, though pharmacologically augmenting the capacity for vivid imagination, by lowering some barrier to it, might sometimes produce better moral judgment, there are circumstances in which it would fail to do so. Consider a case in which, whatever an agent will do, the consequences will be horrific. The agent's task is to select which of two serious atrocities to prevent, say. In this sort of case, one might think that vividly imagining the outcomes would serve only to traumatize the agent in a way that is likely to cloud the agent's moral judgment.

Note, however, that typical *deliberative* conformity enhancements are unreliable in precisely the same way. Consider an agent, Dennis, who has racist beliefs, but on some level recognizes that his beliefs are racist and therefore objectionable. Suppose Dennis seeks to confront his racism through deliberation – say, by reading and reflecting on the *Adventures of Huckleberry Finn*. Such deliberation might well increase his moral knowledge, and might thereby improve his moral conformity. However, even if we accept that moral knowledge reliably produces moral conformity, there is a sense in which Dennis' moral conformity, following his deliberation, is accidental. It is accidental in that his reading and reflecting on *Huckleberry Finn* was not *guaranteed* to produce moral knowledge, at least if Dennis is an ordinary person. *Perfect* moral deliberation might invariably produce moral knowledge. But no ordinary person is a perfect moral deliberator – someone who, when she engages in moral deliberation, always does so perfectly. Moreover, we have particular reasons to doubt Dennis' deliberative abilities: by hypothesis, he has racist beliefs, and we might worry that these beliefs, or emotional reactions that may underpin them, will infect his deliberation. For example, perhaps they will lead him, when deliberating, to selectively read the evidence in a way that helps to maintain those beliefs.

More generally, an agent's moral deliberation will enhance that agent's moral-epistemic resources – her moral knowledge, understanding and judgment – only when it goes well, and this is something that cannot, in an ordinary person, be relied upon. It depends, for example, on the absence of certain external impediments to good deliberation (the absence of temptations and distractions, say). So even where deliberation *does* augment an agent's moral-epistemic resources, and thus leads to greater moral conformity, there is an important sense in which that moral conformity is not reliable. The acquisition of the moral-epistemic resources was itself contingent on favorable circumstances for deliberation.

This is the same problem that we identified in the case of *brute* conformity enhancements. Though such enhancements may augment the moral-epistemic resources, which in turn reliably produce moral conformity, they do not reliably do so. They do so only contingent on favorable circumstances obtaining.

There may, of course, be some individuals – those particularly disposed to sound moral deliberation – in whom moral deliberation will produce more reliable moral conformity than any alternative brute conformity enhancement and will thus, perhaps, be the route to moral conformity most conducive to moral worth. But this does not help the proponent of the Moral Worth Claim, who maintains that *typical* deliberative conformity enhancements are more conducive to moral worth than any brute alternatives that might plausibly be developed in the medium term future. Given that the moral conformity produced by both kinds of conformity enhancement is unreliable in the same sort of way, there seems little reason to suppose that typical deliberative conformity enhancements will produce more reliable moral conformity than brute conformity enhancements of this sort.

8 Conclusions

John Harris holds that we have reason to adopt typical deliberative conformity enhancements in preference to brute conformity enhancements. I granted that the Moral Worth Claim supports this view and considered three attempts to justify that Claim. Each of these attempts drew on ideas that have often been associated with a Kantian approach to moral worth. The first attempt appealed to the standard Kantian view that to have moral worth, the action must be done from the motive of duty. I argued that this attempt fails because, on a standard understanding of acting on the motive of duty, brute conformity enhancements could bring it about that one acts on precisely this motive. The second attempt appealed to the view that moral effort confers moral worth along with the view that brute conformity enhancements would be less effortful than typical deliberative alternatives. I argued that, at best, this attempt succeeds in establishing the Moral Worth Claim only on the assumption that, leaving aside considerations of moral worth, we have more reason to pursue deliberative conformity enhancements than brute alternatives. This assumption deprives the Moral Worth Claim of its interest as a way of defending Harris view. Finally, the third attempt appealed to the view that moral worth requires reliable moral conformity along with the claim that brute conformity enhancements would produce less reliable moral conformity than typical deliberative alternatives. I responded to this attempt by arguing that both kinds of conformity enhancement are unreliable in the same sort of way.

Where does this leave Harris' view that we should prefer deliberative conformity enhancements to brute alternatives? Plainly this depends on (i) whether it is possible to establish the Moral Worth Claim via some route that have not pursued here, and (ii) the persuasiveness of Harris' other concerns about brute conformity enhancements can be plausibly understood in terms other than those specified by the Moral Worth Claim. It also depends on the persuasiveness of Harris' other concerns about brute conformity enhancements.[35] However, the failure of the three attempted justifications for the Moral Worth Claim considered here should, I think, at least significantly lower our credence in Harris' view. This is because I believe that those three attempts are perhaps the most promising means of establishing Harris' view.

---

35 For responses to these concerns, see DeGrazia, "Moral Enhancement, Freedom, and What We (Should) Value in Moral Behavior", Douglas, "Moral enhancement", Douglas, "Moral enhancement via direct emotion modulation", I. Savulescu and J. Persson, "Moral transhumanism," *Journal of Medicine and Philosophy*, Vol. 35, No. 6 (2010), pp. 656–669. and I. Savulescu, T. Douglas and J. Persson, *Autonomy and the ethics of biological behaviour modification*, (unpublished, 2013).

References

Arpaly, Nomy. *Unprincipled Virtue: An Inquiry into Moral Agency.* New York: Oxford University Press, 2003.

Audi, Robert. "Moral Virtue and Reasons for Action." *Philosophical Issues,* Vol. 19, No. 1, 2009, pp. 1–20.

Burnyeat, M. F. "Aristotle on Learning to be Good." In: *Essays on Aristotle's Ethics,* ed. A. O. Rorty. Berkeley, CA: University of California Press, 1980, pp. 69–92.

Cohen, K. R., S. Soskic, T. Iuculano, R. Kanai and V. Walsh. "Modulating Neuronal Activity Produces Specific and Long-Lasting Changes in Numerical Competence." *Current Biology,* Vol. 20, No. 22, 2010, pp. 2016–2020.

Crockett, M. J., L. Clark, M. D. Hauser and T. W. Robbins. "Serotonin Selectively Influences Moral Judgment and Behavior Through Effects on Harm Aversion." *Proceedings of the National Academy of Sciences,* Vol. 107, No. 40, 2010, pp. 17433–17438.

Darley, J. M. "Social Organization for the Production of Evil." *Psychological Inquiry,* Vol. 3, No. 2, 1992, pp. 199–218.

DeGrazia, D. "Moral Enhancement, Freedom, and What We (Should) Value in Moral Behavior." *Journal of Medical Ethics,* Vol. 40, 2014, pp. 361–368.

Douglas, T. "Moral Enhancement." *Journal of Applied Philosophy,* Vol. 25, No. 3, 2008, pp. 228–245.

Douglas, T. "Moral Enhancement via Direct Emotion Modulation: A Reply to John Harris." *Bioethics,* Vol. 27, No. 3, 2011, doi:10.1111/j.1467-8519.2011.01919.x.

Faust, H. S. "Should We Select for Genetic Moral Enhancement? A Thought Experiment Using the MoralKinder (MK+) Haplotype." *Theoretical Medicine and Bioethics,* Vol. 29, No. 6, 2008, pp. 397–416.

Glover, J. *Humanity: A Moral History of the Twentieth Century.* London: Jonathan Cape, 1999.

Harris, J. "Moral Enhancement and Freedom." *Bioethics,* Vol. 25, No. 2, 2011, pp. 102–111.

Harris, J. "What it's Like to be Good." *Cambridge Quarterly of Healthcare Ethics,* Vol. 21, No. 3, 2012a, doi: 10.1017/S0963180111000867.

Harris, J. "'Ethics is for Bad Guys!' Putting the 'Moral' into Moral Enhancement." *Bioethics,* Vol. 27, No. 3, 2012b, doi: 10.1111/j.1467-8519.2011.01946.x.

Harris, J. and S. Chan. "Moral Behavior is not What It Seems." *Proceedings of the National Academy of Sciences*, Vol. 107, No. 50, 2010, E183.

Henson, R. G. "What Kant Might Have Said: Moral Worth and the Overdetermination of Dutiful Action." *Philosophical Review*, Vol. 88, No. 1, 1979, pp. 39–54.

Herman, B. "On the Value of Acting from the Motive of Duty." *Philosophical Review*, Vol. 90, No. 3, 1981, pp. 359–382.

Herman, B. "Making Room for Character." In: *Aristotle, Kant, and the Stoics : Rethinking Happiness and Duty*, eds. Stephen P. Engstrom and Jennifer Whiting. Cambridge: Cambridge University Press, 1996.

Hills, A. "Moral Testimony and Moral Epistemology." *Ethics*, Vol. 120, No. 1, 2009, pp. 94–127.

Hurka, T. *Virtue, Vice, and Value*. New York: Oxford University Press, 2001.

Hursthouse, R. "Virtue Theory and Abortion." *Philosophy & Public Affairs*, Vol. 20, No. 3, 1991, pp. 223–246.

Johnson, R. N. "Kant's Conception of Merit." *Pacific Philosophical Quarterly*, Vol. 77, 1996, pp. 313–337.

Jotterand, F. "'Virtue Engineering' and Moral Agency: Will Post-Humans Still Need the Virtues?" *American Jouranl of Bioethics—Neuroscience*, Vol. 2, No. 4, 2011, pp. 3–9.

Kant, I. *Groundwork of the Metaphysic of Morals*. Trans. H. J. Paton, 1st Harper torchbook ed. New York: Harper & Row, 1964 [1785].

Kant, I. *The Metaphysics of Morals*. Trans. M. Gregor. New York: Cambridge University Press, 1991 [1797].

Karim, A. A., M. Schneider, M. Lotze, R. Veit, P. Sauseng, C. Braun and N. Birbaumer. "The Truth about Lying: Inhibition of the Anterior Prefrontal Cortex Improves Deceptive Behavior." *Cerebral Cortex*, Vol. 20, No. 1, 2010, pp. 205–213.

von Leibniz, G. W. "New System, and Explanation of the New System." In: *Philosophical Writings*, ed. G. H. R. Parkinson, Trans. Mary Morris. London: Dent, 1973 [1695].

Lifton, R. J. *The Nazi Doctors: Medical Killing and the Psychology of Genocide*. New York: Basic Books, 1986.

Markovits, J. "Acting for the Right Reasons." *Philosophical Review*, Vol. 119, No. 2, 2010, pp. 201–242.

Persson, I. and J. Savulescu "The Perils of Cognitive Enhancement and the Urgent Imperative to Enhance the Moral Character of Humanity." *Journal of Applied Philosophy*, Vol. 25, No. 3, 2008, pp. 162–177.

Persson, I. and J. Savulescu "Moral Transhumanism." *Journal of Medicine and Philosophy*, Vol. 35, No. 6, 2010, pp. 656–669.

Sorensen, K. "The paradox of Moral Worth." *Journal of Philosophy*, Vol. 101, No. 9, 2004, pp. 465–483.

Sorensen, K. "Effort and Moral Worth." *Ethical Theory and Moral Practice*, Vol. 13, No. 1, 2010, pp. 89–109.

Sparrow, R. *Better Living Through Chemistry? A Reply to Savulescu and Persson on 'Moral Enhancement'. Journal of Applied Philosophy*, Vol. 31, Issue 1, 2014, pp. 23–32.

Sparrow, R. "(Im)moral Technology? Thought Experiments and the Future of 'Mind Control.'" In: *The Future of Bioethics: International Dialogues*, ed. Akira Akabayashi. Oxford: Oxford University Press, 2014, pp. 113–119.

Wood, A. *Kant's Ethical Thought*. Cambridge, UK: Cambridge University Press, 1999.

Wood, A. "Moral Worth, Moral Merit, and Acting from Duty." In: *The Free Development of Each: Studies on Freedom, Right, and Ethics in Classical German Philosophy*. Oxford: Oxford University Press, 2014.

Adriana Warmbier

Moral Enhancement. Enhancing Motivational Processes and Agent-Based Ethics[1]

Abstract: The philosophical approach to the idea of moral enhancement addresses central ethical issues such as agency, autonomy, self-development, and the normative character of human action. It has been argued that biomedical moral enhancement, which, as its proponents believe, will improve our motivational processes in moral decision making, provide better impulse control, and increase our willingness to cooperate with others, is problematic in both its implications and reasoning. After a short presentation of the concept of moral bioenhancement, I formulate two counter-arguments which aim to demonstrate that in so far as we identify improving one's moral dispositions in terms of a paternalistic intervention, thus reducing agency to a mere form of experience, we have little prospect of success in bringing about a person who would more likely arrive at the right choices. I argue that a plausible conception of moral self-improvement must appeal to the core problem of normativity, which manifests itself in the relation between one's attitude toward reasons and toward the values that are constitutive for one's ends and priorities. The criticism of both the narrow understanding of moral action and the deliberate avoidance of the question of what constitutes "morally better motives" is strengthened through the suggested alternative: an Aristotelian account of the reflective processes of cognition, motivation, and action, which are integrally related to the idea of *phronesis*. I explore the possibility of applying agent-based ethics in the moral enhancement debate, arguing that virtue ethics is not only defensible, but may be also regarded as an attractive alternative to the idea of using biomedical methods for self-improvement.

Keywords: Moral enhancement, Moral improvement, Virtue ethics, Aristotelian *Phronesis*, Kant

1 Introduction: The ethical question of perfecting ourselves

Disagreements about enhancement – and moral enhancement in particular – have persistently remained unresolved. Main controversies do not arise as to whether we should strive for self-development and self-fulfilment, and in doing

---

1 The writing of this article was funded by the Polish National Science Centre (2012/07/D/HS1/01099). I would like to thank Prof. Richard Holton, Prof. Robert Audi and Dr Elizabeth Shaw for reading this paper and for their useful comments and suggestions which helped me to improve this text.

so improve the quality of our life; in fact, we may say that flourishing is a kind of moral obligation. The dilemma occurs when we reflect on the very concepts of human development and perfection. The debate over human enhancement has made it abundantly clear that both the proponents and adversaries of bioenhancement do not share a common conception of what perfection consists in. The concept of human perfection is constitutive for diverse ethical theories such as virtue ethics, Kantian ethics, or utilitarianism, but because of the vast semantic scope of the concept of excellence it cannot serve as a criterion for distinguishing between them. If we focus on the historical framework of different types of human excellence, we will see that it has been identified with various goals and means. For ancient and medieval thinkers, the concept of human perfection was inextricably related to time. They assumed that perfecting oneself has little to do with a state which one may reach instantly as a result of the decision to change. Rather, as they account for, it is a long-term process which requires both a specific attitude of distance toward one's intentions, inclinations, and beliefs, i.e., the ability to call them into question, and an awareness of the grounding of one's choices.[2] Soon thereafter, in modern thought, the idea of human perfection was transformed into a term with pluralistic, diverging meanings.[3] Generally speaking, the concept of human perfection no longer points exclusively to moral development, but it has also acquired a sense which pertains to the biotechnical dimension. However, the tendency to lean toward the latter understanding has not displaced the former one. Some believe that we have been thinking all along in terms of virtue, happiness, and a sense of life satisfaction. If this is true, then ancient ethics bears great relevance to contemporary reflection on human flourishing. Even if there are some aspects of ancient thought that cannot be carried over to the modern world, the central insights of the ancients seem to harmonize with our own perceptions.[4]

The European history of how the ethical question of perfection developed began, naturally, with Plato[5] who believed that people reach perfection when they try to reach the good. The "good life" was a central category for ancient ethics. Aristotle believed that realizing virtuous ends is what conduces one to the

---

2   See Plato, *Charmides*, trans. B. Jowett.
3   For a detailed account of the modern approaches to the idea of human perfection, see the chapter written by Wojciech Załuski.
4   See J. Annas, *The Morality of Happiness* (Oxford: Oxford University Press, 1993), p. 5.
5   See W. Tatarkiewicz, "Moral Perfection," *Dialectics and Humanism*, Vol. 7, No. 3 (1980), pp. 117–124; see also G. Santas, *Goodness and Justice: Plato, Aristotle, and the Moderns*, (Cambridge: Blackwell, 2001).

good life in general. Virtue makes one's end the right one, and *phronesis*, which is the excellence of practical reasoning, allows one to recognize or specify the contents of one's end correctly (NE 1142b28–33, 1144a 7–10, MM I.18).⁶ The final end is the flourishing of one's life. The good life depends on what one choses to do and what sort of person one becomes. As Aristotle recognizes, living a life worth living strongly involves agency. The point of this dependency is to emphasize the autonomous and individual character of the very process of perfecting ourselves. When Aristotle talks about the fundamental goal of living well, he introduces the notion of *eudaimonia* (NE I, 5; I, 7; EE I, 3). The final end of *eudaimonia* is presented as a reflective framework insofar as one can define the content of that end for oneself and can specify the reasons for pursuing it.

The proponents of human bioenhancement justify their demands by appealing to the category of the "best life",⁷ which, for some, might be viewed as a counterpart of the ancient conception of a "good life". But there are reasons for not rushing to establish this similarity too soon. The fundamental reason is that the category of the "best life", which underlies the claims in favor of biomedical interventions aimed at enhancing human capacities, stands for a mere possible state of affairs, disregarding the agent's perspective (perspective of the one who is involved in this state). The problem that arises here goes to the heart of transhumanists' adoption of impersonal reasons for applying the category of the "best life". The ancient idea of the "good life" does not ignore the agent's perspective; on the contrary, agent-relative reasons are constitutive for all the main ancient ethical concepts, starting from the category of the "good life". The Greeks were quite clear on this issue. Aristotle holds that "we intuitively believe that the good is something of our own and hard to take from us" (NE 1095b 26–27). This in turn focuses our attention on important topics such as agency, practical reasoning, and improving the motivational process, which are all relevant to the discussion on whether we should set new limits to self-development considering the radical possibilities of biomedical improvement or, perhaps, as

---

6   Aristotle, *Nicomachean Ethics*, translated with Introduction, Notes, and Glossary by T. Irwin (Indianapolis and Cambridge: Hackett Publishing Company, Inc., 1999), p. 252.
7   See J. Savulescu, "Procreative Beneficence: Why Should We Select the Best Children," *Bioethics*, Vol. 15, No. 5–6 (2001), pp. 413–426; J. Savulescu, G. Kahane, "The Moral Obligation to Create Children with the Best Chance of the Best Life," *Bioethics*, Vol. 23, No 5 (2009), pp. 274–290; J. Harris, *Enhancing Evolution: The Ethical Case for Making Better People* (Princeton and Oxford: Princeton University Press, 2007). For a critical discussion on this issue, see M. Parker, "The Best Possible Child," *Journal of Medical Ethics*, Vol. 33, No. 5 (2007), pp. 279–281.

some argue, we have sound reason to stick to traditional agent-affecting means. Following this question, I attempt to elucidate the role of the agency in the process of perfecting. Since the idea of moral enhancement is regarded by some as an alternative solution to the problem of poor moral decision-making, I consider this idea in light of the complex character of the realm of moral action. I shall turn to the ancients, in particular to the Aristotelian approach to human excellence, which entails the concept of practical wisdom. By appealing to this concept, I defend the view that it may afford harmony and stability for the automatic and reflective processes of cognition and motivation that generate our actions. Aristotelian *phronesis* may be more conducive to moral improvement than any biotechnological means. Of course, it is right to insist that ethicists should pay attention to the relevant empirical data. I will therefore attempt to answer the question of whether the concept of practical wisdom is consistent with experimental findings.

2 The idea of biomedical moral enhancement

Recent developments in biotechnology, neuroscience, and genetic engineering may bring radical and innovative forms of enhancement, and thus will not only force us to confront new types of social structures and life activities but may also influence on our ethical self-understanding as members of our species.[8] The last point ramifies into several issues relevant to the debate on enhancement. The main one has to do with the category of subjectivity and the fact that its normative character has the power of demarcation, enabling us to preserve the crucial differentiation between subjectivity and the world of objects.[9] This argument, which was first articulated by Hans Jonas, has been repeated and developed by Jurgen Habermas, Michael Sandel, and Leon Kass.[10] All four thinkers emphasize that biotechnological intervention is a specific mode of action that involves "self-referentiality and irreversibility of intervention" and leads to

---

8 See J. Habermas, *The Future of Human Nature*, (Cambridge: Polity Press, 2003).
9 Habermas, *The Future of Human Nature*, pp. 44–53.
10 See H. Jonas, *The Imperative of Responsibility. In Search of an Ethics for the Technological Age* (Chicago: The University of Chicago Press, 1984); Habermas, *The Future of Human Nature*; M. Sandel, *The Case Against Perfection. Ethics in the Age of Genetic Engineering* (Cambridge, Massachusetts and London: Harvard University Press, 2007); President's Council on Bioethics, *Beyond Therapy: Biotechnology and the Pursuit of Happiness* (New York: Regan Books, 2003); see also F. Fukuyama, *Our Posthuman Future. Consequences of the Biotechnology Revolution* (New York: Farrar, Straus and Giroux, 2002).

consequences which are no longer in our power to control. To perform genetic engineering means then "to commit something to the stream of evolution in which the producer himself is carried along".[11] Habermas describes this problem in terms of the identity of the species and of self-understanding of moral beings, which in turn provides the framework within which our conception of morality, values, and law are embedded. Thus far, the supporters of enhancement have failed to adequately respond to this argument. Rather, they voice the objection, which takes the form of claiming that in privileging human beings in our ethical thought the adversaries of bioenhancement express a prejudice.[12] However, I endorse Williams' claim that these opponents bring out an important aspect of self-improvement, which must be taken into consideration in the debate over human bioenhancement, namely, the agent-concerning perspective. I shall therefore draw attention to the fact of human agency and to the key notions of reflectiveness, or second-order volition,[13] and practical reasoning, as well as to the quasi-automatic process of cognition, which we associate with the motivational process.

One of several reasons for justifying the demand for moral enhancement is the problem of poor decision-making, lack of self-control, and unwillingness to cooperate with others. There is mutual agreement among both the proponents and the adversaries of biomedical interventions on this point: we must search for possible ways of improving our moral conduct. Some of the proponents claim that some sort of improvement of our moral capacities might be achieved as a result of cognitive enhancement.[14] Others argue that cognitive enhancement would not provide such an outcome, and therefore it needs to be accompanied by a more radical solution of moral enhancement, i.e., interfering in individual agency through the application of pharmacology, neuroscience, and genetic

---

11 Habermas, *The Future of Human Nature*, p. 47.
12 See P. Singer, *Animal Liberation* (London: Pimlico, 1995). For discussion see D. DeGrazia, "Human Animal Chimeras: Human Dignity, Moral Status, and Species Prejudice," *Metaphilosophy*, No. 38 (2007), pp. 309–329. Sound argument to the Singer's objection has been offered by Bernard Williams in his paper: "The Human Prejudice," in: *Philosophy as a Humanistic Discipline* (Princeton: Princeton University Press, 2006), pp. 135–152. See also B. McKibben, *Enough: Staying Human in an Engineered Age* (New York: St. Matin's Press, 2004).
13 The phrase "second-order volition" comes from Harry G. Frankfurt. See his "Freedom of the Will and the Concept of a Person," *The Journal of Philosophy*, Vol. 68, No. 1 (1971), pp. 5–20.
14 See J. Harris, "Moral Enhancement and Freedom," *Bioethics*, Vol. 25, No. 2 (2011), p. 105.

selection or modification.¹⁵ This latter demand rests transparently on specific assumptions about the notion of moral insight. The advocates of extensive moral enhancement of humankind offer a solution that consists in developing a "science of morality" which would enable us to acquire dispositions that make it more likely to arrive at the correct judgment of what the right thing to do is, and also more likely to act on that judgment.¹⁶ Due to this belief, they formulate an "urgent imperative to enhance the moral character of humanity":

> The perils of cognitive enhancement require a vigorous research program on understanding the biological underpinnings of moral behaviour. As Hawking quipped, our future may depend on making ourselves wiser and less aggressive. If safe moral enhancements are ever developed, there are strong reasons to believe that their use should be obligatory, like education or fluoride in the water, since those who should take them are least likely to be inclined to use them. That is, safe, effective moral enhancement would be compulsory.¹⁷

This claim goes with a general approach to human behavior, which, as it stands, is contingent upon many internal and external factors. The conditioned nature of our motivational processes and actions leads, in turn, to another fundamental point. Our ability to act means that we initiate processes without the possibility of turning back, processes whose outcome remains uncertain and unpredictable.¹⁸ This inability to reverse the results of our actions, viewed as a most serious threat in the age of technology, induces some to believe that we need a new framework of ethics, for the former one would no longer contain the "new type of human action" and address the "new type of agency" that performs it. This outlook has been expressed by Hans Jonas, who in response proposes a new imperative based on different premises than the transhumanist principle, namely,

---

15 See I. Persson, J. Savulescu, "The Perils of Cognitive Enhancement and the Urgent Imperative to Enhance the Moral Character of Humanity," *Journal of Applied Philosophy*, Vol. 25, No. 3 (2008), pp. 162–177; A. Buchanan, *Better than Human. The Promise and Perils of Enhancing Ourselves* (Oxford, Oxford University Press, 2011), p. 34.
16 See J. Savulescu and I. Persson, *Unfit for the Future: The Need for Moral Enhancement* (Oxford: Oxford University Press, 2012), p. 403; See also J. Savulescu, I. Persson, "Moral Enhancement, Freedom and the God Machine," *The Monist*, Vol. 95, No. 3 (2012), pp. 399–421.
17 Persson and Savulescu, "The Perils of Cognitive Enhancement and the Urgent Imperative to Enhance the Moral Character of Humanity", p. 174.
18 On this issue, see H. Arendt, *The Human Condition* (Chicago and London: University of Chicago Press, 1998), pp. 230–247.

the imperative of responsibility.[19] Inspired by Max Weber's essay "Politics as a Vocation", Jonas sees the inadequacy of earlier ethical theories, which, as he puts it, form their postulates in the narrow *anthropocentric* framework, and thus are not fit to respond to the challenges of modern technology.[20] This new dimension of responsibility, which opens up both the horizon of the future and the horizon of the space within which we interact with one another, is by all means a significant contribution to the discussion on ethics in the technological age. However, this is not to admit that traditional ethics, with its focus on the individual, has nothing to offer to us directly, and thus should be put aside or replaced. Ethics addresses itself to individuals who have realistic intentions, desires, and goals – not to humanity as such, which is simply an abstraction. Plato makes this point when he says that "for all good and evil, whether in the body or in human nature, originates (…) in the soul, and overflows from thence, as if from the head into the eyes. And therefore if the head and body are to be well, you must begin by curing the soul; that is the first thing".[21]

The effort of appropriately shaping one's moral attitudes requires much more than the process of internalizing moral rules and duties. But this does not mean that rule-following and the notion of duty do not play an important role in the process of improving our moral conduct; it only means that following norms alone, which can doubtlessly help in forming one's moral habits, affords an incomplete account of moral agency, unless it is related to a more fundamental attitude, namely, the individual's approach to norms, values, and reasons, in other words, one's moral engagement.[22] As I shall demonstrate, this aspect of moral agency, which is given its due place within the ancient account of self-development and self-fulfillment, sheds a different light on the imperative to enhance the moral character of humanity. There are reasons to believe that if we are to improve our conduct, we should focus our attention not only on the "negative" aspect of moral development, which would mean, as Jonas and the proponents of bioenhancement claim, pointing out the insufficiency of the framework of traditional ethics, but first we ought to reinstate and account for the primordial condition from which our decisions and actions arise. This, of course, is one's

---

19 See Jonas, *The Imperative of Responsibility*. Jonas modifies Kant's categorical imperative to the form which goes thus: "Act so that the effects of your action are not destructive of the future possibility of such life".
20 Jonas, *The Imperative of Responsibility*, pp. 4–8.
21 Plato, *Charmides*, 156E–157A.
22 See M. Scheler, *Formalism in Ethics*, trans. M. S. Frings, R. Funk (Evanston: Northwestern University Press, 1985).

moral engagement, which seems to me to not have received enough attention in the discussion of moral enhancement. Classical ethics, with its key concepts of practical wisdom and *eudaimonia*, may provide us with a deeper understanding of human behavior and explain to what extent we may influence and correct some of our tendencies that may be destructive for our personal and social well-being. Ancient thought also addresses the difficulties of cultivating one's moral attitude related to the need for rational recognition of the appropriateness of one's decisions and actions. Much of the inner tensions that we experience are generated as a result of the conditioned character of our motivational process. The advocates of the idea of moral enhancement hold that once we manage to eliminate, or at least reduce, that conditioned character of our nature by applying biotechnological means, we will increase the probability of having morally better motives.[23] Those who oppose biomedical interventions viewed as an alternative method for bringing about a better person do not deny the conditioned aspect of our nature. On the contrary, they take this aspect seriously and thus, holding it as their fundamental context, they attempt to explain the idea of the best life, the agent's own good, self-development, and the proper role of emotions. This is to be done within a broad framework of the complex structure of automatic and reflective processes of cognition, motivation, and action. It may be the case that we must take a wider view of the process of decision making if we are to express what really needs to be improved in order to have a person more likely to come to the correct judgment of what is right and to act on that judgment. It is with this framework in hand that we should seek appropriate methods for self-improvement and consider whether the new science of behavioral control might truly increase, as the supporters of bioenhancement believe, the probability of having morally better motives.

3 The alternative: a Kantian and an Aristotelian response to the idea of moral bioenhancement

3.1 Agency and motivation: what rationality may reasonably demand of our motivational process?

The main argument against any forms of moral bioenhancement holds that all of them undermine human autonomy and freedom. This claim, in turn, appeals to the issue that such an intervention will irreversibly affect the overall structure of

---

23 See T. Douglas, "Moral Enhancement," *Journal of Applied Philosophy*, Vol. 25, No. 3 (2008), p. 229.

our moral experience.[24] Our understanding of being persons is primarily based on an agent-centered account of personhood. This active aspect of our nature plays an important role in morality.[25] For those who adopt an Aristotelian or Kantian view of moral decision-making, the agency stands for the capacity to initiate the decision-making process and action, whereas the utilitarians, whose approach stems mainly from Humean empiricism,[26] tend to regard the agency as a mere form of experience. These two different views have far-reaching implications for our understanding of morality and thus the nature of moral choice. The question of whether the use of new techniques of moral bioenhancement results in people better conforming to morality, as the advocates of the idea of bioenhancing moral dispositions assume, depends on the account of what acting morally involves as its constitutive element.[27] This is most evident when they write: "those who had undergone moral bioenhancement would act for the same reasons as those of us who are most moral today do, and the sense in which it is 'impossible' that they do what they regard as immoral will be the same for the morally enhanced as for the garden-variety virtuous person."[28]

The central belief behind this claim holds that the virtue of one's character or the acquisition of a disposition to do the right thing are a matter of mechanical behavioral control achieved through external forces. Hence, as it says, we can

---

24 Habermas reinforces Ronald Dworkin's claim referring mainly to genetic modifications which are aimed at enhancing the human condition. See Habermas, *The Future of Human Nature*, pp. 28–29.
25 For an extensive discussion on this issue, see Ch. M. Korsgaard, "Personal Identity and the Unity of Agency: A Kantian Response to Parfit," *Philosophy and Public Affairs*, Vol. 18, No. 2 (1989), pp. 101–132; J. McDowell, *Mind, Value, and Reality* (Cambridge: Harvard University Press, 2001); H. G. Frankfurt, "Freedom of the Will and the Concept of a Person," *The Journal of Philosophy*, Vol. 68, No. 1 (1971), pp. 5–20; Ch. Taylor, *Human Agency and Language. Philosophical Papers I* (Cambridge: Cambridge University Press, 1985); P. Ricoeur, *Oneself as Another*, trans. K. Blamey (Chicago: The University of Chicago Press, 1995).
26 See Hume's approach to the concept of free will, which he regards rather as the feeling we experience while we use this capacity. In *A Treatise of Human Nature* he holds "I desire it may be observ'd that by the *will*, I mean nothing but *the internal impression we feel and are conscious of, when we knowingly give rise to any new motion of our body, or new perception of our mind*". D. Hume, *A Treatise of Human Nature*, ed. L. A. Selby-Bigge and P. H. Nidditch (Oxford: Clarendon Press, 1978), p. 399.
27 I discussed this issue in "Moral Perfection and the Demand for Human Enhancement," *Ethics in Progress*, Vol. 6, No. 1 (2015), pp. 23–37.
28 Savulescu, Persson, *Moral Enhancement, Freedom and the God Machine*, p. 408.

have biomedical interventions that "produce more moral behaviour that do control the moral agent, subjugating that person to the will of another and removing the freedom to act immorally".[29] The question is whether moral development indeed proceeds in this way. Being a virtuous person requires much more, or rather requires something entirely different, than not having to confront the encountered impulses. It is mystifying why anyone would think morality is a good strategy for achieving such a "liberation". The right conduct in terms of having virtues stands for the capacity for transforming impulses and desires into "morally right" motives on which one chooses to act. Since we are more than animals, we do not merely behave, but first and foremost we *act*, and even if we act on impulse, this is not to say that we act because we have that impulse, but rather because we choose to act on it, having in mind some reason for this choice.[30] Thus, the claims that biomedical interventions will improve the dispositions necessary for morality actually underplay a fundamental aspect concerning those dispositions, which in fact is the entry point for moral development. If we reduce the active aspect of our nature to a merely passive form (to the form of experience), the problem of lack of moral conduct will never be solved. For without regard for the fact that in order to transform bare impulse into the motivation to act appropriately we need the standpoint of practical reason, there is little or no chance to meet the expectations which the proponents of moral bioenhancement keep formulating, and these are: increasing altruism, increasing the willingness to co-operate and sacrifice one's own interests, and improving impulse control. The standpoint of practical reason is the standpoint from which we take our perceptions and desires into full view, as we are not just conscious of their presence, but we refer to them directly. This is where the concept of a moral agent, with a whole framework of intentions, motives, and reasons actually starts. This outlook, namely, this opposition to the utilitarian reduction of agency to a form of experience, has been well illustrated by both Harry G. Frankfurt and Christine Korsgaard. The latter notes that our ability to direct our own intention to our perceptions, impulses, and desires

---

29 Savulescu, Persson, *Moral Enhancement, Freedom and the God Machine*, p. 419.
30 For a discussion on virtues and normativity see D. C. Russell, *Practical Intelligence and the Virtues* (Oxford: Oxford University Press, 2009), pp. 346–348; B. Williams, *Acting as the Virtuous Person Acts*, in: *Aristotle and Moral Realism*, ed. R. Heinaman (London: University College London Press, 1995), pp. 13–23; Annas, *The Morality of Happiness*, pp. 47–131; J. McDowell, "Virtue and Reason," *The Monist*, Vol. 62, No. 3 (1979), pp. 331–350.

sets us a problem no other animal has. It is the problem of the normative. For our capacity to turn our attention on to our own mental activities is also a capacity to distance ourselves from them, and to call them into question. I perceive, and I find myself with a powerful impulse to believe. But I back up and bring that impulse into view and then I have a certain distance. Now the impulse doesn't dominate me and now I have a problem. Shall I believe? Is this perception really a *reason* to believe? I desire and I find myself with a powerful impulse to act. But I back up and bring that impulse into view and then I have a certain distance. Now the impulse doesn't dominate me and now I have a problem. Shall I act? Is this desire really a *reason* to act? The reflective mind cannot settle for perception and desire, not just as such. It needs a *reason*. Otherwise, at least as long as it reflects, it cannot commit itself or go forward.[31]

The capacity to distance ourselves from our own mental activities, desires, and impulses introduces the core problem of normativity that cannot be cut off from a plausible conception of moral enhancement. It is critical distance that makes changes possible in one's inclinations or habits. If we are to influence the goals of our inclinations, to transform impulse into motivation toward morally good action, and to shape our emotional reactions, we must see that there is a significant relation between one's attitude towards reasons and values, which are constitutive for the agent's ends and priorities, and the action performed on those inclinations, impulses, or emotions. It is less probable that a biomedical intervention will effectively influence our moral disposition than that some critical distance from what moves us to act will. This is not to say that the very ability to distance oneself from one's perceptions, desires, and impulses determines the action one takes. Rather, it holds that, unlike animals, whose actions are set off by an impulse, human action requires one's consent to be performed (if we act on the impulse, we act not because we are merely following that impulse, but we act on that impulse because we have consented to act on it). Human action differs in this substantial point: it needs a reason. As Bernard Williams wrote: "Desiring to do something is of course a reason for doing it."[32] A suitable image which elucidates the significance of this distinction is offered by Frankfurt, who introduces the concept of "second-order volition", which stands for the capacity to transcend "first-order desires" and to formulate the former ones. The "second-order volition" is thus the desire from which one acts.[33] In Frankfurt's view, this "second-order volition" is structured in a specific way, namely, we move back from our

---

31 Ch. M. Korsgaard with G. A. Cohen [et al.] *The Sources of Normativity* (Cambridge: Cambridge University Press, 2003), p. 93.
32 B. Williams, *Ethics and the Limits of Philosophy* (London: Routledge, 1993), p. 19.
33 Frankfurt, *Freedom of the Will and the Concept of a Person*, pp. 10–11.

desire to the perspective of our more fundamental ends and preferences, and view the desire from that standpoint (the standpoint from which one's choices are made). This constitutes a radical difference between agency reduced to a form of experience and an agent who raises the question: "Do I really want to be what I now am? (i.e., have the desires and goals I now have?)".[34] The "capacity for reflective self-evaluation" manifests itself in the fact that the ordinary moral agent can desire to be moved by certain desires, that she can revise and shape her behavioral dispositions. Thus, our motivations may be enhanced, strengthened, or altered only from the standpoint of practical reason. And none of the biomedical interventions operates within this standpoint. They may influence our perception or mood, but they have little chance of bringing about a person who reliably makes the right choices. In so far as moral development requires an active agent (with the capability to distance ourselves from our activities), the new science of behavioral control cannot offer us any plausible means of enhancing our moral conduct. As the ancients claim, moral development cannot simply be injected into one through external interference, but it needs to be initiated and proceeded by one's own intention and decision. Otherwise, we risk the possibility of "self-alienation" and of losing one's integrity.

This objection has already been voiced in bioethics literature. In *Beyond Therapy*, Leon Kass and the members of the President's Council on Bioethics point out the problem with mood-altering drugs which, as they argue, will separate us from the actions and experiences that normally accompany those moods.[35] The problem with taking such drugs lies in the fact that this kind of intervention may irreversibly influence our grasp of ourselves. Kass and others claim that in being separated from those drug-mediated experiences, we will also be separated from the notion of who we really are and how the world really is.[36] Hence, we may find ourselves "self-alienated" – disorientated, confounded, and with a sense of loss of our identity.[37] This is to say that if we are not engaged in performing actions fully deliberately, we will be deprived of regarding ourselves as the authors of those actions. Furthermore, the normative relation between the

---

34 Ch. Taylor, "Responsibility for Self," in: *Free Will*, ed. G. Watson (Oxford: Oxford University Press, 1982), p. 111. See also R. Audi, "Responsible Action and Virtues Character," *Ethics* Vol. 101 (1991), p. 309.
35 See President's Council on Bioethics, *Beyond Therapy: Biotechnology and the Pursuit of Happiness* (New York: Regan Books, 2003), p. 238
36 President's Council on Bioethics, *Beyond Therapy*, pp. 253–255.
37 President's Council on Bioethics, *Beyond Therapy*, p. 294.

moral agent and the ends of her actions constitutes her practical identity, and thus her integrity. As Korsgaard points out:

> If I can overcome my cowardice by surgery or medication rather than habituation I might prefer to take this less arduous route. So long as an authentic good will is behind my desire for greater courage, and authentic courage is the result, the mechanism should not matter. But for the Kantian it does matter who is initiating the use of the mechanism. Where I change myself, the sort of continuity needed for identity may be preserved, even if I become very different. Where I am changed by wholly external forces, it is not. This is because the sort of continuity needed for what matters to me in my own personal identity essentially involves my agency.[38]

The point here is that we are identified with actions that we initiate and which flow from the chain of our previous choices, resulting from our attitudes, beliefs, and life projects. One's choice is not an isolated activity but it is determined by one's normative involvement in one's projects and attitudes. This means that our sense of identity and integrity depends on our relationship to our choices and actions, and from the practical standpoint that relationship must be an authorial one. Thus, it does matter whether we are discussing the idea of enhancing one's moral dispositions in terms of a possible paternalistic intervention, or in the context of an agent-centered conception of person. The forms of moral improvement ought to be adjusted to our grasp of ourselves – to us as rational agents who view themselves as authors of their decisions and actions, and who are capable of critical distance from themselves, and are therefore capable of normative engagement with what motives them.

3.2 The cognitive-affective understanding of virtue: in defense of the empirical adequacy of Aristotelian *Phronesis*

What can we learn from the Greeks? What benefit may flow from their moral considerations, in particular, from an ethics of virtue and its core notions of practical wisdom and *eudaimonia*? First and foremost, ancient ethics may offer us useful insight as to why we should strive to perfect ourselves, and what is at stake in the idea of moral development. Since ancient ethical thought emerges from everyday reflection, its entry concepts may not be as remote from the central issues of our own ethical deliberations as some may believe. Much of our concern pertains all the time to the same questions, that is, of the place of certain

---

38 Korsgaard, *Personal Identity and the Unity of Agency*, p. 123.

projects, priorities, and values in one's life taken as a whole, and of the notion of a life which may be regarded as fulfilling. As Julia Annas notes:

> To think about my life as a whole is to ask how I have become the person I now am, how past plans, successes and failures have produced the person who now has the present projects and attitudes that I have. And it is also to think about the future. How do I see my present plans continuing? (…) Ancient ethics gets its grip on the individual at this point of reflection: am I satisfied with my life as a whole, with the way it has developed and promises to continue?[39]

It is not only the end toward which one strives that counts, but the very path that leads to this end plays an important role and affects the one who took it. This conviction lies at the basis of the ancient (mainly Aristotelian) framework within which the Greeks approach the question of motivational processes and advance an explanation. Ancient thought has much to say to those who have reached this point of reflection about their lives. "The arguments and conclusions of ethical philosophy will be effective only with those who have come to them through worrying about real problems: recommendations as to how best to live will have force only with those who have wondered for themselves about how their lives are going."[40] However, for some the Socratic question "How should one live?" may not sound as if it could provide useful and practical guidance about how we can develop ourselves morally. It may be perceived this way because the imperative of reflective life does not instantly afford a ready answer to every moral dilemma and difficult moral situations in which we find ourselves. In fact, the question "How should I live?" or "What kind of person should I be?" ought to be taken, as Bernard Williams urged, as the best place for moral philosophy to start, rather than end.[41] Moreover, the answers to those questions are not meant to be a universal prescription according to which one ought to act. The ancients did not impose such demands on ethical thought, nor did they perceive moral reflection as a sort of method or decision procedure for arriving at morally right conclusions.[42] This demand was put forward by modern moral philosophers (mainly by Henry Sidgwick). During the transition from ancient to modern thought, the Greeks' question "Who ought I to *be*?" became replaced by the Kantian and utilitarian tradition, which is concerned more with the questions

---

39 See Annas, *Morality of Happiness*, pp. 28–29. See also J. Annas, "Should Virtue Make You Happy?," in: *Eudaimonia and Well-Being*, ed. L. Jost and R. Shiner (Kelowna: Academic Printing & Publishing, 2002); H. G. Frankfurt, "On the Usefulness of Final Ends," *Iyyun*, Vol. 41 (1992), pp. 3–19.
40 Annas, *Morality of Happiness*, p. 29.
41 B. Williams, *Ethics and the Limits of Philosophy* (London: Routledge, 1993), p. 4.
42 See Aristotle, *Nicomachean Ethics*, 1094b.

"What ought I to *do*?", "What is our duty?", or "Which deeds are morally wrong and which are morally right?". This transition had two significant consequences. First, it shifted our attention from an agent-centered ethics of excellence to an ethics of obligation focused on the action itself. Secondly, it resulted in a different account of morality, where it is taken to be a set of precepts or principles concerning judgments on whether we should perform, or not perform, a certain action. In the Kantian view of ethics, moral worth of an action is derived not from the intention or the aim, which is to be attained by it, but from the principle which determines it. The utilitarian account has also focused on the act which one should perform according to the one universal standard of utility – affording the greatest overall good.

This legalistic turn in ethics was first criticized by Elizabeth Anscombe in *Modern Moral Philosophy*, published in 1958.[43] Her criticism aimed at deontology and utilitarianism, gave rise to a resurgence of a deep and widespread interest in virtue ethics, as it focuses not only on the character of the agent, on her motivation, beliefs, goals, and emotions, but also offers a positive outline of how virtue can be developed and in what sense the acquisition and exercise of virtue may expand our moral capacities.[44] However, despite the rising interest in virtue ethics, the greatest challenge it currently has to face is to counter the criticism leveled by several social psychologists, labeled "situationists", such as John Doris, Gilbert Harman, Maria Merritt, and Peter Vranas. They claim that some of the findings of behavioral psychology in the recent decades repudiate

---

43 G.E.M. Anscombe, "Moral Modern Philosophy," *Philosophy*, Vol. 33, No. 124 (1958), pp. 1–19.
44 Within the vast ethical tradition centered on virtue, there are a number of various contemporary stances concerning the concept of what virtue is, and in what manner it may be appropriately employed in moral thinking. Amongst the most relevant approaches are: A. MacIntyre, *After Virtue: A Study in Moral Theory*, (Notre Dame: University of Notre Dame Press, 2007); M. Nussbaum, *The Fragility of Goodness: Luck and Ethics in Greek Tragedy and Philosophy* (Oxford: Oxford University Press, 1986); N. Sherman, *The Fabric of Character: Aristotle's Theory of Virtue* (Oxford: Oxford University Press, 1989); N. Sherman, *Making the Necessity of Virtue. Aristotle and Kant on Virtue* (Cambridge: Cambridge University Press, 1997); Annas, *The Morality of Happiness*; J. Annas, *Intelligent Virtue* (Oxford: Oxford University Press, 2013); R. Audi, *Moral Knowledge and the Ethical Character* (Oxford and New York: Oxford University Press, 1997); Ch. Swanton, *Virtue Ethics: A Pluralistic View* (New York: Oxford University Press, 2003); R. Adams, *A Theory of Virtue: Excellence in Being for the Good* (Oxford: Oxford University Press, 2006); R. Hursthouse, *On Virtue Ethics* (Oxford: Oxford University Press, 1999); R. Crisp, M. Slote, ed., *Virtue Ethics* (Oxford: Oxford University Press, 1997).

the existence of moral character and its features as universal.[45] According to their view, these are the situational factors rather than one's alleged dispositions of character that determine human action. However, the most recent empirical research in cognitive and social psychology has given us sound reasons to believe that, contrary to the situationist claims used to support moral bioenhancement, there is an adequate tie between the Aristotelian account of virtue, which relates virtue to *phronesis* and newest experimental findings. To elucidate this connection, let me invoke Walter Mischel's and Yuishi Shoda's conception of the cognitive-affective personality system. It outlines a view that accounts for both the inconsistencies in one's behavior and the relatively stable patterns of variation within the person that generates the former. According to this model, one's personality consists of a number of complex cognitive-affective processes, like one's individualized manner of interpreting the particular situation in which one is to act, along with one's goals, values, emotions, feeling, beliefs, and predictions about the consequences of each of the different behavioral possibilities.[46]

This view thus focuses not so much on the situational factors which influence one's action, as on one's particular mode of perceiving them, along with the various conditions which underpin this mode and the possibilities of shaping it. Although Mischel and Shoda's research does not directly appeal to the question of whether people possess virtues, it may provide better insight into the motivational process pointing out the possibilities of enhancing it, and can thus serve as a valuable contribution to the agent-based approach of the virtue ethics.[47] To put

---

45 J. M. Doris, *Lack of Character*, (Cambridge: Cambridge University Press, 2002); G. Harman, "Moral Philosophy Meets Social Psychology: Virtue Ethics and the Fundamental Attribution Error," *Proceedings of the Aristotelian Society*, Vol. 109 (1999), pp. 316–331; G. Harman, "No Character or Personality," *Business Ethics Quarterly*, Vol. 13 (2003), pp. 87–94; G. Harman, "Skepticism about Character," *Ethics*, Vol. 13 (2009), pp. 234–242; M.W. Merritt, "Aristotelian Virtue and the Interpersonal Aspect of Ethical Character," *Journal of Moral Philosophy*, Vol. 6 (2009), pp. 23–69; P. B. M. Vranas, "The Indeterminacy Paradox: Character Evaluations and Human Psychology," *Nous*, Vol. 39 (2005), pp. 1–42; P. B. M. Vranas, "Against Moral Character Evaluations: The Undetectability of Virtue and Vice," *Ethics*, Vol. 13 (2009), pp. 213–233.

46 See W. Mischel, Y. Shoda, "A Cognitive-Affective System Theory of Personality: Reconceptualizing the Invariances in Personality and the Role of Situations," *Psychological Review*, Vol. 101 (1995), pp. 246–268; see also Y. Shoda, S. L. Tiernan, W. Mischel, "Personality as a Dynamical System: Emergence of Stability and Distinctiveness from Intra- and Interpersonal Interactions," *Personality and Social Psychology Review*, Vol. 6 (2002), pp. 316–325.

47 There are a number of philosophers who advocate the ethics of virtue and in doing so appeal to the conception of cognitive-affective personality system. See Russell, *Practical*

this another way, since the concept of cognitive-affective integration of personality is consistent with Aristotle's account of the reflective processes of cognition, motivation, and action, all integrally related to the idea of *phronesis*, then the agent-based virtue ethics is not only defensible, but it may also be an attractive alternative to the idea of using biomedical methods for self-improvement. What may offer a new direction in the agent-based approach to enhancing motivational processes is the cognitive-affective conception of the virtues.[48] In other words, the key problem of the debate over the adequacy (or inadequacy) of moral psychology on which the Aristotelian approach to virtue is founded lies neither in the empirical research aiming to answer the question of whether there is evidence for the existence of behavioral dispositions and thus of character, nor in the interpretation of that research offered by social psychologists. The problem lies in the framework within which virtue is being defined. The critics of practical wisdom have presented a number of experiments which demonstrate the lack of consistency in a person's behavior. They therefore reject the notion of dispositions of character which determine human action. But in this line of reasoning, the consistency of behavior is sought not in the agent, but in the very behavior itself, i.e., not from the agent's point of view, but from the third-person standpoint. A behavioral account of virtues differs from the virtues as defined in cognitive-affective terms. The former gives us virtues understood as the mindless and automatic habits established from repeated experiences that do not involve deliberation and choices, and which are expected to result in a specific kind of behavior. While the latter, as Julia Annas points out,

> [are] not a habit in the sense in which habits can be mindless, sources of action in the agent which bypass her practical reasoning. A virtue is a disposition to act, not an entity built up within me and productive of behaviour; it is my disposition to act in certain ways and not others. A virtue, unlike a mere habit, is a disposition to act for reasons, and so a disposition which is exercised through the agent's practical reasoning; it is built up by making choices and exercised in the making of further choices. When an honest person decides not to take something to which he is not entitled, this is not the upshot of a causal build-up from previous actions, but a decision, a choice which endorses his disposition to be honest.[49]

Intelligence and the Virtues; N. E. Snow, *Virtue as the Social Intelligence: An Empirically Grounded Theory* (New York: Routledge, 2011).

48 See Annas, *Morality of Happiness*, pp. 53–107; Russell, *Practical Intelligence and the Virtues*, pp. 324–327.

49 J. Annas, "Virtue Ethics," in: *The Oxford Companion to Ethical Theory*, ed. D. Copp, (Oxford: Oxford University Press, 2006). See also J. Annas, "Comments on John Doris'

The main difficulty with the behavioral approach to virtues is that automatic and unthinking actions do not lead one to acquire the practical ability to recognize and know how to act in a wide range of moral situations. This ability is responsible for correctly interpreting a particular moral situation in which one finds oneself, appropriately responding to it, and so having a veridical practical perception of it. Cognition is here not to be understood as scientific knowledge, but it is rather practical in character. While the former is concerned with universals only, the latter appeals to particulars, but it also involves knowledge of universals and of how to bring them to bear appropriately in particular situations – which is *phronesis*.[50] The Aristotelian concept of *phronesis*, being the virtue of practical reasoning, is, in fact, a complex disposition and not mere practical intelligence, for *phronesis* is never morally neutral, but it is always bound up with one's final end (with *eudaimonia*), which determines one's attitude and normative beliefs. Both of these points are stressed in Aristotle's account of the way in which one acquires the disposition to shape one's patterns of action. The key point here is that without one's moral engagement there is little prospect of success in enhancing the processes of cognition, motivation, and action, which would include correcting the patterns of one's emotional reactions, breaking a bad habit and establishing a good one, and, what is most important, understanding the very need for this change. Putting it differently, an effective improvement of one's disposition to arrive at morally right judgments and to act on those judgments might be more likely achieved if we view it from the agent standpoint. With this framework at hand, which involves a cognitive-affective conception of virtue along with its core notions of practical reason and eudaimonistic flourishing, the agent-based virtue ethics may provide true, practical guidance in our moral growth. And this approach, despite its weaknesses,[51] can most definitely be an adequate response to the current debate on moral improvement.

Lack of Character," *Philosophy and Phenomenological Research*, Vol. 71 (2005), pp. 636–642.

50 C. D. C Reeve, *Practices of Reason. Aristotle's Nicomachean Ethics* (Oxford: Oxford University Press, 1992), p. 72. See also R. Hursthouse, "What Does the Aristotelian Phronimos Know?," in: *Perfecting Virtue. New Essays and Virtue Ethics*, ed. L. Jost, J. Wuerth (Cambridge: Cambridge University Press: 2011), pp. 43–47.

51 See R. B. Louden, "On Some Vices of Virtue Ethics," *American Philosophical Quarterly*, Vol. 21 (1984), pp. 227–236; R. B. Louden, "Virtue Ethics and Anti-Theory," *Philosophia*, Vol. 20 (1990), pp. 93–114.

4 Concluding remarks

The problem of developing moral dispositions and the related issue of improving our motivational processes is addressed both by the proponents of the idea of biomedical moral enhancement and by those who oppose it. What we are discussing then are two different responses to the question of whether biotechnological interventions may correct one's character, strengthen one's powers of self-control, and thus results in improving moral decision-making. The supporters of the idea of moral bioenhancement believe that if we manage to fully elaborate the "science of morality", we will be able to influence one's moral dispositions and behavior, and thus to effectively improve them. The adversaries of such interventions claim that moral development should stem from one's own decisions, otherwise we risk the possibility of "self-alienation" – of losing, confounding, and abandoning our identity. Furthermore, as John Harris points out, since the space between knowing the good and doing the good is a region entirely inhabited by freedom, the advocates of enhancing moral competence mistakenly hold that any biomedical or genetic interference will increase the probability of having morally better future motives.[52] I argued that the idea of moral bioenhancement fails to address the question of the normative relation between the moral agent and the goals of her choices and action. The normative relation entails the involvement of the active aspect of our nature (the fact that we understand ourselves as autonomous agents) and if we reduce this active side to the mere passive form of it, there is less likelihood that we will actually come to enhance our motivational processes, for we do not act simply because we have an impulse, but because having a reason for doing so, we choose to act on it. Thus, transforming bare impulses into the motivation to act appropriately means employing the standpoint of practical reason – and none of the biomedical interventions operate within this standpoint. In search of plausible and effective forms of moral improvement I turned to the ancient ethics along with its key concepts of virtue defined in cognitive-affective terms, practical wisdom, and *eudaimonia*. I attempted to demonstrate that if we are to address the challenge posed by recent developments in neuroscience, genetics, and biotechnology, we need to propose an alternative to the idea of moral bioenhancement, one which would be a positive outline of the forms of human flourishing, and which would provide a deeper understanding of human behavior.

---

52 See Harris, *Moral Enhancement and Freedom*, p. 104.

References

Annas, J. *The Morality of Happiness*. Oxford: Oxford University Press, 1993.

Annas, J. "Should Virtue Make You Happy?" In: *Eudaimonia and Well-Being*, eds. L. Jost and R. Shiner. Edmonton: Academic Printing & Publishing, 2002.

Annas, J. "Virtue Ethics." In: *The Oxford Companion to Ethical Theory*, ed. D. Copp. Oxford: Oxford University Press, 2006.

Annas, J. *Intelligent Virtue*. Oxford: Oxford University Press, 2013.

Anscombe, G.E.M. "Moral Modern Philosophy." *Philosophy*, Vol. 33, No. 124, 1958, pp. 1–19.

Arendt, H. *The Human Condition*. Chicago, London: University of Chicago Press, 1998.

Aristotle. *Nicomachean Ethics*. Translated with Introduction, Notes, and Glossary by T. Irwin. Indianapolis, Cambridge: Hackett Publishing Company, Inc., 1999.

Audi, R. "Responsible Action and Virtues Character" *Ethics*, Vol. 101, 1991, pp. 304–321.

Audi, R. *Moral Knowledge and the Ethical Character*. Oxford, New York: Oxford University Press, 1997.

Buchanan, A. *Better than Human. The Promise and Perils of Enhancing Ourselves*. Oxford: Oxford University Press, 2011.

DeGrazia, D. "Human Animal Chimeras: Human Dignity, Moral Status, and Species Prejudice." *Metaphilosophy*, Vol. 38, 2007, pp. 309–329.

Doris, J. M. *Lack of Character*. Cambridge: Cambridge University Press, 2002.

Douglas, T. "Moral Enhancement." *Journal of Applied Philosophy*, Vol. 25, No. 3, 2008, pp. 228–245.

Fukuyama, F. *Our Posthuman Future. Consequences of the Biotechnology Revolution*. New York: Farrar, Straus and Giroux, 2002.

Frankfurt, H. G. "Freedom of the Will and the Concept of a Person." *The Journal of Philosophy*, Vol. 68, No. 1, 1971, pp. 5–20.

Frankfurt, H. G. "On the Usefulness of Final Ends." *Iyyun*, Vol. 41, 1992, pp. 3–19.

Habermas, J. *The Future of Human Nature*. Cambridge: Polity Press, 2003.

Harman, G. "Moral Philosophy Meets Social Psychology: Virtue Ethics and the Fundamental Attribution Error." *Proceedings of the Aristotelian Society*, Vol. 99, 1999, pp. 315–332.

Harris, J. *Enhancing Evolution: The Ethical Case for Making Better People*. Princeton, Oxford: Princeton University Press, 2007.

Harris, J. "Moral Enhancement and Freedom." *Bioethics*, Vol. 25, No. 2, 2011, pp. 102–111.

Hursthouse, R. "What Does the Aristotelian Phronimos Know?" In: *Perfecting Virtue. New Essays and Virtue Ethics*, ed. L. Jost and J. Wuerth. Cambridge: Cambridge University Press, 2011.

Jonas, H. *The Imperative of Responsibility. In Search of an Ethics for the Technological Age.* Chicago: The University of Chicago Press, 1984.

Korsgaard, Ch. M. "Personal Identity and the Unity of Agency: A Kantian Response to Parfit." *Philosophy and Public Affairs*, Vol. 18, No. 2, 1989, pp. 101–132.

Korsgaard, Ch. M. with G. A. Cohen [et al.] *The Sources of Normativity*. Cambridge, Massachusetts: Cambridge University Press, 2003.

Louden, R. B. "On Some Vices of Virtue Ethics." *American Philosophical Quarterly*, Vol. 21, 1984, pp. 227–236.

McDowell, J. "Virtue and Reason."*The Monist*, Vol. 62, No. 3, 1979, pp. 331–350.

McDowell, J. *Mind, Value, and Reality*. Cambridge: Harvard University Press, 2001.

McKibben, B. *Enough: Staying Human in an Engineered Age*. New York: St. Matin's Press, 2004.

Merritt, M. W. "Aristotelian Virtue and the Interpersonal Aspect of Ethical Character." *Journal of Moral Philosophy*, No. 6, 2009, pp. 23–69.

Mischel, W. and Y. Shoda. "A Cognitive-Affective System Theory of Personality: Reconceptualizing the Invariances in Personality and the Role of Situations." *Psychological Review*, Vol. 101, 1995, pp. 246–268.

Nussbaum, M. *The Fragility of Goodness: Luck and Ethics in Greek Tragedy and Philosophy*. Oxford: Oxford University Press, 1986.

Parker, M. "The Best Possible Child." *Journal of Medical Ethics*, Vol. 33, No. 5, 2007, pp. 279–281.

Persson, I. and J. Savulescu. "The Perils of Cognitive Enhancement and the Urgent Imperative to Enhance the Moral Character of Humanity." *Journal of Applied Philosophy*, Vol. 25, No. 3, 2008, pp. 162–177.

President's Council on Bioethics, *Beyond Therapy: Biotechnology and the Pursuit of Happiness*. New York: Regan Books, 2003.

Reeve, C.D.C. *Practices of Reason. Aristotle's Nicomachean Ethics*. Oxford: Oxford University Press, 1992.

Russell, D. C. *Practical Intelligence and the Virtues*. Oxford: Oxford University Press, 2009.

Sandel, M. *The Case Against Perfection. Ethics in the Age of Genetic Engineering.* Cambridge, Massachusetts and London: Harvard University Press, 2007.

Santas, G. *Goodness and Justice: Plato, Aristotle, and the Moderns.* Cambridge: Blackwell, 2001.

Savulescu, J. "Procreative Beneficence: Why Should We Select the Best Children." *Bioethics*, Vol. 15, No. 5–6, 2001, pp. 413–426.

Savulescu, J. and G. Kahane. "The Moral Obligation to Create Children with the Best Chance of the Best Life." *Bioethics*, Vol. 23, No. 5, 2009, pp. 274–290.

Savulescu, J. and I. Persson. *Unfit for the Future: The Need for Moral Enhancement.* Oxford: Oxford University Press, 2012.

Savulescu, J. and I. Persson. "Moral Enhancement, Freedom and the God Machine." *The Monist*, Vol. 95, No. 3, 2012, pp. 399–421.

Scheler, M. *Formalism in Ethics.* Trans. M. S. Frings and R. Funk. Evanston: Northwestern University Press, 1985.

Sherman, N. *The Fabric of Character: Aristotle's Theory of Virtue.* Oxford: Oxford University Press, 1989.

Singer, P. *Animal Liberation.* London: Pimlico, 1995.

Snow, N. E. *Virtue as the Social Intelligence: An Empirically Grounded Theory.* New York: Routledge, 2011.

Tatarkiewicz, W. "Moral Perfection." *Dialectics and Humanism*, Vol. 7, No. 3, 1980, pp. 117–124.

Taylor, Ch. "Responsibility for Self." In: *Free Will*, ed. G. Watson. Oxford: Oxford University Press, 1982.

Taylor, Ch. *Human Agency and Language. Philosophical Papers I.* Cambridge: Cambridge University Press, 1985.

Vranas, P.B.M. "The Indeterminacy Paradox: Character Evaluations and Human Psychology." *Nous*, Vol. 39, 2005, pp. 1–42.

Warmbier, A. "Moral Perfection and the Demand for Human Enhancement." *Ethics in Progress*, Vol. 6, No. 1, 2015, pp. 23–37.

Williams, B. *Ethics and the Limits of Philosophy.* London: Routledge, 1993.

Williams, B. "Acting as the Virtuous Person Acts." In: *Aristotle and Moral Realism*, ed. R. Heinaman. London: University College London Press, 1995.

Williams, B. "The Human Prejudice." In: *Philosophy as a Humanistic Discipline*, ed. B. Williams. Princeton: Princeton University Press, 2006, pp. 135–152.

Jacek Jaśtal

Human Flourishing and the Coherence of Knowledge. An Aristotelian Perspective

Abstract: In this paper I discuss human flourishing in comparison with bioenhancement methods. I turn to the ancient idea of *eudaimonia*, which was the key concept in philosophical practice, and pose the question whether the efforts put towards *eudaimonia* can be strengthened by using modern technologies which try to improve our natural functions. The rationale for the legitimacy of this kind of inspiration is either an indication of some important analogies between the ancient world – as we describe it from our perspective – and the modern world, or an indication of a fundamental error of modern philosophy, which has made our ethical reflection significantly indigent. In the case of substantiating the return to eudaimonic ethics, attention is drawn to such shortcomings of modern ethics as the excessive concentration on action itself, an unjustified rejection of theologism in practical philosophy, overt individualism, and the reductionist concept of entity.

Keywords: Virtue ethics, Aristotle, Eudaimonia, Coherence, Human flourishing

Over the recent decades, modern paradigms of thought concerning man, his life activity and, as a result, the apprehension of his obligations and ethical values, have clearly been exhausted. The reasons for this are numerous and one of them is undoubtedly the need to set new limits to individual self-development and self-fulfilment, which have been pushed back by new forms of social structures, new forms of life activity, and new technologies. The dispute about the meaning of these reasons and the search for a way to resolve key dilemmas encourages us to reach for seemingly outmoded categories and theoretical models. Among them are those whose lineage goes beyond modern tradition, in particular, the ancient thought, in which *eudaimonia* was a key concept in philosophical practice. The rationale for the legitimacy of this kind of inspiration is either an indication of some important analogies between the ancient world – as we describe it from our perspective – and the modern world,[1] or an indication of a fundamental error in modern philosophy, which has made our ethical reflection significantly

---

1 Similarities can be found, for example, in certain aspects of globalisation, with the formlessness of the social world typical for them, in the concept of community, or in social time structures (the analogy of an ancient concept of cyclical time and the modern concepts of timelessness).

indigent. In the case of substantiating the return to eudaimonic ethics, attention is drawn to such shortcomings of modern ethics as the excessive concentration on action itself, an unjustified rejection of *telos* in practical philosophy, overt individualism, or a reductionist concept of entity. The latter will be of particular importance in the following considerations, especially in view of the problematic aspects of emotions.

The search for ancient inspirations for a new concept of man based on the category of *eudaimonia* leads to two – to all appearances, related – traditions: Stoic philosophy and Aristotelian ethics. Stoicism has been variously in vogue throughout the history of European culture. The program of concentrating on privacy and one's inner side, as well as an encouragement to suppress passion and limit desires, always gains attractiveness when existing ways of conceptualizing our social existence become inadequate under the influence of rapid and complex changes, and the socially accepted goals are beginning to be seen as unrealistic. Stoic philosophy was a response to man losing himself in a multicultural world after the conquests of Alexander the Great, the confusion experienced in the face of the stagnation of the Roman Empire or during the time of barbarian invasions, as well as the fall of the medieval vision of the world on the threshold of modernity; therefore, there is no surprise that it also attracts interest in the era of "the end of the world as we know it".[2]

The case of the renaissance of Aristotelian ethics, however, is different. First of all, contrary to common opinion, it was never in the mainstream of practical philosophy: its knowledge was always brief, at best, limited to far-reaching reinterpretation and, at worst, to flashy slogans, such as the "golden mean" principle. Many elements of this philosophy, for various reasons, did not continue in the thought of Aristotle's successors, they did not affect essential philosophical debates of the following centuries, and today, newly rediscovered, they may prove useful in conceptualizing our own intuition concerning an active life in the world of the post-modern era. In fact, the anthropology of Aristotle seems to harmonize with the image of man drawn by contemporary psychology much better than other currents of ancient thought. Moreover, it is also considered as a model that provides the tools for the philosophical development of what modern human sciences are telling. Thus, Aristotle's practical philosophy has the charm of a long-lost road that has been unexpectedly rediscovered and promises

---

2 The title of I. Wallerstein's book *The End of the World As We Know It. Social Science for the Twenty-first Century* (Minnesota: University of Minnesora Press, 1999) perfectly summarizes a quite common assessment of the historical point we are currently at.

great expectations. Secondly – and, from our perspective, most importantly – the Aristotelian program cannot be considered a form of escapism: it tries to redefine and reorder the fields of man's practical activity in terms of reaching a sense of life satisfaction in complete acceptance of the real conditions we live in here and now, and not with regard to categories of imposed obligations to be fulfilled in a world that is alien, hostile and incomprehensible to us. The greatest strengths of this program are its general optimism and complete affirmation of man and his life in the conditions of fortune's contingency and the uncertainty of knowledge, which make it seem so relevant today. Aristotle's ethics continue to fascinate many with their flexibility, anti-dogmatism, and sensitivity to human relationships, and these attitudes, in terms of this theory alone, can be easily identified with the highest form of life wisdom (*phronesis*) and achieving true *eudaimonia*. It is difficult to find a more optimistic program of practical philosophy today, especially in the face of growing pessimism about the possibilities of philosophy in general.

The infatuation with the general undertone of Aristotle's theory or the analogies possible to be explicated in the symbolic sphere is, however, something different than a systematic elaboration of the reinterpretations that can be accepted today. Some of these reinterpretations come from individual issues raised by Aristotle and are only loosely associated with the entirety of his system. The catalogue of such problems is quite vast and, for obvious reasons, the choice of the starting point depends on the goals set by a contemporary philosopher. To some, the legacy of Aristotle will be limited to uncodifiability[3] or exhibiting the community context[4]; to others, ethical perfectionism[5] or the ability to take into account such elements as constitutive luck[6] or weakness of will. As a result, we may receive entirely different concepts, which can be labelled as neo-Aristotelian only in general, although in many of them we will easily find not only references

---

3   Cf. e.g., S.G. Clarke, E. Simpson, ed., *Anti-Theory in Ethics and Moral Conservatism* (New York: State University of New York Press, 1989).
4   This aspect is highlighted especially by A. MacIntyre, as well as by communitarists. It is also very important in today's reinterpretations of the Aristotelian rhetoric.
5   Cf. e.g., T. Hurka, *Perfectionism* (Oxford: Oxford University Press, 1993).
6   This term comes from T. Nagel (T. Nagel, *Mortal Questions* (Cambridge: Cambridge University Press, 1979), p. 28), who it is rather hard to consider a neo-Aristotelian in modern ethics. However, it also appears in the context of the debate on *moral luck* induced by B. Williams, who shows the clearly visible influence of Aristotelian practical philosophy. Cf. also e.g., M. Nussbaum, *The Fragility of Goodness* (Cambridge: Cambridge University Press, 1986).

to some of Aristotle's ideas, but also a fascination with the general nature of his practical philosophy.

The problem of Aristotelian inspirations becomes considerably more complicated when we want to start with a systematic, comprehensive interpretation of his practical philosophy. Perhaps the greatest achievement of the last fifty years of research on the practical philosophy of Aristotle is a greater awareness of the extent to which we do not understand it. Typical interpretations of *Nicomachean Ethics*, made using ethical categories ultimately developed in the 19th century, proved to be very simplistic and distorting. However, the better we know the culture of Antiquity, the better we understand the mechanisms underlying the society of that time, and the more aspects and nuances of Aristotle's ethics we uncover, the more difficult it is to assemble all its elements into a coherent whole. Therefore, while synthesizing, we must, out of necessity, make selections and far-reaching interpretations whose starting point are our contemporary theoretical models. It is obvious, therefore, that no representation of Aristotle's practical philosophy is entitled to be the only correct one. However, if it is to be Aristotelian from the bottom up, it must remain within the framework set by the cultural context of Antiquity and the general nature of Aristotle's philosophy. Trying to understand his practical philosophy is different from using it to solve contemporary philosophical problems. Even if both constitute our constructions, they are made at different levels, although sometimes with the use of similar tools.[7]

In this article I will focus on the category of *eudaimonia*, as it seems to be the essence of disputes over both the interpretation of Aristotle's philosophy, and the reinterpretations that can be accepted today. In the first part I will limit myself to a brief presentation of some currents of Aristotelian thought concerning *eudaimonia*, without going too deeply into the disputes over the interpretation conducted by historians of philosophy. Therefore, it will be a very synthetic presentation, which even though maintaining a foothold in the source texts, in line with what was stated above, may certainly be considered highly questionable in many of its elements. In the second part I will propose a certain far-reaching reinterpretation of the concept of *eudaimonia*, which may allow taking an in-depth look at man's objectives and his development in the modern world, which I will address in part three. It is clear that today it is impossible to return to Aristotle's

---

7 Therefore, it is difficult to settle to what extent and scope such dissimilar philosophers as E. Anscombe, I. Murdoch, Ph. Foot, A. MacIntyre, B. Williams, J. McDowell or M. Nussbaum may be considered neo-Aristotelian in ethics.

practical philosophy, as if the two thousand years (or at least the last five centuries) of history of philosophy were a mere mistake.

1 Eudaimonia and human nature

The term *eudaimonia*, usually translated into modern languages as "happiness", in the beginnings of Greek tradition was related to well-being experienced thanks to the care of a good deity.[8] This condition essentially remains beyond human control and, at most, may be requested through various forms of worship. When the Greek philosophers concentrated their efforts on searching for permanent principles of the natural world, soon also a natural need to describe human life in similar categories emerged. The process of converting *eudaimonia* into a philosophical category was therefore an answer to the question about the scope and methods of controlling one's own life, regardless of the twists of capricious fate. In this way, the category of soul (*psuche*), *eudaimonia*, as its objective or full bloom, and virtue (*arete*) were combined as exemplary dispositions leading to this goal. The starting point, of course, is the concept of soul, as it expresses the essence of the concept of man and his role in the world.

The period of creating a comprehensive anthropological concept became completed thanks to Socrates. Socrates considered *psuche* a medium of intellectual processes, i.e., the processes which determine the uniqueness of man. To him, only cognitive processes are a source of action, and the guarantee of their correctness is found in virtue, perceived as knowledge. Thus, virtue ceased to be understood as proficiency associated with civic activity (as it was understood by the Sophists) and obtained the dimension of a fundamental category expressing the adequate state of the human soul (cf. e.g., *Fedon* 66b-67a), i.e., the state of *eudaimonia*. The explanation of motivational processes in terms of intellectual needs adopted by Socrates, however, was unable to clarify the phenomenon of the divergence between desires and knowledge, which was caused by placing emotions beyond *psuche*. To correct Socrates's shortcomings, it was necessary to expand the concept of the human soul with aspects related to desires and emotions, and then show how these emotions can be organized and constrained by reason. When approaching this task, Plato adopted the assumption that particular types of desires must be related to different parts of a soul – the same part of the soul cannot feel opposing desires (cf. e.g., *Rep.* 436b8). Such an assumption has led to three parts of the soul being delineated: the base

---

8   *Eudaimon* means "good spirit", therefore a happy man is a blessed man.

or appetitive (*epithymetikon*), related to biological needs, the spiritual (*thymos*), oriented toward prestige and social respect, and the rational (*logistikon*) related to cognitive needs. As a result, we obtained a picture of the soul as a merger of three homunculi, each of them being humanoid, but disadvantaged in essential aspects.[9] This kind of concept creates completely new problems. Firstly, it turns out to be very arbitrary, since it is not clear why, for example, only three types of desires are distinguished, why desires of the same kind cannot be contrary, or how the desires delineated are associated with different emotions, e.g., fear or regret. Secondly, the principle of controlling the three types of desires through virtues, which are assigned to individual parts of the soul, but have to be – in some mysterious way – remain in close relation, remains unclear.[10]

Aware of the problems of his predecessors, Aristotle approached the question of the soul, emotions, and virtue in a completely different way. First and foremost, in his approach, the question of *psuche* is not a question of the uniqueness of man, but a general question of the principle of life in all its complexity. Therefore, there is no reason to assume a priori that man is a radically different being from other living organisms. To Aristotle, soul is a general principle that explains the workings of each living organism – i.e., an organism capable of change (DA 412a-b) – and not a separate being linked to corporeality and unique only to man. When explaining human nature, one has to show how it functions as an integral whole, able to carry out complex life functions. The category of soul that describes it has to be uniform (DA 411b), because individual aspects of human activities overlap and generate – but sometimes also modify and restrict – each other. Different types of the soul, or rather of what has authority over it, which have to be distinguished due to various dominant forms of activity, are merely components of a theoretical model explaining the complexity of human functioning, therefore they should not be imagined as separate, self-contained parts (as Plato did), but rather as the "concavity and convexity on the circumference of a circle" (EN 1102b26–31): they can be separated only by cognitive means.

One such authority is the emotional sphere, associated with desires (*orexis*), which is responsible for motivations – Aristotle does not see the possibility of assigning motivation to reason (*dianoia, nous*), because it is incapable of setting something in motion (EN 1139a, DA 433a-30).[11] All desires emerge from the

---

9 Cf. e.g., R. de Sousa, *The Rationality of Emotion* (Cambridge, Mass., and London: MIT Press, 1987), pp. 24–26.
10 Cf. e.g., T. Irwin, *Plato's Ethics* (Oxford: Oxford University Press, 1995).
11 Here I assume the interpretation according to which *boulesis* (rational desire) is in some way a corrected *alogos orexis* (irrational desires), and, as the latter, belongs to

orectic part, thus some of them remain in conflict, which – to a certain extent – is a natural state. Ethics understood in terms of character, and not only as correct action, must therefore be associated with some sort of arrangement of emotional reactions. Aristotle assumes that these reactions can generally remain in harmony, and this is how personal development ought to look like. This harmonization cannot, however, be achieved by reason simply containing emotions, as it is – in fact – powerless (suppressing emotions, *apatheia*, is therefore a road to nowhere, and it only means a rejection of activity). Organizing desires must, nonetheless, rely on controlling them in some way (searching for "proper measures"), as it is clear that they cannot be harmonized while assuming their simultaneous satisfaction to the maximum extent. However, this restriction has to stem from the very nature of emotions, so it is rather a self-correction on the level of the orectic soul, a modification of some emotional responses by matching them to other emotional responses, and not forcing them into an imposed order. To make this possible, the process of ordering emotions must start from the disposition for experiencing them, and not concern only their particular manifestations in given circumstances. Aristotle defended this assumption, because he rejected the possibility of postulating a system of rules relating only to action itself, due to the variety of specific circumstances, impossible to systematize, in which we need to make decisions. Ultimately, it is assumed that the target state is to obtain a certain characterological trait (understood as a set of dispositions), through which emotional responses are adequate to the circumstances of the act put in a long-term perspective covering the entirety of life's efforts. This adequacy is an effect of combining cognitive, motivational, and affective processes, and it results from possessing constant dispositional attributions.

How are those characterological dispositions shaped? The explanation of this process requires a significant adjustment of the Platonic model on two levels. On the first – one may say, psychological – level, it is necessary to significantly expand the conceptualization of emotional responses. Admittedly, Aristotle never presented a coherent interpretation of his concept of emotions, but on the basis of the comments he made in various places (primarily in *Rhetoric*), we can attribute to him a very complex picture of emotions as reactions, functionally

---

the emotional, rather than rational, realm. The difference between them lies in the degree of correctness of cognitive processes related to them and the general purpose toward which they are directed (cf. e.g., J. M. Cooper, "Some Remarks on Aristotle's Moral Psychology," in: *Reason and Emotion*, ed. J. M. Cooper (Princeton: Princeton University Press, 1999), pp. 237–252; a different position is defended by T. Irwin, "Who Discovered the Will," *Philosophical Perspective* Vol. 6, (1992), pp. 453–472.

necessary for survival, directly linked with simple forms of cognition and valuation.[12] The adequacy of emotional responses is possible to obtain through the correction of cognitive and evaluative processes integral to emotions, i.e., by expanding them outside the immediate context of action, taking into account the vast causal relationships and far-reaching goals. On the second – ethical – level, it is necessary to revise the entire science of virtues as permanent dispositions relating to emotions, which allow us to obtain a long-term coherence of life choices. A significant portion of *Nicomachean Ethics* is dedicated to this aspect of practical philosophy program, especially the sections explaining the category of virtue itself, specific virtues and the rules concerning them, such as the principle of the unity of the virtues, or the concept of the mean. Teaching these things is the objective of the educational program, in which the main roles will not be played by state institutions (as in Plato's concept), but by informal, interpersonal relations, which are much more sensitive to the emotional side of human nature.

A complete integration of both these levels, however, requires the indication of a single purpose – traditionally named *eudaimonia* – whose achievement would be the full realization of human potential, a kind of peak achievement on the path of self-development or prosperity. The pursuit of this objective is the starting point of the process of ordering emotional responses, therefore, for the sake of motivation, it must be somehow related to the orectic soul, which has to perceive this goal as an internal compulsion, and not a task imposed from the outside. This type of fulfilment should, therefore, be practically satisfying and involve a sense of subjective contentment.[13]

Unfortunately, all of the inaccuracies of Aristotle's program accumulate at this point, because the problems resulting from exposing the categories of character as a layout of fixed dispositions begin to reveal themselves. Dispositions are not manifested directly, but only throughaction, however, specific action is always a resultant between disposition and given situational factors. What is more, each action partially results from the disposition, but it may also modify them.

---

12 On the matter of Aristotle's take on emotions, cf. e.g., J. M. Cooper, "Aristotle on Emotion," in: *Reason and Emotion*, ed. Cooper; W. W. Fortenbaugh, *Aristotle on Emotion* (London: Duckworth, 2002); G. Striker, "Emotions in Context," in: *Essays on Aristotle's Rhetoric*, ed. A. O. Rorty (Berkeley and Los Angeles: University of California Press, 1996).

13 Aristotle does not say, however, that it is possible in the case of every individual fortune. Sometimes specific external circumstances may prove to be insurmountable, which makes human life constantly marked by a hint of tragedy, which can reveal itself contrary to all our efforts (cf. e.g., *EN* 1100a8).

The key turns out to be the time factor imposed, on the one hand, on unique circumstances, and on natural psychological (e.g., fatigue) and biological (e.g., ageing) processes we undergo, on the other. This constant balancing between act and possibility, cause and effect, and repetition and exceptionality, makes it incredibly difficult to specify not only how *eudaimonia* can be reached, but also what it is ultimately about.[14]

Aristotle tries to solve these problems by referring to the order of the universe in which man is perceived only as a species and not as an individual. Only such a perspective allows to accept the function and purpose of man as *objectively* provided. Therefore, the description of *eudaimonia* as the target condition of human development must correspond with the general scheme governing all of reality, in which every being has its place in a hierarchically organized, perfect, and non-developing universe. As everything that exists, man pursues an ultimate, final goal, which must be complete and self-sufficient (*EN* 1097a15–b21). This goal must be related – as in the case of any other species – to the best function he can perform (the so-called "function argument" *EN* 1097b-1098a18) and, at the same time, remain consistent with the objective hierarchy of beings. As this function is "the act of an intellectual element" (*EN* 1098a), man's highest good is therefore related to theoretical cognition, which allows man to objectively recognize their place in the world and strive towards its perfect accomplishment.

Aristotle's undoubted intention was to merge *eudaimonia* with contemplation as the perfect human function. However, the way he does so turns out to be completely inconsistent with the earlier conclusions. Therefore, the entire effort of modern interpreters seeks to show that it is possible to defend Aristotle's practical philosophy by aligning contemplation and the orethic perfection of a part of the soul, whether it is by showing that they are relatively independent, or by showing that ethical virtues are necessary to exercise intellectual virtues[15] – either

---

14 In modern ethics these problems have been considered so fundamental that they led to the abandonment of the project of ethics based on luck and characterological traits of man, in favour of focusing only on actions.
15 Of course, there are some commentators who think that *eudaimonia* consists only of contemplation. On the interpretation of *eudaimonia*, cf. e.g., W.F.R. Hardie, "The Final Good in Aristotle's Ethics," *Philosophy*, Vol. 40 (1965), pp. 277–295; J. L. Ackrill, "Aristotle on Eudaimonia," in: *Essays on Aristotle's Ethics*, ed. A. O. Rorty (Berkeley: University of California Press, 1980); J. M. Cooper, *Reason and Human Good in Aristotle* (Cambridge, Mass.: Harvard University Press, 1975) and "Contemplation and Happiness: a Reconsideration," in: *Reason and Emotion*, ed. Cooper; G. R. Lear, *Happy Lives and the Highest Good* (Princeton: Princeton University Press, 2004).

in the sense that they are a prerequisite, or in the sense that orethic perfection and contemplation can only be achieved *together*. The basis for the adoption of the latter concept can be found in the recognition that contemplation is also an *action*, and therefore it is associated with motivation, it is based on emotional needs and, ultimately, it is an *experience*, which must evoke a subjective state of satisfaction and fulfilment, and in turn the relevant dispositions relating to choices must be based on the dispositions relating to theoretical cognition[16] – the act of contemplation simply cannot be treated as a separate type of activity in relation to activity associated with the practical side of life.

No matter how we handle the numerous vulnerabilities that appear in these parts of Aristotle's concept, there is no doubt that his ethical program does not lead to the elimination of emotion or its suppression by rational functions, but to the integration of all functions of the soul. Aristotle put particular emphasis on the pursuit of the coherence of emotional reactions defined in terms of permanent dispositions (*hexis*) manifesting themselves in long periods of human life. Aristotle's optimism lies in the fact that – starting from the general assumption about the unity of soul – he assumes that all functions of a human being are essentially in harmony, and the state of their permanent mismatch, which we constantly experience, is ultimately an accidental state.

2 Eudaimonia in postmodern era

The brief recapitulation of the Aristotelian program of practical philosophy presented above places an emphasis on three elements: the holistic psychological concept outlining the rational and cognitive role of emotions, the ethical concept focused on developing appropriate dispositions, and the metaphysical concept defining the purpose of man in the context of the overall concept of the universe. Each of these elements has a lot of shortcomings and gives rise to a lot of interpretative problems. Some of the solutions adopted by Aristotle, however, arouse so much controversy that it is difficult to expect that his practical philosophy may become more widely recognized today without significant modification. Let us investigate these controversies in greater detail.

The first element, that is, the theory of emotion, is without a doubt the strongest part of Aristotle's concept. Its general character was completely innovative, which was probably why it was initially misunderstood and almost immediately

---

16 Cf. e.g., T. Tuzzo, "Contemplation, the Noble, and the Mean," in: *Aristotle, Virtue, and the Mean.Apeiron XXV*, ed. R. Bosley, R. Shiner and J. Sisson (Edmonton: Academic Printing and Publishing, 1995).

became abandoned. It was not until the second half of the 20th century that psychology rediscovered Aristotle's theory of emotion, which to a large extent determines the current attractiveness of his ethics and is seen as a normative formulation of those psychological assumptions. Aristotle's psychology is nowadays also considered to be the archetype of the whole paradigm of research, recognizing the key role of emotions for the functioning of the human being, and the heart of this paradigm is the so-called cognitive theory of emotion, which considers them to be cognitive-motivational-affective states with a high degree of complexity. Naturally, today the categories of this theory are also used to fill in the gaps in the concept created by Aristotle himself, as well as in the eudaimonic ethics.[17]

The second element, the theory of virtue, raises significantly more questions. Some of them are, one could say, of a technical nature and are the subject of common philosophical disputes led by ethicists. These disputes relate to issues such as specific relationships of virtues to particular emotions, ways of understanding the principle of the golden mean,[18] or the recognition of virtues as dispositions. Some concerns, however, are of a more general nature. The most important one relates to the recognition of the process of shaping virtues as a process of socialization. As we have seen above, Aristotle's concept does not take into account the individuality of the entity and its autonomy at all – development is not described in terms that leave room for the distinctness of a person as a moral entity recognized from the perspective of 20th century philosophy.[19]

---

17 On the Aristotelian sources of contemporary psychology of emotion, cf. e.g., R. C.Solomon, "The Politics of Emotion," *Midwest Studies in Philosophy*, Vol. 22 (1998), pp.1–4; N. Frijda, A.S.R. Manstead and S. Bem, *Emotions and Beliefs: How Feelings Influence Thoughts* (Cambridge: Cambridge University Press, 2000), p. 1. On the issue of using the cognitive theory of emotions to reinterpret Aristotle's thought cf. e.g., N. Shermann, *Making a Necessity of Virtue* (Cambridge: Cambridge University Press, 1997), esp. pp. 26, 55, 57; Summary overview of this issue: J. Jaśtal, *Natura cnoty. Problematyka emocji w neoarystotelesowskiej etyce cnót* [*Nature of Virtue. Emotions in neoaristotelian virtue ethics*] (Kraków: Księgarnia Akademicka, 2009) [in Polish].
18 Cf. e.g., J. Urmson, "Aristotle's Doctrine of the Mean," in: *Essays on Aristotle's Ethics*, ed. Rorty; R. Hursthouse, "A False Doctrine of the Mean," in: *Aristotle's Ethics*, ed. N. Sherman (Lanham: Rowman & Littlefield Publishers, 1999); Ch. Young, "The Doctrine of the Mean," *Topoi*, Vol.15 (1996).
19 It is very difficult to find references to the individualised entity in Aristotle's writings, a conscious "Me" – even the theory does not submit to such an interpretation easily. Perhaps the most differentiating element of his concept would be the vices (the pursuit of virtues would, therefore, erase this individualism), although they are also discussed in terms of human types, and not as a result of experiences related to the individual

For the Ancients, the process of shaping dispositions primarily consisted of the complete subordination of an individual to the community, and a socialization oriented towards making them accept their social role and socially accepted vision of the world. If Aristotle's ethics is to be treated today as a serious alternative to modern ethical and anthropological concepts, it must be able to take into account the individuality of the subject, firstly, as a unique person with all the uniqueness of their life experience, mental structure and value system, and, secondly, as an entity capable of making independent, free decisions, for which they can assume moral responsibility.

Similar doubts are raised by the concept of *eudaimonia* as an objective of humanity, which has been determined by the nature of the universe. The contemporary postulate to take into account the uniqueness of life experience is not just a matter of displaying an individual set of dispositions, but it also raises the question of the individual fulfilment of life or subjective aspects of *eudaimonia*. In its essence, *eudaimonia* in Aristotle's system arises from the purpose-bound vision of the world and the related concept of knowledge, which was rejected by modern philosophy and is virtually impossible to defend today. The rejection of teleologism and the replacement of the concept of rationality as the manifestation of the objective order of nature with the concept of procedural rationality causes *eudaimonia* to be perceived as a subjective, particular goal.[20] Therefore, anything that can lead to it, including behavioral dispositions, virtues, and emotional reactions is subject to relativization. What is more, the belief that the diversity of these reactions and happiness understood differently and allowing the possibility of an objective arrangement falters.[21] Any improvement of the function of man is an improvement *to something* – if this *something* becomes relative, and stands in opposition to *something else*, which can also be carried out by the same functions, then improvement has no objective value, it cannot be a base of evaluating actions in a moral sense. Similar techniques of supporting our functions also become purely instrumental due to currently preferred goals.

---

way of life or individual personality differences. Cf. e.g., R. Sorabji, *Self. Ancient and Modern Insights about Individuality, Life and Death* (Chicago: The University of Chicago Press, 2006).

20 The first philosopher who recognized the relativity of happiness and virtue, and consequently rejected the entire eudaimonic and aretolgical tradition in ethics, was William Ockham; he accepted the moral law given by God and our will to execute it as the basis for ethics.

21 It is one of the main arguments raised by I. Kant against eudaimonic ethics.

The fundamental questions are as follows: If there are no clear social roles and no common vision of the world pretending to be objective, what then might be considered to be the ultimate goal? How to describe individual goals an entity is to pursue autonomously, if their implementation requires the modification of disposals through activities that are determined by those dispositions? Is it possible to rephrase the Aristotelian concept of *eudaimonia* to allow it to take into account philosophical beliefs that are crucial to us?[22]

To answer these questions it is worth to adopt the Cartesian starting point and ask, what can be considered, from the perspective of one's internal experience – according to the Aristotelian description of *eudaimonia* – as an ultimate, complete, and self-sufficient goal, that is, a goal one cannot intentionally not want and strive towards.[23] All external goals (no matter how we describe them in detail) do not meet this criterion. One can feasibly consider whether species interests, social interests (related to one's own community or humankind in general), the interests of one's loved ones, or personal interests such as, for example, improving one's physical capabilities or broadening knowledge, should be realized always and everywhere. Even if these interests were realized knowingly and intentionally (and not only under the influence of internal, unconscious factors), many people would certainly not consider them to be absolute and paramount. In turn, the category of happiness understood as contentment or life satisfaction is so vague and so dependent on external circumstances that it does not bring much to the description of *eudaimonia*. It seems that today the only need that can lay claim to being the goal in an Aristotelian sense, is the need for a consistent, cognitive experience. It is difficult to imagine that anyone intentionally pursues the increase of cognitive dissonance and internal disharmony as the ultimate goal (although, of course, such actions may be intended as part of the process of building the ultimate consistency).

If we now extend, in the spirit of Aristotle's psychology, the concept of cognitive consistency with cognitive elements related to emotions, and we assign the basic role of motivation from cognitive activities (in the sense mentioned on the occasion of discussing the Aristotelian concept of contemplation) to emotions, then it becomes clear that achieving such a consistency must entail the shaping

---

22 A negative answer to this question does not have to lead to the rejection of the Aristotelian program, but can become the starting point for a comprehensive critique of the whole *acquis* of modern philosophy, including its individualism and epistemological bias. It seems that this way was chosen, for example, by A. MacIntyre.
23 Here I present a concept which I have discussed in detail in: Jaśtal, *Natura cnoty*, esp. pp. 362–387.

of emotional responses. The rationality of emotion is understood here as the effectiveness of achieving the current goals of a subject under the conditions of limited knowledge and a lack of time to make a decision. The shaping of rationality understood in such a way consists in taking into account a wider palette of immediate goals, including long-term ones, on the one hand, and improving cognitive elements, on the other. In this way one can talk about objective adequacy of responses and the circumstances of the action, recognized from the perspective of the sum of life experiences.

Such a formulation of the process of achieving *eudaimonia* is assumed by metaphysical realism: the world around us is a world partly independent from us, it resists us and sometimes it is completely indifferent to us. Our cognitive structures and attitudes related to them are limited by the reality of the world, therefore, they can only be assessed as more or less accurate (which is verified in practice). At the same time, however, our responses are a part of the objective world, they have an impact on its processes, which – in turn – secondarily affect us. Coherence understood in such a way is never given to us as something finite and sustainable, because also, along with life experience and changes in the emotional sphere, our long-term goals may change. Therefore, ultimately, our view of the world – taking into account our own experience, with all of its uniqueness – always remains a narrative structure of a more or less wavering nature, which can only conditionally be described as adequate to reality. We must also remember that sometimes we take shortcuts, that is, we use all sorts of simplifying mental mechanisms, which are clearly intended for immediate reduction of tensions. The dependence of our image of the world on social factors, which provide patterns to build our own narrative, influence the selection of certain problem-solving techniques, or emphasize certain values, also cannot be ruled out.

One more assumption, acting as a kind of a regulatory rule to this kind of Aristotelian constructivism, is needed in order to avoid suspicion that *eudaimonia* understood in such a way will ultimately serve as a goal that can be used to justify any relatively consistent set of cognitive and emotional responses.[24] This assumption stems from the acceptance of ourselves as parts of the biological world, the psycho-corporeal whole, whose vital functions were developed while confronting the world around us. This means granting the primacy of practical

---

24 On the matter of Aristotelian version of metaethical constructivism, cf. M. LeBar, "Aristotelian Constructivism," *Social Philosophy and Policy*, Vol. 25, No. 1 (2008), pp. 182–213; J. Jaśtal, "Konstruktywizm w metaetyce – perspektywa Arystotelesowska" ["Constructivism in metaethics – Aristotelian point of view"], *Diametros* Vol. 45 (2015), pp. 122–143 [in Polish].

reason over theoretical reason. We assume that ultimately, we are able to construct an image of the world that will be the most similar to the isomorphic image, thanks to the constant verification of our actions, related to the resistance of the world around us. At the same time, this image will be linked to the satisfaction of our emotional needs, such as the feeling of accomplishment, contentment, a sense of inner harmony, that is, generally speaking, happiness. Even if this happens to only some of us, rarely and only for a moment (as the states of Aristotelian contemplation), the efforts related to the constant adjustment of the set of cognitive and emotional dispositions are not pointless. We are reminded of it by the sense of absence and failure that we experience – and that we are constantly trying to break free from.

3 Human flourishing and transhuman technologies

Can the efforts put towards reaching thus understood *eudaimonia* be strengthened by using modern technologies that try to improve our natural functions? In accordance with the adopted model, the state of *eudaimonia* can be achieved by enhancing two elements. The first possibility is to improve our natural intellectual abilities, forming the base of emotional responses, for example, memory, the capacity to associate facts, or analytical skills. These traits, however, do not guarantee the improvement of the coherence of life experience, and might as well impede its achievement by complicating, for example, the decision making process through an excessive extension of the information at our disposal, or the overdevelopment of the hypothetical chains of cause and effect that our actions become tangled in. As a result, the internal tension connected with the emotional and cognitive dissonance may increase along with our efforts to take into account all available information and apply all the techniques of their analysis known to us.

The second way is to directly improve emotional responses. Of course, it may not consist in their elimination, as in this way we can, in fact, deprive ourselves of the capacity to function properly in certain situations. It seems that this rule applies even to responses such as anger: limiting the excessive tendency to aggression should be rather done through reinforcing self-control, and not through preventing the emergence of aggressive responses (although in some cases, the restriction of certain emotions may be appropriate, e.g., when a person's anxiety completely destabilises their life; the elimination or reduction of fear, however, cannot be the basis for a permanent solution). Even the principle of moderation (*medio tutissimus ibis*) does not work in practical life if it is understood as the determinant of the desired, sustainable state of internal balance, and not only as an immediate rule saying that violent manifestations of emotions are rarely the appropriate action. Paradoxically, therefore, the improvement of emotional responses should rather intensify them, for example by increasing sensitivity, empathetic capacities, or expanding the palette of emotional relations and making them more flexible. This recommendation corresponds to a program which, in modern psychology, is related to the development of the so-called emotional intelligence. It is difficult to say whether it can be achieved by the direct modification of our capacities, for example, through genetic engineering or pharmacological intervention. Probably, to a certain extent, yes – but this kind of treatment surely cannot guarantee positive results, similarly, as the pursuit of *eudaimonia* through improving cognitive capabilities may turn out entirely counter-productive.

Similar concerns arise from the conclusion that the process of valuation is an important component of an emotional reaction.[25] Although it seems that a specific feeling, e.g., fear, is based on a relatively simple evaluation of something as a threat, the ability to evaluate what and to what extent is a threat in a given moment requires some training. Due to the complexity of interpersonal relations, learning about the current social context, the socially valid conventions, patterns, lifestyles etc. is a part of this training, which always takes time and practice. Even if it comes more easily to some people, thanks to their capacities, than to others, the ultimate effect always remains the resultant of inborn characteristics and the tedious process of their improvement.

As demonstrated in the article, the primary problem associated with the pursuit of *eudaimonia*, understood as the coherence of emotional/cognitive reactions, lies in the fact that these reactions must be adequate for the *current* circumstances of life, and that their proper course depends equally on our internal dispositions and the state of the world in which we operate.[26] The formation of these dispositions must take place under the influence of particular

---

25 As one can guess, Aristotle was fully aware of the evaluative nature of emotional responses. It is reflected both by the descriptions of specific feelings in *Rhetoric* (e.g., shame is associated with *defamation*, and resentment is characterised as a feeling caused by experiencing *undeserved* happiness by someone [Rhet 1383b, 1386b]; both emphasized categories are heavily dependent on social interpretation), and the general characteristic of virtue as *hexis*, whose vital element is a reference to an unspecified system of values (cf. e.g., D. S. Hutchinson, *The Virtues of Aristotle* (London: Routledge Kegan & Paul, 1986), pp. 8–36).

26 Aristotle assumed that virtues cannot be inherent, because they must undergo a shaping process (EN 1103a–b).

circumstances and the associated opportunities and risks which update these dispositions. The starting point of the Aristotelian program of practical philosophy is the assumption that the world we live in is so complex that there is no way to indicate any solid rules pertaining to actions. Therefore, these rules cannot be used as indicators of initial, permanent modification of our responses. Any adaptation of human nature to preconceived behavior patterns would ultimately prove to be a form of reduction of human capabilities. The pursuit of *eudaimonia* is always contextual and can only be effectively supported by certain heuristic techniques. These techniques, however, are designated by a different central analogy than in Classical culture. A model referring to the process of psychotherapy, in which the subjects are actively interested in internal change due to the psychological tension they are experiencing and their factual knowledge, has no decisive role in this process, in place of the model of *paideia*, exhibiting the meaning of oppressive socialization and education, understood as improving the knowledge relevant to the performance of tasks designated by the assigned social role. This change involves a certain paradox in the concept of *eudaimonia* understood as a coherent image of the world. The paradox lies in the fact, that after crossing a certain unclear – and probably very relative, both generationally and individually – threshold of accepting the complexity of the social world in which we live. Coherence is even more difficult to achieve the more we know and the richer our life experience is; therefore, how we deal with reducing the widely understood cognitive dissonance is equally dependent on the degrees of natural abilities we possess, and the conditions under which we live – it may be easier to live happily when one's simple image of the world is sufficient for normal functioning, and life's challenges are repetitive and predictable. Therefore, it should not come as a surprise that nowadays some people seek such conditions, perceiving changes in their way of life as the best way to achieve inner peace. However, the problem lies in the fact that the state in which I *cognitively* do not want to know or experience something is not equivalent to a state, in which *I am not aware* that I can know or experience something. Therefore, it is always a form of escapism, and maintaining the state of internal cohesion in such conditions requires recoursing to certain psychological mechanisms which help us believe that the world we are withdrawing from is, in some regards, worthless. These escapist mechanisms, however, usually bring serious inconsistencies to our lives and cause us to give up on some forms of action in the world. The need for restriction appears when we do not focus on experiencing emotions. It is impossible to determine, in the form of rules or objective laws of nature discovered by empirical sciences, where the golden mean of internal states lies. The only thing Aristotle assures us of is that such a state is essentially attainable by all of us in the environment in which we operate if only we put enough effort in developing ourselves.

References

Ackrill, J. L. "Aristotle on Eudaimonia." In: *Essays on Aristotle's Ethics*, ed. A. O. Rorty. Berkeley: University of California Press, 1980.

Aristotles. *Nicomachean Ethics*. Trans. T. Irwin. Indianapolis and Cambridge: Hackett Publishing Company, 1999.

Clarke, S. G. and E. Simpson, eds. *Anti-Theory in Ethics and Moral Conservatism*.New York: State University of New York Press, 1989.

Cooper, J. M. *Reason and Human Good in Aristotle*. Cambridge, Massachusetts: Harvard University Press, 1975.

Cooper, J. M. *Reason and Emotion*. Princeton: Princeton University Press, 1999.

Cooper, J. M. "Contemplation and Happiness: A Reconsideration." In: *Reason and Emotion*, J. M. Cooper. Princeton: Princeton University Press, 1999.

Cooper, J. M. "Aristotle on Emotion." In: *Reason and Emotion*, J. M. Cooper. Princeton: Princeton University Press, 1999.

Cooper, J. M. "Some Remarks on Aristotle's Moral Psychology." In: *Reason and Emotion*, J. M. Cooper. Princeton: Princeton University Press, 1999.

Fortenbaugh, W. W. *Aristotle on Emotion*. London: Duckworth, 2002.

Frijda, N., A.S.R. Manstead and S.Bem. *Emotions and Beliefs: How Feelings Influence Thoughts*. Cambridge: Cambridge University Press, 2000.

Hardie, W.F.R. "The Final Good in Aristotle's Ethics." *Philosophy*, Vol. 40, 1965, pp. 277–295.

Hurka, T. *Perfectionism*. Oxford: Oxford University Press, 1993.

Hursthouse, R. "A False Doctrine of the Mean." In: *Aristotle's Ethics*, ed. N. Sherman. Lanham: Rowman & Littlefield Publishers, 1999.

Hutchinson, D. S. *The Virtues of Aristotle*. London: Rowman & Littlefield Publishers, 1986.

Irwin, T. *Aristotle's First Principles*, Oxford: Oxford University Press, 1988.

Irwin, T. "Who Discovered the Will." *Philosophical Perspective*, Vol. 6, 1992, pp. 453–472.

Irwin, T. *Plato's Ethics*. Oxford: Oxford University Press, 1995.

Jaśtal, J. *Natura cnoty. Problematyka emocji w neoarystotelesowskiej etyce cnót [Nature of Virtue. Emotions in neoaristotelian virtue ethics]*. Kraków: Księgarnia Akademicka, 2009 [in Polish].

Jaśtal, J. "Konstruktywizm w metaetyce – perspektywa Arystotelesowska" [Constructivism in metaethics – Aristotelian point of view]. *Diametros*, Vol. 45, 2015, pp. 122–143 [in Polish].

Lear, G. R. *Happy Lives and the Highest Good*. Princeton: Princeton University Press, 2004.

LeBar, M. "Aristotelian Constructivism." *Social Philosophy and Policy*, Vol. 25, Issue 1, 2008, pp. 182–213.

Nagel, T. *Mortal Questions*. Cambridge: Cambridge University Press, 1979.

Nussbaum, M. *The Fragility of Goodness. Luck and Ethics in Greek Tragedy and Philosophy*. Cambridge: Cambridge University Press, 1986.

Shermann, N. *Making a Necessity of Virtue*. Cambridge: Cambridge University Press, 1997.

Solomon, R. C. "The Politics of Emotion." *Midwest Studies in Philosophy*, Vol. XXII, 1998, pp. 1–20.

Sorabji, R. *Self. Ancient and Modern Insights about Individuality, Life and Death*. Chicago: The University of Chicago Press, 2006.

de Sousa, R. *The Rationality of Emotion*. Cambridge and London: MIT Press, 1987.

Striker, G. "Emotions in Context." In: *Essays on Aristotle's Rhetoric*, ed. A. O. Rorty. Berkeley and Los Angeles: University of California Press, 1996.

Tuzzo, T. "Contemplation, The Noble, and The Mean." In: *Aristotle, Virtue, and the Mean*. Apeiron XXV, eds. R. Bosley, R. Shiner and J. Sission. Edmonton: Academic Printing and Publishing, 1995.

Urmson, J. "Aristotle's Doctrine of the Mean." In: *Essays on Aristotle's Ethics*, ed. A. O. Rorty. Berkeley: University of California Press 1980.

Wallerstein, I. M. *The End of the World As We Know It. Social Science for the Twenty-first Century*. Minnesota: University of Minnesota Press, 1999.

Young, Ch. "The Doctrine of the Mean." *Topoi*, Vol. 15, 1996, pp. 89–99.

Wojciech Lewandowski

Intrinsic and Instrumental Values in the Assessment of Human Enhancement[1]

Abstract: One of the most discussed issues in the debate concerning human enhancement is the problem of balancing competing reasons for undergoing some enhancing interventions. It is often claimed that reasons based on the instrumental value of some enhanced trait should be balanced against reasons based on the risk of eliminating the existing intrinsic value related to engagement in traditional human practices. This kind of argumentation serves Nicholas Agar as a justification for rejecting radical enhancement, while at the same time accepting moderate enhancement, which joins the instrumental values connected with the higher level of human capacities with the preservation of intrinsic values as seen from the individual and anthropocentric point of view. The main challenge to this view is that it requires a justification of the primacy of the anthropocentric point of view over the objective impartial one. Such a manner of reasoning therefore requires some model of comparing outcomes in respect of the increase and decrease in intrinsic and instrumental values. In this paper I analyze the possibilities of building such a model based on the combination of two criteria: the criterion of positive and negative outcomes of using a given capacity, and the criterion of compatibility with current human standards of valuation related to the intrinsic value of ground projects or interpersonal relationships. I argue that the application of the category of weak superiority between such outcomes may help to resolve some problems with the assessment of human enhancement, though it cannot serve as the only basis for the moral evaluation of enhancing interventions.

Keywords: Bioconservatism, Intrinsic value, Instrumental value, Transformative change

One of the objections against bioconservatism is that it insufficiently distinguishes between the various degrees of human enhancement. Taking drugs to improve memory or concentration is regarded in the same way as embryo selection for cognitive enhancement, and this, in turn, in the same way as creating an enhanced brain-machine interface. Advocates of human enhancement claim that there are no reasons strong enough to justify the prohibition of all enhancing interventions. According to them, the reasons for and against various

---

1 The writing of this article was funded by the Polish National Science Centre (Dec-2013/10/E/HS5/00157).

interventions should be balanced against each other to formulate a moral judgment about the permissibility or impermissibility of a given intervention. The debate on the strength of reasons in the assessment of human enhancement is usually based on one of the following distinctions of reasons. The first one is a distinction between person-affecting and impartial reasons. It concerns the effects of actual decisions about using enhancing technologies on the existence, identity, and number of future people. Predominantly, it is assumed that in situations in which it is possible to show that a given intervention would be beneficial or harmful for a future person, the person-affecting reasons are stronger than impersonal ones.[2] The main problem is that many actions taken in the context of human enhancement, e.g., procreative decisions and embryo selection, can hardly be viewed as person-affecting harm.[3] The second distinction is between agent-relative and agent-neutral reasons. The former are based on the agent's perspective – they concern her own well-being, her relations with other people, and obligations which only that individual agent has. Agent-neutral reasons abstract from the agent's perspective and concern general well-being and obligations to all those with moral status. The latter are viewed as stronger in theories in which impartiality is an essential element of morality, however, many authors are trying to justify human enhancement by reference to both kinds of reasons.[4] The third distinction divides reasons into those based on intrinsic and instrumental values. The common understanding of these values is that the proper way of referring to instrumental values is to use them for the realization of further ends, and the proper way of referring to intrinsic values is to promote or respect them. From this point of view, it would be simple to see reasons based on intrinsic values as essentially stronger than those based on instrumental values. It should be noted, however, that not all intrinsic values are seen to have absolute superiority. The best example are the attempts at balancing both kinds of values in environmental ethics. Analogically, if it is possible to justify that actual human nature is intrinsically valuable, then the risk of loss of this value may draw the line beyond which the reasons for increasing instrumental values through the enhancement of human capacities may cease to prevail. This

---

2   J. Savulescu, "The Nature of the Moral Obligation to Select the Best Children," in: *The Future of Bioethics. International Dialogues*, ed. A. Akabayashi (Oxford: Oxford University Press, 2014), pp. 170–182.

3   B. Saunders, "Why Procreative Preferences May Be Moral – And Why it May Not Matter if They Aren't," *Bioethics* Vol. 29, No. 7 (2015), pp. 503–504.

4   J. Savulescu, I. Persson, *Unfit for the Future: The Need of Moral Enhancement* (Oxford: Oxford University Press, 2012).

kind of argumentation is formulated by Nicholas Agar in his book *Truly Human Enhancement*, in which he shows that some ways of enhancing our cognitive capacities or prolonging life are undesirable as they raise these traits to an excessively high level.[5] What differs his argumentation from the bio-conservative view is an attempt to formulate and justify a model of balancing instrumental and intrinsic values based on the category of "transformative change" which allows for a differentiated approach to various kinds of human enhancement. In this paper I am going to analyze the possibilities of building a general model of balancing these values in the context of human enhancement.

1 Transformative change as the boundary for enhancing interventions

Nicholas Agar uses the criterion of degree of enhancing interventions to distinguish between radical enhancement, which improves significant attributes and abilities to levels greatly exceeding what is currently possible for human beings, and moderate enhancement, improving significant attributes and abilities to levels within or close to what is currently possible for human beings.[6] These significant attributes and abilities may be defined by their prudential value, i.e., how they can promote our interests or well-being. Prudential value may be viewed from one of two points of views: objective or anthropocentric. From the objective perspective, prudential values are instrumental values, dependent on the results of actions taken with the use of human attributes and abilities. Enhanced cognitive capacities have greater instrumental value because they enable us to solve increasingly more difficult and complex problems, and enhanced strength increases efficiency in physical activities. Prudential value, seen from the anthropocentric perspective, is an intrinsic value, independent of the results of the use of our capacities, and corresponding to the engagement in the exercise of these capacities.[7]

The distinction proposed by Agar is inspired by Alasdair MacIntyre's conception of goods internal and external to the practice:

> By 'practice' I am going to mean any coherent and complex form of socially established cooperative human activity through which goods internal to that form of activity are realized in the course of trying to achieve those standards of excellence which are

---

5   N. Agar, *Truly Human Enhancement: A Philosophical Defence of Limits*, Cambridge, Mass.: MIT Press, p. 2.
6   Agar, *Truly Human Enhancement*, p.2
7   Agar, *Truly Human Enhancement*, pp. 17–18, 28.

appropriate to, and partially definitive of, that form of activity, with the result that human powers to achieve excellence, and human conceptions of the ends and goods involved, are systematically extended.[8]

According to this definition, the arts, sciences, complex games, politics, or the making and sustaining of family life are practices, and such activities as playing tic-tac-toe or bricklaying are not. Goods external to practices are those which may be achieved differently than by a given practice. When they are achieved, they remain individual property and may constitute the object of competition. On the other hand, goods internal to a practice may be defined only by the concepts taken from the given practice and may be recognized only through participation in this practice. The achievement of these kinds of goods is a good for the whole community.[9]

It may be noted that the possibility of enhancement of future people's capacities is a challenge for MacIntyre's view. Raising the IQ level of a handicapped child, who so far could not participate in the practice of chess-playing, would enable him to achieve both goods internal and external to this practice. Further improvement of his intelligence – allowing him to play on Samuel Reshevsky's level – would increase his chance of achieving more goods external to the practice, while maintaining the same level of goods internal to the practice. Problems arise when the child's intelligence is raised to a level so high that playing chess would not offer him any challenge, being comparable in difficulty to tic-tac-toe. Continuous victories over world's best players would not constitute a good for the whole community and may even entail the loss of some internal goods. From the point of view of advocates of human enhancement, in order to avoid negative consequences such as the alienation of enhanced individuals, as many people as possible should actually undergo enhancement, which would lead to the formation of new practices and provide opportunities to achieve new goods internal to them. The prospect of a future society consisting of individuals whose capacities significantly exceed the current abilities of actual people, and who are engaged in practices completely different from the current ones, including a different approach to morality, may be compatible only with the objective point of view, as it collides with the anthropocentric perspective. *Our own* practices and goods internal to them matter *to us* more than any future practices of enhanced people. How to balance both kinds of values? According to Agar, "We arrive at a final

---

8   A.MacIntyre, *After Virtue. A Study in Moral Theory* (Notre Dame: Notre Dame Press, 2007), p. 187.
9   MacIntyre, *After Virtue*, pp. 190–191.

reckoning of the prudential value of an enhancement by appropriately balancing the instrumental and intrinsic value of the resulting capacity."[10] The key criterion is the threshold set by "transformative change":

> A transformative change alters the state of an individual's mental or physical characteristics in a way that causes and warrants a significant change in how that individual evaluates a wide range of their own experiences, beliefs, or achievements.[11]

This proposition of a criterion for the limit of human enhancement presented by Agar is based on the individual's point of view. The main threat for the individual is a fundamental change in the way in which he/she evaluates his/her life.

From this point of view, the main argument against interventions leading to transformative change concerns the potential threat for our identities, as these interventions may affect the psychological connections between our past and actual selves by altering our assessment of our autobiographical memories. If before enhancing our physical abilities our participation in a marathon was memorable as a valued achievement, then after enhancement the same achievement may be unimpressive and not worth remembering.[12] It should be noted that such argumentation would only concern the individuals who actually have some autobiographical memories from a time before the transformative change, and hence it would not help to resolve problems concerning procreative decisions, embryo selection, or interventions performed on fetuses or infants. Furthermore, this argumentation requires a justification of the concept of identity based on autobiographical memories.[13]

The second argument against radical enhancement is based on the fact that a transformative change may be irreversible. Radical cognitive or moral enhancement may lead to an outcome in which the enhanced person possesses extremely different beliefs and moral dispositions as compared with those from before the transformation. Changing the basis for moral valuations may have the effect that the person who decided to implement this change will not have a chance to recognize a possible error and regret her decision.[14] According to Agar, this constitutes a good reason to avoid such modifications. Objections to this argument consist primarily of attempts to demonstrate that the individual may have reasons for irreversible transformative change, even beyond the context of

---

10 Agar, *Truly Human Enhancement*, pp. 27–28.
11 Agar, *Truly Human Enhancement*, pp. 5–6.
12 Agar, *Truly Human Enhancement*, p. 64.
13 Agar, *Truly Human Enhancement*, pp. 60–62.
14 Agar, *Truly Human Enhancement*, p. 14.

human enhancement. According to Nicholas Bostrom and Toby Ord, growing up is such an irreversible process.[15] Although this transformative change is not discontinuous, there is a clear difference between beliefs and valuing standards possessed by the person in childhood and the same person after reaching adulthood. The fact of this difference cannot be a reason for a child to be against this kind of transformative change. Agar's reply is based on the thesis that radical enhancement differs from growing up in that the child may assign significant value to his/her practices, but he/she cannot assign a value to the fact that these practices are valuable for him/her. In contrast, adult people can connect their practices into the whole project of their lives and assign value to being a human adult.[16] It seems that this argument can be defended only from the "human point of view" understood as the point of view of rational human beings participating in practices considered as valuable. In this case, however, the previous argument based on individual perspective and focused on the threat for individual identity loses its strength, and at the forefront stands the species-relativism view, according to which "certain experiences and ways of existing properly valued by members of one species may lack value for the members of another species."[17] The main argument against radical enhancement then turns out to be based on the risk of loss of intrinsic values significant from the "human point of view".

The argumentation presented above is not exposed to the speciesism objection, as it does not assume that the intrinsic values of members of the *homo sapiens* species are objectively superior to the possible values shared by potential rational members of other species. It also does not say that the participation in the common valuable practices of rational beings belonging to different species is impossible. It only says that rational beings of one species, having different ideals, may not want to participate in some practices of rational members of other species.

The main challenge for species-relativism is justifying why the way of assigning value would depend primarily on biological species membership and not, for example, on having certain cognitive capacities or ways of experiencing the world. Another possible objection against basing intrinsic values on species-relativism is that it requires the abandonment of the universalist approach in moral theory. It is significant that one of the most famous defenses of human

---

15 N. Bostrom, T. Ord, "The Reversal Test: Eliminating Status Quo Bias in Applied Ethics," *Ethics*, Vol. 116, No. 4 (2006), p. 671.
16 Agar, *Truly Human Enhancement*, pp. 74–75, N. Agar, *Humanity's End. Why We Should Reject Radical Enhancement* (Cambridge, Mass: MIT Press), pp. 188–189.
17 Agar, *Humanity's End*, p. 12.

Tab. 1: The comparison of the objective and anthropocentric ideals

Objective ideal		Anthropocentric ideal	
Instrumental value of human capacity		Intrinsic value of human capacity	
Goods external to practices		Goods internal to practices	
The criterion of positive or negative outcomes of the use of a given capacity		The criterion of compatibility with current human standards of valuation	
Q^-	Q^+	I^-	I^+
Decrease of the quality of life	Increase of the quality of life	Lesser possibility to engage in practices compatible with the current human standards of valuation	Greater possibility to engage in practices compatible with the current human standards of valuation

values against enhancement has been formulated by a prominent representative of the anti-theory approach.[18] MacIntyre's view, on which Agar's argumentation is based, is also far from universalism, which makes an attempt to create an objective model of evaluating potential enhancing interventions by balancing intrinsic and instrumental values even more difficult.

2 Balancing instrumental and intrinsic values

On the basis of the idea described above, the following criteria and categories for comparing instrumental and intrinsic values can be distinguished (see Table 1).

How could the four above categories be compared? One of the possibilities is to apply the concept of lexical superiority of values. The strong lexical superiority of value A over value B means that *any* amount of A value is better than *any* amount of B value. The weak lexical superiority of A over B means that some amount of A is better than *any* amount of B.[19] The most popular illustration of the concept of lexical superiority is John Stuart Mill's famous quote: "it is better to be a human being dissatisfied than a pig satisfied."[20] Abstracting from

---

18 B. Williams, "The Human Prejudice," in: B. Williams, *Philosophy as a Humanistic Discipline* (Princeton, N.J.: Princeton University Press, 2006), pp. 135–152.
19 G. Arrhenius, "Superiority in Values," in: T. Rønnow-Rasmussen, M.J. Zimmerman, *Recent Work on Intrinsic Value* (Dordrecht, London: Springer, 2011), p. 291.
20 J. S. Mill, *Utlilitarianism* (Oxford: Oxford University Press, 1998), p 57. The problem of whether Mill's thesis may be interpreted as applying lexical superiority is discussed by B. Saunders, "J. S. Mill Conception of Utility," *Utilitas* Vol. 27, No. 1 (2010), pp. 52–69. Analysis of relationships between strong and weak superiority: G. Arrhenius, W. Rabinowicz, "Millian Superiorities," *Utilitas*, Vol. 14, No. 2 (2005), pp. 127–146.

hedonistic axiology, if applied to the problem of human enhancement, this sentence may be interpreted in following way:

$I^+Q^- > I^-Q^+$

> – strong lexical superiority

Applying this to the problem of human enhancement, it would mean that, with respect to the actual human condition, every enhancing intervention which would increase the quality of life while simultaneously decreasing the possibilities of achieving goods internal to valued practices would be worse than an intervention which increases the opportunities to participate in these practices despite the reduction of the quality of life. Paraphrasing Mill's words: it is better to be a human being dissatisfied than a post-human satisfied.

Assuming that I^0Q is the actual state of intrinsic and instrumental values, the model for evaluating decisions concerning the enhancement of human capacities based on strong superiority of intrinsic values would be as follows:

$I^+Q^+ > I^+Q^0 > I^+Q^- > I^0Q^+ > I^0Q^0 > I^0Q^- > I^-Q^+ > I^-Q^0 > I^-Q^-$

The assumption that compatibility with current human standards of valuation sets the lexical threshold makes it possible to claim that no matter how great the potential growth of instrumental value would be, the choice of a given intervention is better than refraining from it only when it is associated with the maintenance or increase of intrinsic values. Using this model might be advantageous, as it may lead to solutions of some common problems when assessing enhancing interventions. Furthermore, it could help to avoid the "status quo bias" objection, as lexical superiority of intrinsic value may provide a good basis for the preservation of actual human capacities on the current level.

There are, however, many problems which should be resolved before the adoption of this model. First of all, it is necessary to answer the question of whether it can support deontological constraints against performing enhancing interventions which lead to the loss of intrinsic values. On the one hand, the strong lexical superiority view seems to be perfect ground for these constraints, because it makes it possible to justify the claim that all options leading to a loss of intrinsic values are impermissible, and at the same time, it also allows for a gradation of the other options that set the intrinsic value at least at the initial level. The task of authors defending the bio-conservative position would be to justify the connection between

---

Possible interpretations of lexical superiority in the models of social choice are analyzed by J. H. Moldau, "On the Lexical Ordering of Social States According to Rawls' Principles of Justice," *Economics and Philosophy* Vol. 8, No. 1 (1992), pp. 142–143.

the axiological thesis that decisions promoting intrinsic values are better than others and the normative thesis that these values should be respected and protected.[21]

The second problem is constituted by the possible implausible implications of a strong superiority of intrinsic values. The classic example is a situation when a huge loss of values from the lower level is accompanied by only a slight increase in the values of the higher level.[22] In the context of human enhancement, this situation would occur in the case of a rapid increase of suffering with a slightly improved possibility of achieving goods internal to practices. According to Agar, such a view would be bio-conservative extremism. Mankind has developed many practices which allow it to face pain, suffering, and death, however, it does not seem that the prospective increase of such virtues as compassion or heroism during a war or a natural disaster would justify allowing these catastrophes.[23] Furthermore, there are possible situations in which the prevention of extreme social injustice or of the extinction of human species would justify giving up some intrinsic values.[24] A possible answer to this objection is that the first situation concerns the necessity of balancing two different kinds of intrinsic values and the second is a situation in which both intrinsic and instrumental values are endangered. According to Francis Fukuyama, "no one can make a brief in favor of pain and suffering," but "the highest and most admirable human qualities (…) are often related to the way that we react to, confront, overcome, and frequently succumb to pain, suffering, and death."[25] This claim is accompanied by the thesis that "our good characteristics are intimately connected to our bad ones" and that violence and aggression are often needed to defend ourselves, the feelings of exclusivity are needed to be loyal, without jealousy we could never feel love and our mortality allows us as a species to survive and adapt.[26] This

---

21 One of the possible ways of justifying the universal normative character of intrinsic values is basing them on virtues. The difficulties of this approach are analyzed by E. D. Pellegrino, "Toward a Virtue-Based Normative Ethics for Health Professions," *Kennedy Institute of Ethics Journal* Vol. 5, No. 3 (1995), pp. 253–277.
22 J. Griffin, *Well-being. Its Meaning, Measurement and Moral Importance* (Oxford: Clarendon Press, 1989), pp. 83–89.
23 N. Agar, "Radical Human Enhancement, And What's Wrong with It," in: *Designer Biology. The Ethics of Intensively Engineering Biological and Ecological Systems*, ed. R.L. Sandler and J. Basl (Lenham: Lexington Books, 2013), pp. 93–94.
24 J. Pugh, G. Kahane and J. Savulescu, "Cohen's Conservatism and Human Enhancement," *Journal of Ethics*, Vol. 14, No. 4 (2013), p. 343.
25 F. Fukuyama, *Our Posthuman Future. Consequences of the Biotechnology Revolution* (New York: Farrar, Straus and Giroux 2002), p. 173.
26 Fukuyama, *Our Posthuman Future*, p. 173

argument, however, seems to miss the point as it assumes the radical impossibility of the human condition which excludes aggressive behavior and discrimination without excluding defense against aggression, loyalty, and love. It should rather be said that human virtues should work in good times and in bad, which means that there should be the possibility of achieving internal goods both in favorable conditions and in the face of pain, suffering, and death. Furthermore, the earlier mentioned combination of this model with deontological constraints would permit claiming that the necessity of avoiding wars and natural disasters is justified not only by balancing values, but also by human rights.[27] Another answer to the problem of conflict situations might be to apply this model only to decisions assuming a positive quality of life. The argument for this solution is that most of the human enhancement projects assume a positive quality of life at the starting point, and conflict situations of the kind described earlier would occur only in therapeutic decisions which need more complex models of ethical assessment. This answer requires the assumption that participation in practices is to some extent dependent on quality of life. If so, then the option in which quality of life falls below the minimal level needed to participate in a given practice should never be preferable.

The third challenge for this model is a matter of clarifying how to define the current level of intrinsic values and the criteria of growth or loss of these values. If this model is to be applied to political decisions concerning the implementation of technologies improving human capacities, and not only to individual decisions whether or not to use these technologies, then the obvious criterion would be the risk of elimination of a given valuable practice from social life. The easiest way to estimate these changes is by referring to the number of people with the capacities needed to participate in this practice. The downside of this model is "anthropocentric eugenics", because on this basis we could claim that the existence of enhanced people whose capacities and traits are not compatible with actual human standards of valuation is undesirable. More plausible would be a reference to the existence or absence of a given practice. The problem here, however, is the lack of precision in identifying those practices which should be preserved. It seems that many examples given by Agar or MacIntyre would hardly be seen as intrinsic values lexically superior to the values of improving the quality of life. Playing chess or running a marathon certainly have intrinsic

---

27 S. Caney, "*Climate* Change, Human Rights and Moral Thresholds," in: *Human Rights and Climate Change*, ed. S. Humphrey (Cambridge: Cambridge University Press, 2009), p. 165.

value, however, it is difficult to consider them to be common values from the anthropocentric point of view. The main task, then, is clarifying and narrowing down the concept of intrinsic values related to participation in a practice.

In the next part of this paper I will analyze two possibilities of doing that: basing intrinsic value on human commitment to ground projects, and on interpersonal relationships.

3 Ground projects and interpersonal relationships after transformative change

According to Roduit, Baumann, and Heilinger, a good framework for assessing enhancing interventions is Martha Nussbaum's list of ten capabilities fundamental for human beings: life; bodily health; bodily integrity; senses, imagination, and thought; emotions; practical reason; affiliation; other species; play; and control over one's environment.[28] The prognosis about how human enhancement may affect human life in these respects can play both a constraining and guiding role in the assessment of such interventions, since the Capability Approach provides us both with ten thresholds which demarcate the levels under which human life would be seriously impoverished, and with a holistic account of human life.[29] The problem is that decisions about individual or species transformative changes are made from an agent-relative perspective. In order to assign intrinsic value to these capabilities and compare them with instrumental values, they need to be referred to some more fundamental values in the agent's perspective.

Consider an example of transformative change of sensual experience given by Laurie Paul:

> Imagine that neuroscientists and engineers invent a microchip that, when implanted in the brain, gives humans a new sensory ability, a sixth sense in addition to the usual five. If this sense is truly new, rather than a combination of more familiar senses, before getting the chip, we cannot know what it would be like to experience the new, sixth, sense. With respect to this sensory ability, anyone without the chip would be in the same position as a person who is blind from birth, or a person who has never been able to hear: there would be a human sensory ability that some people have and others don't, but those without the chip would not know what it is like to have it until they experience it.[30]

---

28 M. Nussbaum, *Creating Capabilities: The Human Development Approach* (Cambridge, Mass: Belknap Press, 2011), p. 25.
29 J.A.R. Roduit, H. Baumann, J. C. Heilinger, "*Ideas* of Perfection and the Ethics of Human Enhancement," *Bioethics*, Vol. 29, No. 9 (2015), p. 629.
30 L. A. Paul, *Transformative Experience* (Oxford: Oxford University Press, 2014), p. 6.

The Capability Approach needs a justification of the role which having five senses instead of six or more plays in our lives. That means that assessment of transformative change should be first personal.[31] In contemporary ethical debate there are two widely discussed sources of intrinsic values which could serve as a holistic view on the value of capabilities: ground projects and special relationships.

The first proposition which supports strong lexical priority of intrinsic values over instrumental ones is inspired by Bernard Williams' reflections on the agent's integrity.[32] If the possibility of satisfying one's categorical desires is a condition of integrity, then it is a natural consequence to accept the claim that commitment to ground projects has value which is lexically superior to instrumental values:

$$I_p^+Q^+ > I_p^+Q^0 > I_p^+Q^- > I_p^0Q^+ > I_p^0Q^0 > I_p^0Q^- > I_p^-Q^+ > I_p^-Q^0 > I_p^-Q^-$$

I_p – intrinsic value of commitment to ground projects

The main argument for this proposition is that the reduction of value of ground projects to values based on maximization of utility leads to counter-intuitive conclusions. Objections against consequential theories are often illustrated by cases in which the agent has an obligation to sacrifice his own projects in order to maximize general well-being. Williams' view on the connection between integrity and identity seems to support strong lexical superiority, as there is no level of quality of life high enough which could outweigh our commitment to everything which makes us *us*. In the context of human enhancement, a great increase in the quality of life is never worth abandoning our human values.[33]

The problem of this justification is the difficulty of determining the nature of lexical superiority of ground projects' value. Not all projects seem to have such a value. Technological advances have led to the elimination of many practices and thus – many intrinsic values have been abandoned in order to give way to the growth of instrumental values. More and more human practices are carried out through information technologies. In the future, new practices will appear and future people will have the possibility of choosing old and new life projects and achieving them without effort.[34] A model based on strong lexical priority

---

31 According to Paul, the main criterion for this kind of decisions is the subjective desirability of revelation (Paul, *Transformative Experience*, pp. 124–126). It seems, however, that also the desirability of revelation should be balanced against the possibility or the lack of possibility of participating in practices compatible with the current human standards of valuation.
32 B. Williams, "A Critique of Utilitarianism," in: *Utilitarianism. For and Against*, ed. B. Williams, J.J.C. Smart (Cambridge: Cambridge University Press, 1973), pp. 116–117.
33 Williams, "The Human Prejudice," pp. 149–152.
34 M. Schermer, "Enhancements, Easy Shortcuts and the Richness of Human Activities," *Bioethics* Vol. 22, No. 7 (2008), p. 363.

of intrinsic values of ground projects should assume the possibility of mutual comparison of these values. The criterion of compatibility with current human standards of valuation is insufficient on its own. According to Joseph Raz, even now it is difficult to compare the intrinsic value of a career of a lawyer and of an artist, because both of these practices contain some subjectively chosen project.[35] For the same reason it would be difficult to justify the obligation to preserve a specific profession in the society, or the demand for the capacities or traits related to this profession.

An attempt to defend the strong superiority of ground projects' intrinsic value could be based on a classic conception of human nature, in which the agent's ground project is pursued to its ultimate end. In this perspective, perfection is the fullest development of the agent's traits according to natural human inclinations. This approach could include human enhancement, provided that the interventions preserve the ability to act in harmony with natural human inclinations and do not modify these inclinations themselves.[36] According to MacIntyre, the rejection of the claim that people have their ultimate end is the main source of the failure of the Enlightenment Project to create universal moral standards for resolving practical problems.[37] In the context of the human enhancement problem, the lack of consideration about the end towards which human practices are aimed, or the reduction of this end to the category of well-being, causes consecutive, contradictory proposals to arise for new directions of enhancing human capacities, and the arguments formulated by each faction appear to be equally compelling. Furthermore, the lack of the category of an ultimate end would make it impossible to differentiate between the practices which require commitment and make life meaningful, and mere consumption.[38] The main foundation for lexical superiority of the ground projects' value would be, therefore, the fact that any attempt to balance one's projects against the quality of life would constitute self-corruption and a threat to one's integrity.

It should be noted that the above arguments can be applied not only to a defense of the model involving the lexical ordering of values, but also against it, as they could support the view opposing any attempt to build a theoretical

---

35 J. Raz, *The Morality of Freedom* (Oxford: Clarendon Press, 1986), p. 342.
36 This approach can be found in J.T. Eberl, "A Thomistic Appraisal of Human Enhancement Technologies," *Theoretical Medicine and Bioethics* Vol. 35, No. 4 (2014), pp. 289–310.
37 MacIntyre, *After Virtue*, p. 52.
38 A. Borgmann, *Technology and the Character of Contemporary Life: A Philosophical Enquiry* (Chicago: Chicago University Press, 1984), p. 219.

framework for comparing different kinds of values. Such a framework would require conceiving the ultimate end of human life as clearly as possible in order to determine which practices may lead to this end. If the commonsense view on perfection is that it should be based on practical wisdom rather than on theoretical models, then building these models for the optimal balance between intrinsic and instrumental values should be viewed as another doomed-to-failure consequence of the Enlightenment Project described by MacIntyre. It could also be said that there is no possibility of even formulating a comprehensive project of human life, the implementation of which would allow for the achievement of perfection.[39] Human life may rather be viewed as a creative search for a way to realize a whole range of different kinds of practices, with the possibility of abandoning some projects, starting others, or returning to the previous ones.

A compromise between these positions would be the acceptance of the weak superiority of intrinsic values:

$$I_p^+Q^+ > I_p^+Q^0 > I_p^+Q^- > I_p^0Q^+ > I_p^0Q^0 > I_p^0Q^- > I_p^-Q^+ > I_p^-Q^0 > I_p^-Q^-$$

› – weak lexical superiority

According to this version, the possibility of a sufficient increase of instrumental values or the prevention of their decrease might outweigh the growth of intrinsic values or the prevention of their decrease. The degree of the superiority would not be defined once and for all, but it would be dependent on individual practical wisdom. If technological progress could help avoid some catastrophic decrease of instrumental values at the expense of the elimination of some practices which were sources of intrinsic values, then it would only depend on the agents' view concerning the significance of that practice whether a future without this value would be better than with it.[40]

It seems that the best candidate for such a significant practice which could generate values that outweigh a great decrease of the quality of life are the practices related to some interpersonal relationships, such as family ties or friendship. This would be more consistent with deontological constraints than the integrity view, since special relationships seem both to have intrinsic value and to generate obligations.[41] Furthermore, it can serve as an easy justification of

---

39 M. Slote, *The Impossibility of Perfection: Aristotle, Feminism and the Complexity of Ethics* (New York: Oxford University Press, 2011).
40 This case abstracts from other considerations concerning obligations or duties which may affect the ultimate judgment.
41 J. Raz, "Liberating Duties," *Law and Philosophy* Vol. 8, No. 1 (1989), p. 19; S. Scheffler, "Projects, Relationships and Reasons," in: *Reason and Value: Themes From the Moral*

why some transformative changes are undesirable. If cognitive enhancement led to a great increase of the quality of life, but at the cost of the lack of possibility to engage in practices developing previous relationships and to act in accordance with one's obligations generated by these relationships, then one would have strong reasons not to undergo this change. From a general point of view, the possibility of engaging in these relations can be seen as a fundamental condition of the acceptance of enhancing interventions.

Given that previous arguments against strong lexical superiority may also be applied to intrinsic values from practices related to interpersonal relationships, the new version of the model would be based on weak lexical superiority:

$$I_R^+E^+ > I_R^+E^0 > I_R^+E^- > I_R^0E^+ > I_R^0E^0 > I_R^0E^- > I_R^-E^+ > I_R^-E^0 > I_R^-E^-$$

I_p – intrinsic value of commitment to interpersonal relationships

> A possible objection to this model is similar to the previous one: how to formulate a list of special relationships which are sources of the intrinsic value of a practice? The answer here may be more plausible than previously: though it is impossible to define special relationships, there is almost universal agreement that at least some of them, like family or friendship, should be taken into account in all ethical considerations. This claim, however, does not resolve the second problem: whether special relationships are a source of intrinsic values irreducible to the quality of life or other values. The value of a relationship may be dependent on, e.g., its quality or the degree of commitment to it. According to Simon Keller, these relationships can have only extrinsic value, even though they are agent-relative, irreplaceable, immense, and not straightforwardly dependent upon some other value:[42] Suppose we discover a society in which a certain kind of special relationship is entirely absent. Imagine that children in this society are raised communally and that there are no special loving bonds between children and their parents. Suppose that the communal arrangement appears to work very well and that the society functions no less successfully than ours. People in this society, as compared to ours, seem just as happy, just as virtuous, just as secure, just as respected, and so on.[43]

If we imagine that the society described by Keller is a consequence of some kind of human enhancement, then according to the proponents of these interventions it is hard to maintain that the model of lexical superiority may be based on rational assumptions. It should be noted, however, that from the anthropocentric point of view the question is not about the objective value of possible future

---

Philosophy of Joseph Raz, ed. R. J. Wallace (Oxford: Oxford University Press, 2004), p. 260.
42 S. Keller, *Partiality* (Princeton, NJ: Princeton University Press, 2013), p. 55.
43 Keller, *Partiality*, pp. 58–59.

societies, but about what kind of society we have most reason to pursue. If the future society is built through individual procreative decisions, then one of the most important reasons for having children is to build a special relationship with them. The model of weak lexical superiority may be one of the possible ways of conceptualizing these reasons.

4 Conclusion

Assessing human enhancement with the models for comparing intrinsic and instrumental values makes it possible to avoid some problems in contemporary dispute. Using these models indicates that at least in theory some types of interventions may have positive effects for the preservation of intrinsic values. This also makes it possible to show that bioconservative arguments against human enhancement may rely not on an incorrect estimation of the future value, but on an assumption of the primacy of intrinsic values over instrumental ones. Finally, it makes it possible to justify the claim that a predicted increase in the quality of life shouldn't be the only criterion for the assessment of human enhancement. At the same time, these models cannot constitute an independent criterion for the assessment of human enhancement, as they should also be accompanied by deontological constraints. Using these models also requires an axiological framework to explain the relation between the intrinsic value of capacity and the intrinsic value of relationships and individuals. Furthermore, the superiority of these intrinsic values is justified in this model only by assuming the priority of the agent-relative and anthropocentric perspective over the agent neutral. Finally, it should be still justified that the strength of reasons to pursue or refrain from pursuing some end through enhancing interventions is proportional to the lexical superiority of intrinsic values over the instrumental ones.

References

Agar, N. *Humanity's End. Why We Should Reject Radical Enhancement*. Cambridge, Massachusetts: MIT Press, 2010.

Agar, N. "Radical Human Enhancement, And What's Wrong with It."
In: *Designer Biology. The Ethics of Intensively Engineering Biological and Ecological Systems*, eds. R. L. Sandler and J. Basl. Lenham: Lexington Books, 2013, pp. 87–104.

Agar, N. *Truly Human Enhancement: A Philosophical Defence of Limits*. Cambridge: Massachusetts: MIT Press, 2014.

Arrhenius, G. "Superiority in Values." In: *Recent Work on Intrinsic Value*, eds. T. Rønnow-Rasmussen and M. J. Zimmerman. Dordrecht, London: Springer, 2011, pp. 291–304.

Arrhenius, G. and W. Rabinowicz. "Millian Superiorities." *Utilitas*, Vol. 14, No. 2, 2005, pp. 127–146.

Borgmann, A. *Technology and the Character of Contemporary Life: A Philosophical Enquiry*. Chicago: Chicago University Press, 1984.

Bostrom, N. and T. Ord. "The Reversal Test: Eliminating Status Quo Bias in Applied Ethics." *Ethics*, Vol. 116, No. 4, 2006, pp. 656–679.

Caney, S. "Climate Change, Human Rights and Moral Thresholds." In: *Human Rights and Climate Change*, ed. S. Humphrey. Cambridge: Cambridge University Press, 2009, pp. 163–177.

Eberl, J. T. "A Thomistic Appraisal of Human Enhancement Technologies." *Theoretical Medicine and Bioethics*, Vol. 35, No. 4, 2014, pp. 289–310.

Fukuyama, F. *Our Posthuman Future. Consequences of the Biotechnology Revolution*. New York: Farrar, Straus and Giroux, 2002.

Griffin, J. *Well-being. Its Meaning, Measurment and Moral Importance*. Oxford: Clarendon Press, 1989.

Keller, S. *Partiality*. Princeton, New Jersey: Princeton University Press, 2013, pp. 58–59.

MacIntyre, A. *After Virtue. A Study in Moral Theory*. Notre Dame: Notre Dame Press, 2007.

Mill, J. S. *Utlilitarianism*. Oxford: Oxford University Press, 1998.

Moldau, J. H. "On the Lexical Ordering of Social States According to Rawls' Principles of Justice." *Economics and Philosophy*, Vol. 8, No. 1, 1992, pp. 141–148.

Nussbaum, M. *Creating Capabilities: The Human Development Approach*. Cambridge, Massachusetts: Belknap Press, 2011.

Paul, L. A. *Transformative Experience*. Oxford: Oxford University Press, 2014.

Pellegrino, E. D. "Toward a Virtue-Based Normative Ethics for Health Professions." *Kennedy Institute of Ethics Journal*, Vol. 5, No. 3, 1995, pp. 253–277.

Pugh, J., G. Kahane and J. Savulescu. "Cohen's Conservatism and Human Enhancement." *Journal of Ethics*, Vol. 14, No. 4, 2013, pp. 331–354.

Raz, J. *The Morality of Freedom*. Oxford: Clarendon Press, 1986.

Raz, J. "Liberating Duties." *Law and Philosophy*, Vol. 8, No. 1, 1989, pp. 3–21.

Roduit, J.A.R., H. Baumann and J. C. Heilinger. "Ideas of Perfection and the Ethics of Human Enhancement." *Bioethics*, Vol. 29, No. 9, 2015, pp. 622–630.

Saunders, B. "J.S. Mill Conception of Utility." *Utilitas*, Vol. 27, No. 1, 2010, pp. 52–69.

Saunders, B. "Why Procreative Preferences May Be Moral – And Why it May Not Matter if They Aren't." *Bioethics*, Vol. 29, No. 7, 2015, pp. 499–506.

Savulescu, J. "The Nature of the Moral Obligation to Select the Best Children." In: *The Future of Bioethics. International Dialogues*, ed. A. Akabayashi. Oxford: Oxford University Press, 2014, pp. 170–182.

Savulescu, J. and I. Persson. *Unfit for the Future: The Need of Moral Enhancement.* Oxford: Oxford University Press, 2012.

Scheffler, S. "Projects, Relationships and Reasons." In: *Reason and Value: Themes From the Moral Philosophy of Joseph Raz*, ed. R. J. Wallace. Oxford: Oxford University Press, 2004, pp. 247–269.

Schermer, M. "Enhancements, Easy Shortcuts and the Richness of Human Activities." *Bioethics*, Vol. 22, No. 7, 2008, pp. 355–363.

Slote, M. *The Impossibility of Perfection: Aristotle, Feminism and the Complexity of Ethics.* New York: Oxford University Press, 2011.

Williams, B. "A Critique of Utilitarianism." In: *Utilitarianism. For and Against*, eds. B. Williams and J.J.C. Smart. Cambridge: Cambridge University Press, 1973, pp. 77–150.

Williams, B. "The Human Prejudice." In: *Philosophy as a Humanistic Discipline*, ed. B. Williams. Princeton, New Jersey: Princeton University Press, 2006, pp. 135–152.

Beata Płonka

Excellence and Its Biological Limitations[1]

Abstract: The idea of excellence is very popular nowadays, but also frequently overused. Different concepts of excellence are used in various areas of human activity, mainly based on the norm-referenced or the criterion-referenced definitions. The present paper addresses the main problems of the concept of excellence and its limitations from the biological perspective, and discusses the important distinctions between genotype and phenotype, as well as various types of phenotypic traits. It also examines the difficulty with estimating the influence of genetic and environmental factors on the phenotypic diversity of individuals in populations, such as missing heritability or epigenetics, and discusses the concept of normality and the definitions of "health" and "disease". Finally, the paper analyzes the concept of excellence in the transhumanism movement from the evolutionary perspective, presenting the ambiguity of this concept, as well as the difficulties and threats of implementing it in the human population.

Keywords: Excellence, Genotype, Phenotype, Heritability, Epigenetics, Transhumanism

1 Introduction

The idea of excellence is an old philosophical concept which is now extremely popular in many areas of human activity. We apply the concept to individuals as well as groups of people and organizations in every situation connected to striving for perfection. We think about excellence in health care, governance or management, artistic performance, sport, research and education, etc. We also think about excellence when we make long-term plans for the development of societies or new technologies, as the idea of progress implies striving for excellence. It is obvious to many people that we should try to become better in some aspects if we want to achieve success, especially in our professional life. We respect and sometimes worship those who are able to achieve – or come close to – perfection in certain areas, such as great writers, artists, scientists, or sport champions. Since scientists want to determine the basis of the ability to achieve excellence, they look for genes, environmental factors or psychological traits necessary for the successful development of talent. There are many scientific papers dealing with empirical data obtained during the survey of groups of successful

---

[1] The writing of this article was funded by the Polish National Science Centre (2012/07/D/HS1/01099).

individuals (e.g., elite musicians or athletes) and seeking practical solutions to the problem of the effective realization of one's potential.[2] A significant effort has been put into genomic research in order to find a direct link between certain genes and physical performance (endurance, speed, power, etc.), leading to the discovery of more than 200 genetic variations that may be involved.[3]

The search for genetic variants common among athlete champions revealed about 20 genetic variants specifically associated with their elite status: only about

---

2 As an example, see the study of psychological traits of successful musicians (A. MacNamara, P. Holmes, D. Collins, "Negotiating transitions in musical development: the role of psychological characteristics of developing excellence," *Psychology of Music*, Vol. 36, No. 3 (2008), pp. 335–352) or athletes studied by the Psychological Characteristics of Developing Excellence Questionnaire (PCDEQ) (A. MacNamara, D. Collins, "Development and initial validation of the Psychological Characteristics of Developing Excellence Questionnaire,"*Journal of Sports Sciences*, Vol. 29, No. 12 (2011), pp. 1273–1286; A. MacNamara, D.Collins, "Do mental skills make champions? Examining the discriminant function of the psychological characteristics of developing excellence questionnaire," *Journal of Sports Sciences*, Vol. 31, No. 7 (2013), pp. 736–744).

3 Several genes from those that are suspected to contribute to elite performance have been identified and confirmed, mainly *ACE* (angiotensin-converting enzyme), *ACSL1* (acyl coenzyme A synthetase long-chain 1), *ACTN3* (α-actinin-3), *ADG* (angiotensinogen gene), *AMPD1* (adenosine monophosphate deaminase), *EPAS1* (endothelial PAS domain protein 1), *GDF-8* (myostatin (growth and differentiation factor)), *HFE* (hereditary haemochromatosis gene), *MCT1* (monocarboxylate (lactate/pyruvate) transporter 1), *NOS3* (nitric oxide (NO) synthase 3), *PAPSS2* (3'-phosphoadenosine-5'-phosphosulfate synthase 2), *PPARGC1A* (peroxisome proliferator-activated receptor-γ, coactivator 1, α), *PPARA* (peroxisome proliferator-activated receptor α gene) (N. Eynon, E. D. Hanson, A. Lucia, P. J. Houweling, F. Garton, K. N. North and D. J. Bishop, "Genes for elite power and sprint performance: ACTN3 leads the way," *Sports Medicine*, Vol. 43, No. 9 (2013), pp. 803–817; N. Eynon, J. R. Ruiz, J. Oliveira, J. A. Duarte, R. Birk, A. Lucia, "Genes and elite athletes: a roadmap for future research," *The Journal of Physiology*, Vol. 589, No. 13 (2011), pp. 3063–3070; V. Ginevičienė, A. Jakaitienė, A. Pranculis, K. Milašius, L. Tubelis and A. Utkus, "AMPD1 rs17602729 is associated with physical performance of sprint and power in elite Lithuanian athletes," *BMC Genetics*,Vol. 15, (2014), p. 58; M. Sawczuk, L. K. Banting, P. Cięszczyk, A. Maciejewska-Karłowska, A. Zarębska, A. Leońska-Duniec, Z. Jastrzębski, D. J. Bishop and N. Eynon, "MCT1 A1470T: a novel polymorphism for sprint performance?," *Journal of Science and Medicine in Sport*, Vol. 8, No. 1 (2015), pp. 114–118; S. Voisin, P. Cieszczyk, V. P. Pushkarev, D. A. Dyatlov, B. F. Vashlyayev, V. A. Shumaylov, A. Maciejewska-Karlowska, M. Sawczuk, L. Skuza, Z. Jastrzebski, D. J. Bishop, N. Eynon "EPAS1gene variants are associated with sprint/power athletic performance in two cohorts of European athletes," *BMC Genomics*, Vol. 5 (2014), p. 382).

66% of the variance could be explained by additive genetic factors, while about 34% was contributed by non-shared environmental factors. In order to distinguish elite athletes from the general population, as well as between endurance and power athletes, a concept of a "genotype score" has been introduced.[4] Scientific attempts to quantitatively estimate the "potential" of some genotypes to form desired traits characteristic for sport champions may, however, unfortunately be misused for commercial purposes. This danger has been recognized by some researchers, who address the problem of rapidly developing supply of direct-to-consumer marketing (DTC) tests aimed at the identification of children's sport talents and support the idea of abandoning DTC genetic testing of individuals in order to select future champions or alter their training methods according to the current state of knowledge.[5]

The idea of excellence is probably one of the most commonly used concepts in papers, regulations, policies, etc., as well as in our personal way of thinking. As stated by Fred Botting: "Excellence is everywhere. At the heart of the performance-centered, technocratic language of business, sport and the university, the ubiquity of excellence is accompanied by a transparency of effects that mean nothing but nevertheless manage to transform practices within and between all cultural

---

4 The "genotype score" quantifies combined contribution of certain genetic variants to athlete performance, attributing numbers from 0 to 100 to various genotype combinations. The "genotype score" of most champions is not of maximal level. It is interesting that the calculated probability of a "perfect" genotype score for a Caucasian individual is remarkably low (0.0005%) (M. H. De Moor, T. D. Spector, L. F. Cherkas, M. Falchi, J. J. Hottenga, D. I. Boomsma and E. J. De Geus, "Genome-wide linkage scan for athlete status in 700 British female DZ twin pairs," *Twin Research and Human Genetics*, Vol. 10, No.6 (2007), pp. 812–820; Eynon, Ruiz, Oliveira, Duarte, Birk and Lucia,"Genes and elite athletes"; A.G. Williams and J. P. Folland,"Similarity of polygenic profiles limits the potential for elite human physical performance," *The Journal of Physiology*, Vol. 586, No. 1 (2008), pp. 113–121).
5 It should be stressed that the main objection of scientific community to DTC tests comes from the empirical foundation not being strong enough to support the claim. Another problem concerns the poor quality of many DTC laboratories, general lack of efficient genetic counseling for consumers (parents and coaches), as well as the lack of legislation and universally accepted guidelines for DTC testing (N. Webborn, A. Williams, M. McNamee, C. Bouchard, Y. Pitsiladis, I. Ahmetov, E. Ashley, N. Byrne, S. Camporesi, M. Collins, P. Dijkstra, N. Eynon, N. Fuku, F. C. Garton, N. Hoppe, S. Holm, J. Kaye, V. Klissouras, A. Lucia, K. Maase, C. Moran, K. N. North, F. Pigozzi and G. Wang, "Direct-to-consumer genetic testing for predicting sports performance and talent identification: Consensus statement," *British Journal of Sports Medicine*, Vol. 49, No. 23 (2015), pp. 1486–1491).

and corporate institutions".[6] The term "excellence" has become an indicator of development, improvement and progress, and is used as a project motto of various institutions and organizations in creating the so-called "culture of excellence". In this context, "excellence" is treated as a technical tool to achieve a high quality of performance in medical services, research (mainly research productivity) or education.[7] This overwhelming urge to achieve and measure institutional excellence is extremely common nowadays and leads to devising policies and procedures allowing to achieve this goal. This prevailing trend stems from competitiveness and leads to a substantial neglect of the concept of equality, although some measures are taken to implement it to some extent (e.g., gender issues). The problem of proper balance in achieving both excellence and equality is especially acute in the higher education system. It is symptomatic that striving for excellence currently seems to dominate higher education institutions. Karl Dittrich briefly describes the situation in the preface to the report *The Concept of Excellence in Higher Education*:

---

6   F. Botting, "Culture and excellence," *Cultural Values*, Vol. 1, No. 2 (1997), p. 139.
7   It is worth mentioning, that the term "excellence" is frequently used in scientific papers and various consortia reports, and this frequency may be treated as a rough measure of its popularity. For example, the search in one of the most recognized databases for biomedical research – PubMed database (US National Library of Medicine National Institutes of Health) returned 39291 results for excellence, 1876 for centers of excellence, 10758 for clinical excellence, and 2497 for nursing excellence (access 21.11.2015, 10pm). Excellence is usually treated as an evidence of high quality and a lot of effort is put into preparing procedures to achieve it in research and education (J. A. Douglass, "Profiling the Flagship University Model: An Exploratory Proposal for Changing the Paradigm From Ranking to Relevancy," *ROPS.CSHE*, Vol. 5, No.14 (2014); M. R. Rao, "International Centers of Excellence for Malaria Research," *The American Journal of Tropical Medicine and Hygiene*, Vol. 93, No. 3 (2015), Suppl., pp. 1–4) or health care (A. R. Campbell, "Building a culture of excellence from the ground up," *Radiology Management*, Vol. 35, No.3 (2013), pp. 14–18; quiz 19–20; L. Coulon, M. Mok, K. L. Krause, M. Anderson, "The pursuit of excellence in nursing care: what does it mean?," *Journal of Advanced Nursing*, Vol. 42, No. 4 (1996), pp. 817–826; P. A. Hickey, "Building a culture of excellence in Boston and beyond," *World Journal for Pediatric & Congenital Heart Surgery*, Vol. 1, No. 3 (2010), pp. 314–320; A. Kabcenell, K. Luther, "Creating a culture of excellence. It's not as difficult as you might think," *Healthcare Executive*, Vol. 27, No. 4 (2012), pp. 68–71; C. Laserre, "Fostering a culture of service excellence," *The Journal of Medical Practice Management*, Vol. 26, No. 3 (2010), pp. 166–169; P. G. Tropello, "Magnet status as a competitive strategy of hospital organizations: marketing a culture of excellence in nursing services," *Journal of Hospital Marketing & Public Relations*, Vol. 14, No. 2 (2003), pp. 53–57).

Excellence and Its Biological Limitations 241

During the past few years, politicians and higher education institutions (HEIs) have discovered the concept of excellence. Rankings undoubtedly have stimulated this concept, both in positive and negative ways. One positive effect of rankings is the latest drive for enhancement. In addition, the rankings have removed the fiction of "equality" between and within HEIs. The reality is much more complex than a one-dimensional structure (…) As a general tendency, more and more attention will be given to the differences in the student population and the student experience. The concept of equality appears to be losing ground with students and staff. Numerous initiatives have been taken to promote excellent tracks, honours degrees and more challenging educational environments for students who are willing and who are capable of achieving higher levels of attainment.[8]

There is a long lasting discussion about a possible conflict between equity and excellence, and numerous definitions of excellence. W. J. Smith and C. Lusthaus propose a model of the non-linear relationship between equality and excellence (quality) in which the two are compatible.[9] According to K. A. Strike, it is logically impossible to simultaneously realize excellence and equity. He states (p. 410): "That only some can be excellent is true for the same reason that not everyone can be better than average."[10] However, according to Strike, this rule applies only to the norm-referenced notion of excellence, which defines it in relation to the norm (normalized distribution of performance in the reference population). At the same time, a criterion-referenced definition understands excellence in relation to a standard that allows one to erase competition.[11]

---

8 M. Brusoni, R. Damian, J. R. Sauri, S. Jackson, H. Komurcugil, M. Malmedy, O. Matveeva, G.Motova, S. Pisarz, P. Pol, A. Rostlund, E. Soboleva, O. Tavares, L. Zobel, *The Concept of Excellence in Higher Education* (Brussels: European Association for Quality Assurance in Higher Education AISBL, 2014), p. 5. Dittrich also mentions the negative aspect of the current trend. "A negative effect of the concept of excellence is the ease with which politicians use the word and the idea that excellence can be quickly and easily achieved. Universities play a role by asserting in their strategic plans 'that they strive for excellence in research and teaching, thereby challenging those who have to evaluate them on whether they actually deliver what they promise. HEIs might make themselves vulnerable in this way if they do not deliver outstanding quality." (Brusoni et al., The Concept of Excellence in Higher Education)
9 See W. J. Smith, C. Lusthaus, "The nexus of equality and quality in education: a framework for debate," *Canadian Journal of Education*, Vol. 20, No. 3 (1995), pp. 378–391.
10 See K. A. Strike, "Is there a conflict between Equity and Excellence?," *Educational Evaluation and Policy Analysis*, Vol. 7, No. 4 (1985), pp. 409–416.
11 Strike, "Is there a conflict between Equity and Excellence?", pp. 409–416. As Strike puts it: "Let me summarize the argument to this point. I have claimed that we should distinguish between norm-referenced and criterion-referenced conceptions of excellence. A norm-referenced conception will define excellence relative to the performance of others. Thus people will be in competition for excellence and, as a matter of logic, not

These practical issues deal with different concepts of excellence that seem to be closely connected to human biology. In what follows, I will discuss the basis and limitations of the notion of excellence from population biology and the evolutionary perspectives. At the beginning I will introduce a fundamental distinction between genotype and phenotype, as well as various types of phenotypic traits of living organisms. This topic will be followed by a discussion of the problem of biological normality and genotype-phenotype relations, leading to the great diversity and plasticity of individuals in the population. Finally, I will look at the concept of excellence in the transhumanism movement and evolutionary thinking in order to show the ambiguity of this concept as well as difficulties and threats connected with implementing it in the human population.

2 Genotype and phenotype

The distinction between genotype and phenotype is crucial for understanding the structure and function of living organisms. It is necessary, however, to define the meaning of "biological trait" first. In biology, "trait" is understood as a feature (characteristics, attribute) of an organism. M. J. West-Eberhard describes it in the following way:

> A 'trait' is simply a somewhat discrete characteristic of an organism. It could be an aspect of morphology, a physiological state, a behavior, a molecule, or a disease, but the implication is that it is a product of development that is qualitatively distinct relative to other aspects of the organism (…) In addition to the discrete on-off qualitative traits of organisms, there are other traits, such as body size or longevity, that are 'quantitative traits' – features that are described in terms of their numerically measurable (quantifiable) values (e.g., weight, mass, or life span). Discrete, qualitative traits have dimensions (for example, the length of a bone, the duration of a behavior) that can be measured as quantitatively variable traits.[12]

---

everyone can attain it. A criterion-referenced view, however, will define excellence in relation to a standard such that people are not in competition for it and, in principle, if not necessarily in fact, everyone can attain it. Second, I suggested that our conception of excellence will tend to be defined by our purposes. If our purposes are to develop human capital, we are going to have a norm-referenced conception of excellence. If our purposes are those of the Jeffersonian ideal, we may be able to have a criterion-referenced concept of excellence" (p. 412).

12 M. J. West-Eberhard, "Are Genes Good Markers of Biological Traits?," in: *Biosocial Surveys, National Research Council (US) Committee on Advances in Collecting and Utilizing Biological Indicators and Genetic Information in Social Science Surveys*, ed. M. Weinstein, J. W. Vaupel and K. W. Wachter (Washington, DC: The National Academies Press, 2008), p. 178. She adds: "Some authors use the term 'module' to describe a

These traits are examples of the so-called phenotype. The term "phenotype" refers to the observable physical and/or biochemical properties (characteristics, traits) of an organism. It must be noted that the term may have a more general meaning and refer to all traits (observable characteristics) of an organism, but it may also be used to describe a particular trait, like eye color or blood type. Phenotype is usually contrasted with genotype, which means, in a broad sense, the genetic constitution (genetic makeup, entire set of genes) of an organism. The term, in a more narrow sense, can also be applied to the alleles (variants) of a specific gene. For diploid organisms (such as humans), the genotype of a particular gene is represented by a pair of alleles at a specific genetic position (locus). Two identical alleles in a particular locus form a homozygous genotype, while two different alleles lead to a heterozygous genotype. The distinction between genotype and phenotype is important in biology, but there is a fundamental relation between them, as an organism's phenotype depends strongly on its genotype (a genotype may be treated as the genetic contribution to the phenotype). It is sometimes stated that the phenotype of an individual is determined by its genotype, but such a statement can be easily misunderstood. One must remember that phenotype (traits) is a result of the interaction between genotype (usually a number of genes) and the environment. The influence of environmental factors varies among traits, with some traits being determined mainly by the genotype, and other traits being more environment-dependent. The strong dependence of a trait formation on the influence of environment leads to a common phenomenon that phenotypes of individuals with identical genotypes (e.g., identical twins) are not identical. Moreover, the unique history of every individual also shapes the trait, as products from previous gene expression events become part of the internal environment of the multi-cellular organism that influences subsequent gene expression.

There is also another important aspect that should be mentioned here. It has already been stated that phenotype is the observable physical and/or biochemical characteristics of an organism. However, the term may be applied to the more "macroscopic" properties, such as the eye color or body height, but also to the "microscopic" (or even "molecular") ones, like particular mRNAs or enzymes (and their levels). The phenotypes (such as mortality) that are a

---

discrete trait. In operational terms, a discrete or modular trait can be defined as a product of a separate developmental pathway. But it is more accurate to say that a trait is 'somewhat discrete' rather than 'discrete', or that it is 'modular' rather than 'a module' because no trait is completely independent of all other traits in an integrated individual organism" (West-Eberhard, "Are Genes Good Markers of Biological Traits?," p. 178).

result of the influence (or competition) of a huge number of genetic and non-genetic factors are referred to as distant phenotypes. The phenotypes that are directly dependent on genes (such as mRNA or protein levels) are called proximal phenotypes or endophenotypes. The substantial difference in the number of determinants for both types of phenotype means that particular genetic variants typically have a smaller effect on the distant phenotypes than in the case of proximal phenotypes.[13]

It must be stressed that the term "phenotypic trait" can also be applied to our mental (or behavioral) abilities. Relations between genome and behavioral traits seem to be extremely complex and are the focus of behavioral genetics. However, the link is often perceived in a simplified way (even by scientists), as described by Y. Levy and R. P. Ebstein:

> There have been quite a few articles in which a plea has been made to behavioral scientists to revise their misconceptions about gene-behavior correlates if they hope to 'untangle the webs that link genes to cognition' (Fisher, 2006, p. 270). A frequent misunderstanding concerns talk about 'smart genes', 'language genes' or 'aggressive genes'. Such talk implies direct pathways from genes to complex behaviors, whereas biology tells us that those routes are multifaceted and nonlinear (Marcus & Fisher, 2003). Furthermore, such discourse neglects the role played by the intricate sets of ontogenetic factors, environments, developmental timing and stochastic events on the behavioral outcome (Rutter, Moffitt, & Caspi, 2006).[14]

Scientists are becoming increasingly aware of the level of complexity of genotype-phenotype relations that lead to a great phenotypic diversity in populations. This diversity has two main sources, namely, environmental influences and genetic variation among individuals. The occurrence of various genetic variants within a population is called "genetic polymorphism", and the term "epigenetic effect"

---

13 See A. J. Marian, "Molecular genetic studies of complex phenotypes," *Translational Research: The Journal of Laboratory and Clinical Medicine*, Vol. 159, No. 2 (2012), pp. 64–79

14 See Y. Levy, R. P. Ebstein, "Research review: crossing syndrome boundaries in the search for brain endophenotypes," *Journal of Child Psychology and Psychiatry, and Allied Disciplines*, Vol. 50, No. 6 (2009), pp. 657–668 (p. 657); S. E. Fisher, "Tangled webs: tracing the connections between genes and cognition," *Cognition*, Vol. 101, No. 2 (2006), pp. 270–297; G. F. Marcus, S. E. Fisher, "FOXP2 in focus: what can genes tell us about speech and language?," *Trends in Cognitive Science*, Vol. 7, No. 6 (2003), pp. 257–262; M. Rutter, T. E. Moffitt, A. Caspi, "Gene-environment interplay and psychopathology: multiple varieties but real effects," *Journal of Child Psychology and Psychiatry, and Allied Disciplines*, Vol. 47, Nos. 3–4 (2006), pp. 226–261.

is applied to the mechanism of the environmental influence. I will discuss both phenomena in detail in the next section.

3 Origin of variation

In biology, variation refers to naturally occurring differences (phenotypic and genotypic) among individuals in a population. Genetic variation is an extremely important phenomenon, as it allows populations to adapt to the changing environment conditions much better than populations that are not genetically variable. It may be assumed that the genetic variation of a population depends on the number of individuals and the mutation rate. The average rate of single basepair substitutions has been estimated to be 10^{-8} per nucleotide per generation (rates of mutation are different among species). Such a rate means that in the genome of an organism with at least 100 million base pairs, the probability of some spontaneous single base-pair substitutions in every generation is pretty high.[15] However, even the full description of a genetic variation in our species (which is currently far from being complete) does not mean finding direct links with appropriate phenotypic traits. It is interesting that the extent of genetic variation is interpreted in a different way by various researchers. For example, according to L. Feuk, A. R. Carson, and S. W. Scherer.

> A striking observation from the analysis of the human genome is the extent of DNA-sequence similarity among individuals from around the world: any two humans are thought to be about 99.9% identical in their DNA sequence. It is therefore through studies of a small fraction of the genome – which constitutes the genetic variation

---

15 Studies of human genome revealed a variation, on average, in one out of every 1,000 bases (approximately 3.3 million single base-pair changes in the whole genome). The total number of changes in the genome may reach 15 million base pairs due to insertions and deletions (so-called indels) of longer DNA fragments (also whole genes), although these events are comparatively rare (a few hundred thousand per genome) in relation to single base-pair changes (a few million events per genome). Some types of changes in the genome have a higher probability of occurrence than other sorts, e.g., particular types of substitutions, replication errors of homopolymers (DNA fragments of eight or more identical bases) or microsatellites (sequences of few nucleotides repeated many times). See E. E. Harris, D. Meyer, The molecular signature of selection underlying human adaptations, *American Journal of Physical Anthropology*, Vol. Suppl. 43 (2006), pp. 89–130; D. M. Kingsley, "From atoms to traits," *Scientific American*, Vol. 300, No. 1 (2009), pp. 52–59; T. Strachan, A. P. Read, *Human Molecular Genetics*, second edition (Oxford: BIOS Scientific Publishers Ltd., 1999), pp. 210–235.

between individuals – that insights into phenotypic variation and disease susceptibility can be gained.[16]

However, the opposite opinion concerning the same data may also be found: "Humans are genetically very diverse. They differ in approximately 0.1% of their genomes".[17]

Nevertheless, genetic diversity as the basis of various human phenotypic traits (especially those causing diseases) is the aim of extensive genomic research. In recent years the genome-wide association study (GWAS) has provided a lot of valuable data, but has been focused mainly on the so-called single nucleotide polymorphism SNP – a variation among individuals at a single position in the genome.[18] It was assumed that SNP variants play the major role in genetic

---

16 See L. Feuk, A. R. Carson, S. W. Scherer, "Structural variation in the human genome," *Nature Reviews. Genetics*, Vol. 7, No. 2 (2006), pp. 85–97. Average nucleotide diversity was estimated to be 0.051% for a worldwide sample and 0.046% for a European sample, but it is suggested that more complex models may be necessary to explain all data. Moreover, there are many regions in genome with different rates of sequence variation (See M. Cargill, D. Altshuler, J. Ireland, P. Sklar, K. Ardlie, N. Patil, N. Shaw, C. R. Lane, E. P. Lim, N. Kalyanaraman, J. Nemesh, L. Ziaugra, L. Friedland, A. Rolfe, J. Warrington, R. Lipshutz, G. Q. Daley, E. S. Lander, "Characterization of single-nucleotide polymorphisms in coding regions of human genes," *Nature Genetics*, Vol. 22, No. 3 (1999), pp. 231–238. Erratum in: *Nature Genetics*, Vol. 23, No. 3 (1999), p. 373; M. Przeworski, R. R. Hudson, A. Di Rienzo, "Adjusting the focus on human variation," *Trends in Genetics*, Vol. 16, No. 7 (2000), pp. 296–302; D. E. Reich, S. F. Schaffner, M. J. Daly, G. McVean, J. C. Mullikin, J. M.Higgins, D. J.Richter, E. S. Lander, D. Altshuler, "Human genome sequence variation and the influence of gene history, mutation and recombination," *Nature Genetics*, Vol. 32, No. 1 (2002), pp. 135–142).
17 See Marian, "Molecular genetic studies of complex phenotypes," p. 65.
18 See L. A. Hindorff, P. Sethupathy, H. A. Junkins, E. M. Ramos, J. P. Mehta, F. S. Collins, T. A. Manolio, "Potential etiologic and functional implications of genome-wide association loci for human diseases andtraits," *Proceedings of the National Academy of Sciences of the United States of America*, Vol. 106, No. 23 (2009), pp. 9362–9367. SNP variants for many species (including humans) are collected in the free public Single Nucleotide Polymorphism Database (dbSNP) maintained by the National Center for Biotechnology Information (NCBI) in collaboration with the National Human Genome Research Institute (NHGRI). The dbSNP encompasses not only SNP variants, but also other forms of genetic variation – short deletion and insertion polymorphisms, multinucleotide polymorphisms (MNPs), heterozygous sequences, and microsatellites or short tandem repeats (STRs). New data are revised and made available in irregular intervals as a series of "builds". For example, the build 146 (released on Nov 24, 2015) for *Homo sapiens* contains 150,482,731 reference SNP clusters (refSNP), whereas build 132 (released on Sep 23, 2010) 30,442,771 refSNP. Database available online

Excellence and Its Biological Limitations   247

variation, but there is also a growing body of evidence of the importance of structural variations.[19] Structural variants are defined operationally as alterations in the genome involving DNA segments larger than 1 kb (microscopic or submicroscopic). There are many types of structural variants, but the most important ones seem to be inversions, translocations, and especially copy number variants (CNVs).[20] The situation is aptly summarized by L. Feuk, A. R. Carson. and W. Scherer:

> The first wave of information from the analysis of the human genome revealed SNPs to be the main source of genetic and phenotypic human variation. However, the advent of genome-scanning technologies has now uncovered an unexpectedly large extent of what we term 'structural variation' in the human genome. This comprises microscopic

---

http://www.ncbi.nlm.nih.gov/SNP. See S. T. Sherry, M. H. Ward, M. Kholodov, J. Baker, L. Phan, E. M. Smigielski, K. Sirotkin, "dbSNP: the NCBI database of genetic variation," *Nucleic Acid Research*, Vol. 29 (2001), pp. 308–311. The genome-wide association study data concerning SNPs and SNP-trait associations that are statistically significant are collected in the online GWAS Catalog provided by the NHGRI and the European Bioinformatics Institute (EMBL-EBI) (See D. Welter, J. MacArthur, J. Morales, T. Burdett, P. Hall, H. Junkins, A. Klemm, P. Flicek, T. Manolio, L. Hindorff, H. Parkinson, "The NHGRI GWAS Catalog, a curated resource of SNP-trait associations," *Nucleic Acids Research*, Vol. 42 (2014) (Database issue), pp. D1001–D1006).

19 See A. C. English, W. J. Salerno, O. A. Hampton, C. Gonzaga-Jauregui, S. Ambreth, D. I. Ritter, C. R. Beck, C. F. Davis, M. Dahdouli, S. Ma, A. Carroll, N. Veeraraghavan, J. Bruestle, B. Drees, A. Hastie, E. T. Lam, S. White, P. Mishra, M. Wang, Y. Han, F. Zhang, P. Stankiewicz, D. A. Wheeler, J. G. Reid, D. M. Muzny, J. Rogers, A. Sabo, K. C. Worley, J. R. Lupski, E. Boerwinkle, R. A. Gibbs, "Assessing structural variation in a personal genome-towards a human reference diploid genome," *BMC Genomics*, Vol. 16 (2015), p. 286; Feuk, Carson and Scherer, "Structural variation in the human genome"; H. Kahrer-Sawatzki, "What a difference copy number variation makes," *Bioessays*, Vol. 29, No. 4 (2007), pp. 311–313; C. L. Usher, S. A. McCarroll, "Complex and multi-allelic copy number variation in human disease,"*Briefings in Functional Genomics*, Vol. 14, No. 5 (2015), pp. 329–338.

20 Copy-number variant (CNV) is a DNA segment (1 kb or larger) that is present in a variable number of copies as compared with a reference genome. This definition includes deletions, insertions, duplications and large-scale copy number variants (variants that involve DNA segments ≥50 kb). Inversion is a segment of DNA with orientation that is reversed with respect to the rest of the chromosome. Translocation (intra- or interchromosomal) means a change in position within a genome of a DNA segment, without any change of the total DNA content. Other types of structural variations include: heteromorphisms, fragile sites, ring and marker chromosomes, isochromosomes, double minutes, segmental uniparental disomy, and gene-conversion products (See Feuk, Carson and Scherer, "Structural variation in the human genome").

and, more commonly, submicroscopic variants, which include deletions, duplications and large-scale copy-number variants – collectively termed copy-number variants or copy-number polymorphisms – as well as insertions, inversions and translocations. Rapidly accumulating evidence indicates that structural variants can comprise millions of nucleotides of heterogeneity within every genome, and are likely to make an important contribution to human diversity and disease susceptibility.[21]

It has been estimated, that up to 13% of the human genome may be subject to structural variations.[22] Moreover, the surprising large-scale copy number variations were discovered in the human genome in 2004, broadening the view of our genetic and phenotypic variation.[23] Further studies revealed the link between copy number variants (usually several copies, but even up to 50) and complex traits, but the full extent of CNV still remains relatively under-ascertained.[24]

---

21 Feuk, Carson and Scherer, "Structural variation in the human genome", p. 85.
22 See D. F. Conrad, D. Pinto, R. Redon, L. Feuk, O. Gokcumen, Y. Zhang, J. Aerts, T. D. Andrews, C. Barnes, P. Campbell, T. Fitzgerald, M. Hu, C. H. Ihm, K. Kristiansson, D. G. Macarthur, J. R. Macdonald, I. Onyiah, A. W. Pang, S. Robson, K. Stirrups, A. Valsesia, K. Walter, J. Wei; Wellcome Trust Case Control Consortium, C. Tyler-Smith, N. P. Carter, C. Lee, S. W. Scherer, M. E. Hurles, "Origins and functional impact of copy number variation in the human genome," *Nature*, Vol. 464, No. 7289 (2010), pp. 704–712.
23 See N. P. Carter, "As normal as normal can be?," *Nature Genetics*, Vol. 36, No. 9 (2004), pp. 931–932; A. J. Iafrate, L. Feuk, M. N. Rivera, M. L. Listewnik, P. K. Donahoe, Y. Qi, S. W. Scherer, C. Lee, "Detection of large-scale variation in the human genome," *Nature Genetics*, Vol. 36, No. 9 (2004), pp. 949–951; J. Sebat, B. Lakshmi, J. Troge, J. Alexander, J. Young, P. Lundin, S. Månér, H. Massa, M. Walker, M. Chi, N. Navin, R. Lucito, J. Healy, J. Hicks, K. Ye, A. Reiner, T. C. Gilliam, B. Trask, N. Patterson, A. Zetterberg, M. Wigler, "Large-scale copy number polymorphism in the human genome," *Science*, Vol. 305, No. 5683 (2004), pp. 525–528.
24 See Conrad et al., "Origins and functional impact of copy number variation in the human genome;" E. R. Gamazon, B. E. Stranger, "The impact of human copy number variation on gene expression," *Briefings in Functional Genomics*, Vol. 14, No. 5 (2015), pp. 352–357; Kahrer-Sawatzki, "What a difference copy number variation makes;" R. Redon, S. Ishikawa, K. R. Fitch, L. Feuk, G. H. Perry, T. D. Andrews, H. Fiegler, M. H. Shapero, A. R. Carson, W. Chen, E. K.Cho, S. Dallaire, J. L. Freeman, J. R. González, M. Gratacòs, J. Huang, D. Kalaitzopoulos, D. Komura, J. R. MacDonald, C. R. Marshall, R. Mei, L. Montgomery, K. Nishimura, K. Okamura, F. Shen, M. J. Somerville, J. Tchinda, A. Valsesia, C. Woodwark, F. Yang, J. Zhang, T. Zerjal, J. Zhang, L. Armengol, D. F. Conrad, X. Estivill, C. Tyler-Smith, N. P. Carter, H. Aburatani, C. Lee, K. W. Jones, S. W.Scherer, M. E. Hurles, "Global variation in copy number in the human genome," *Nature*, Vol. 444, No. 7118 (2006), pp. 444–454; Usher and McCarroll, "Complex and multi-allelic copy number variation in human disease."

Copy number variants may encompass particular genes, affecting their regulation and amount of product (in a dosage-specific manner), or the regions outside of genes, but may alter the expression of proximal genes.[25] One of the most famous examples of a trait associated with multiple copies of a particular gene is the ability to digest the starch present in food. There is a substantial variation in the number of salivary amylase gene copies among humans (up to 10 copies), with more (average) copies and higher amylase levels populations with high-starch diets.[26]

It has been evident for decades that changes of DNA sequence within genes are extremely important. A large number of mutations in protein-coding genes substantially alter the protein function, either by changing its structure, or expression (inhibition, increase or decrease in expression). The non-synonymous mutations in the gene change the amino acid sequence of the protein encoded by this gene, and it has been proven that some of them can cause protein to be totally ineffective. The other kind of mutations, the so-called synonymous mutations, does not alter the sequence of the protein despite changing the sequence of its gene and has been considered completely unimportant (silent mutations). However, some notions concerning the influence of various kinds of changes in the gene on its product have undergone major changes with the discovery of the possible impact of synonymous mutations. These mutations can lead to substantial changes in the protein folding resulting in structural differences that can, in turn, cause important phenotypic changes, even contributing to the development of cancer.[27] As aptly stated by Z. E. Sauna & C. Kimchi-Sarfaty,

---

25 See D. F.Conrad et al., "Origins and functional impact of copy number variation in the human genome;" R. R. Haraksingh, M. P. Snyder, "Impacts of variation in the human genome on gene regulation," *Journal of Molecular Biology*, Vol. 425, No. 21 (2013), pp. 3970–3977.

26 See Kingsley, "From atoms to traits;" G. H. Perry, N. J. Dominy, K. G. Claw, A. S. Lee, H. Fiegler, R. Redon, J. Werner, F. A. Villanea, J. L. Mountain, R. Misra, N. P. Carter, C. Lee, A. C. Stone, "Diet and the evolution of human amylase gene copy number variation," *Nature Genetics*, Vol. 39, No. 10 (2007), pp. 1256–1260.

27 See J. V. Chamary, J. R. Parmley, L. D. Hurst, "Hearing silence: non-neutral evolution at synonymous sites in mammals," *Nature Reviews. Genetics*, Vol. 7(2006), pp. 98–108; A. Dana, T. Tuller, "Mean of the typical decoding rates: a new translation efficiency index based on the analysis of ribosome profiling data," *G3: Genes - Genomes – Genetics*, Vol. 5, No. 1 (2014), pp. 73–80; K. L. Fung, M. M. Gottesman, "A synonymous polymorphism in a common MDR1 (ABCB1) haplotype shapes protein function," *Biochimica et Biophysica Acta*, Vol. 1794 (2009), pp. 860–871; G. Kudla, A. W. Murray, D. Tollervey, J. B. Plotkin, "Coding-sequence determinants of gene expression in *Escherichia coli*," *Science*, Vol. 324, No. 5924 (2009), pp. 255–258; F. Supek, B. Miñana, J. Valcárcel, T.

Synonymous mutations – sometimes called 'silent' mutations – are now widely acknowledged to be able to cause changes in protein expression, conformation and function. The recent increase in knowledge about the association of genetic variants with disease, particularly through genome-wide association studies, has revealed a substantial contribution of synonymous SNPs to human disease risk and other complex traits.[28]

Another well-known and important example of a uniquely human trait that may have a different genetic background is lactase persistence: all mammals use milk as a significant source of nutrition, but only some groups of humans are capable of doing it during adulthood. This ability is dependent on a special (mutant) form of the enhancer (regulatory) region of the intestinal lactase gene, which is usually active (in most mammals and most humans) only during the infant nursing period. However, in populations with a long tradition of dairy herding and the extensive use of milk in their diet, most individuals carry a variant of the lactase gene that is active beyond childhood. What is more interesting, though, is that lactase persistence has occurred independently in various human populations in Europe, Africa, and the Middle East, in regions strongly dependent on milk (from cattle, goats and camels). The traces of selection that have taken place within the past 5,000-10,000 years are among the strongest observed for any gene in the human genome.[29] It is, as accurately described by D. M. Kingsley, "a striking example of the repeated evolution of a similar trait by independent

---

Gabaldón, B. Lehner, "Synonymous Mutations Frequently Act as Driver Mutations in Human Cancers," *Cell*, Vol. 156, No. 6 (2014), pp. 1324–1335; Ch.-J. Tsai, Z. E. Sauna, Ch. Kimchi-Sarfaty, S. V. Ambudkar, M. M. Gottesman, R. Nussinov, "Synonymous mutations and ribosome stalling can lead to altered folding pathways and distinct minima,"*Journal of Molecular Biology*, Vol. 383 (2008), pp. 281–291; S. Zheng, H. Kim, R. G. Verhaak, "Silent mutations make some noise," *Cell*, Vol. 156, No. 6 (2014), pp. 1129–1131.

28 See Z. E. Sauna, C. Kimchi-Sarfaty, "Understanding the contribution of synonymous mutations to human disease," *Nature Reviews. Genetics*, Vol. 12, No. 10 (2011), pp. 683–691. This is another example of the increasing complexity of the genome–phenotypic traits relations. The phenomenon that for a long time had been considered to be unimportant has suddenly been revealed as a "tip of an iceberg". The possible implications may be staggering if we take into account that only about 10,000 out of total number of 3.5×10^6 SNPs in the human genome are non-synonymous SNPs. There are many other mechanisms involved in the regulation of gene expression this affect phenotype, such as long non-coding RNAs, microRNAs, histone modifications, splice variants, post-translational modifications of the proteins and epigenetics (see Marian, "Molecular genetic studies of complex phenotypes"). Some of these phenomena will be discussed later.

29 See Kingsley, "From atoms to traits."

changes affecting one gene (…) Its retention in milk-dependent societies also illustrates how culture can reinforce the forces of evolution".[30]

All of the facts presented here make it clear that the relations between genome and phenotypic traits are extremely complex.[31] However, it had become apparent that the basis of human diversity is even more complicated when we realized that

---

30 Kingsley, "From atoms to traits," pp. 58–59. Lactase persistence is connected with the presence of specific variants of regulatory regions (enhancer regions), outside the *locus* of the lactase gene (*LCT*), that enhance the activity of the promoter of the gene. Three different variants of regulatory regions that are embedded in the neighboring *MCM6* gene have been discovered to date. The most frequent allele T_{-13910} has been found in population of Central and Northern Europe (also among US inhabitants of European origin), while the G_{-13907} variant is prevalent in Eastern Africa and G_{-13915}-C_{-3712} allele is characteristic for the Middle East population (See T. Bersaglieri, P. C. Sabeti, N. Patterson, T. Vanderploeg, S. F. Schaffner, J. A. Drake, M. Rhodes, D. E. Reich, J. N. Hirschhorn, "Genetic signatures of strong recent positive selection at the lactase gene," *American Journal of Human Genetics*, Vol. 74, No. 6 (2004), pp. 1111–1120; N. S. Enattah, T. Sahi, E. Savilahti, J. D. Terwilliger, L. Peltonen, I. Järvelä, "Identification of a variant associated with adult-type hypolactasia," *Nature Genetics*, Vol. 30, No. 2 (2002), pp. 233–237; N. S. Enattah, T. G. Jensen, M. Nielsen, R. Lewinski, M. Kuokkanen, H. Rasinpera, H. El-Shanti, J. K. Seo, M. Alifrangis, I. F. Khalil, A. Natah, A. Ali, S. Natah, D. Comas, S. Q. Mehdi, L Groop., E. M. Vestergaard, F. Imtiaz, M. S. Rashed, B. Meyer, J. Troelsen, L. Peltonen, "Independent introduction of two lactase-persistence alleles into human populations reflects different history of adaptation to milk culture," *American Journal of Human Genetics*, Vol. 82, No. 1 (2008), pp. 57–72; Kingsley, "From atoms to traits,"; M. Kuokkanen, J. Kokkonen, N. S. Enattah, T. Ylisaukko-oja, H. Komu, T. Varilo, L. Peltonen, E. Savilahti, I. Jarvela, "Mutations in the translated region of the lactase gene (LCT) underlie congenital lactase deficiency," *American Journal of Human Genetics*, Vol. 78, No. 2 (2006), pp. 339–344).

31 It should also be noted that various variants (or genes) may have different impact on the phenotype, as described by A. J. Marian: "The focus of human genetic studies of complex traits is primarily on the DNA sequence variants (DSVs) among individuals, which contribute to susceptibility to a disease, clinical outcomes or response to therapy. (p. 65) Single gene disorders are caused by rare variants with large effect sizes. In addition to the main causal variant, which typically exhibits a Mendelian pattern of inheritance, several other non-Mendelian variants contribute to expression of the phenotype. On the opposite end of the spectrum are the common complex traits, which are caused, in part, by the cumulative effects of a very large number of DSVs, each imparting a modest effect size. In oligogenetic phenotypes, several alleles with moderate size effects and a large number of alleles with small effect sizes contribute to the phenotype". (See Marian, Marian, "Molecular genetic studies of complex phenotypes," pp. 65–68)

genes actually occupy only a small part of our genome (about 5%).[32] Moreover, it has been estimated that only about 1.5% of the 3.2 billion nucleotides of our genome are directly involved in the formation of phenotypic traits. All exons (protein-coding segments), comprising the so-called exome, account for about 1% while the inter-gene regions (serving as template for non-coding RNAs) and the regulatory regions together occupy another 0.5% of the genome.[33] Although for many years the genetic studies of complex phenotypes have been gene-centric (or exon-centric), it has become obvious that other segments of our genome must play an important role as they are also under evolutionary selection pressure.[34] Still, about 98.5% of our genome is sometimes referred to as "the dark matter of the genome", as its function is virtually unknown.[35] In recent years, these findings have led to the shift of attention to the non-coding segments of the human genome.[36] The huge ENCODE (Encyclopedia of DNA Elements) project has already revealed in its initial stage that the vast majority (80%) of human genome undergoes the transcription process.[37] Most of the transcripts

---

32 Genes contain the protein coding regions (exons), introns (between exons) and the regulatory regions (Marian, "Molecular genetic studies of complex phenotypes," pp. 65–68). It has been estimated that human genome contains approximately 180,000 exons in about 23,500 genes (See E. S. Lander, L. M. Linton, B. Birren, C. Nusbaum, M. C. Zody, J. Baldwin, K. Devon, K. Dewar, M. Doyle, W. FitzHugh, R. Funke, D. Gage, K. Harris, A. Heaford, J. How land, L. Kann, J. Lehoczky, R. LeVine, P. McEwan, K. McKernan et al., International Human Genome Sequencing Consortium, "Initial sequencing and analysis of the human genome," *Nature*, Vol. 409, No. 6822 (2001), pp. 860–921, Erratum in: *Nature*, Vol. 411, No. 6838 (2001), p. 720; Marian, "Molecular genetic studies of complex phenotypes," pp. 65–68; J. C. Venter, M. D. Adams, E. W. Myers, P. W. Li, R. J. Mural, G. G. Sutton, H. O. Smith, M. Yandell, C. A. Evans, R. A. Holt, J. D. Gocayne, P. Amanatides, R. M. Ballew, D. H. Huson, J. R. Wortman, Q. Zhang, C. D. Kodira, X. H. Zheng, L. Chen, M. Skupski et al., "The sequence of the human genome," *Science*, Vol. 291, No. 5507 (2001), pp. 1304–1351; Erratum in: *Science*, Vol. 292, No. 5523 (2001), p. 1838).
33 See M. Blaxter, "Genetics. Revealing the dark matter of the genome," *Science*, Vol. 330, No. 6012 (2010), pp. 1758–1759; E. S. Lander, "Initial impact of the sequencing of the human genome," *Nature*, Vol. 470, No. 7333 (2011), pp. 187–197; E.S. Lander et al., op. cit.; J.C. Venter et al., op. cit..
34 See E.S. Lander et al., "Molecular genetic studies of complex phenotypes."
35 See Blaxter, "Genetics. Revealing the dark matter of the genome."
36 See R. P. Alexander, G. Fang, J. Rozowsky, M. Snyder, M. B. Gerstein, "Annotating non-coding regions of the genome," *Nature Reviews. Genetics*, Vol. 11, No. 8 (2010), pp. 559–571.
37 See E. Pennisi, "Genomics. ENCODE project writes eulogy for junk DNA," *Science*, Vol. 337, No. 6099 (2012), pp. 1159–1161.

Excellence and Its Biological Limitations   253

belong to highly heterogeneous groups of short non-coding RNAs (sncRNAs) which may have regulatory functions.[38] It is now evident that genetic variants of non-coding regions (including single-nucleotide and copy number variants) may influence complex traits and diseases.[39] These surprising discoveries, as well as other findings (e.g., the existence of a gene within another gene) have even challenged our notion of the gene. It has been noted that a new gene definition is necessary to accommodate the recent advances in genomics (study of genome), ribonomics (study of RNAs), and proteomics (proteome study).[40] M. B. Gerstein et al. suggested the following definition: "a gene is a union of genomic sequences encoding a coherent set of potentially overlapping functional products".[41] All these unexpected discoveries have been "paradigm shifting" and have opened our eyes to the real complexity of living organisms. However, nature has had in store more shocks for scientists in the form of "missing heritability", which I will discuss it in the next section.

---

38 See G. V. Glinsky, "Phenotype-defining functions of multiple non-coding RNA pathways," *Cell Cycle*, Vol. 7, No. 11 (2008), pp. 1630–1639.
39 See F. Zhang, J. R. Lupski "Non-coding genetic variants in human disease," *Human Molecular Genetics*, Vol. 24, No. R1 (2015), pp. 102–110.
40 See Y. Jia, L. Chen, Y. B. Ma, J. Zhang, N. Xu, D.J. Liao, "To Know How a GeneWorks, We Need to Redefine It First but then, More Importantly, to Let the Cell Itself Decide How to Transcribe and Process Its RNAs," *International Journal of Biological Sciences*, Vol. 11, No. 12 (2015), pp. 1413–1423.
41 See M. B. Gerstein, C. Ruce, J.S. Rozowsky, D. Zheng, J. Du, J. O. Korbel, O. Emanuelsson, Z. D. Zhang, S. Weissman, M. Snyder, "What is a gene, post-ENCODE? History and updated definition," *Genome Research*, Vol. 17, No. 6 (2007), pp. 669–681. There have been many gene definitions for the past century – from the first concept (abstract elements of heredity that act as discrete, distinct units in the work of Gregor Mendel 1866), through gene as a distinct locus (Thomas Morgan 1915), to the gene as an open reading frame (ORF) and annotated genomic entity, enumerated in the sequence databanks (Gerstein et al., "What is a gene, post-ENCODE?"). In 2006 the Sequence Ontology Consortium formulated a tentative definition of a gene as "a locatable region of genomic sequence, corresponding to a unit of inheritance, which is associated with regulatory regions, transcribed regions and/or other functional sequence regions" (See H. Pearson, "Genetics: what is a gene?," *Nature*, Vol. 441, No. 7092 (2006), pp. 398–401 p. 401). It is worth mentioning, that it took two days of heated discussion for 25 scientists involved in the consortium to reach a consensus, as described by the consortium coordinator K. Eilbeck (Pearson, "Genetics: what is a gene?").

4 Missing heritability and epigenetics

It has already been mentioned that a complex phenotype results from a large number of interactions of various genetic and non-genetic factors. Genome-wide association studies (GWAS) of complex traits and diseases have revealed that for the great majority of them the identified variants only account for the small portion of previously estimated heritability.[42] To address this discrepancy, the term "missing heritability" is used by most scientists who search for factors that contribute to it. According to some researchers, the main reason of "missing heritability" is the inability of the standard GWAS analyses to detect small-effect variants or the quality of statistical techniques applied, increasing the efforts to refine them.[43]

One of the most surprising discoveries concerned genetic variants influencing human height, one of the classic polygenic traits. Previous studies had estimated 80-90% heritability for the trait, but the GWAS studies that associated over 40 variants with height differences have only been able to account for about 5% of phenotypic variance.[44] Fortunately, Yang J et al. considerably improved the score

---

42 See B. Maher, "Personal genomes: The case of the missing heritability," *Nature*, Vol. 456, No. 7218 (2008), pp. 18-21; R. Makowsky, N. M. Pajewski, Y. C. Klimentidis, A. I. Vazquez, C. W. Duarte, D. B. Allison, G. de los Campos, "BeyondMissing Heritability: Prediction of Complex Traits," *PLoS Genetics*, Vol. 7 (2011), p. e1002051; T. A. Manolio, F. S. Collins, N. J. Cox, D. B. Goldstein, L. A. Hindorff, D. J. Hunter, M. I. McCarthy, E. M. Ramos, L. R. Cardon, A. Chakravarti, J. H. Cho, A. E. Guttmacher, A. Kong, L. Kruglyak, E. Mardis, C. N. Rotimi, M. Slatkin, D. Valle, A. S. Whittemore, M. Boehnke, A. G. Clark, E. E. Eichler, G. Gibson, J. L. Haines, T. F. Mackay, S. A. McCarroll, P. M. Visscher, "Finding the missing heritability of complex diseases," *Nature*, Vol. 461 (2009), pp. 747-753; M. Trerotola, V. Relli, P. Simeone, S. Alberti, "Epigenetic inheritance and the missing heritability," *Human Genomics*, Vol. 9 (2015), p. 17.
43 See S. Berger, P. Pérez-Rodríguez, Y. Veturi, H. Simianer, G. de los Campos, "Effectiveness of Shrinkage and Variable Selection Methods for the Prediction of Complex Human Traits using Data from Distantly Related Individuals," *Annals of Human Genetics*, Vol. 79, No. 2 (2015), pp. 122-135; G. de los Campos, A. I. Vazquez, R. Fernando, Y. C. Klimentidis, D. Sorensen, "Prediction of Complex Human Traits Using the Genomic Best Linear Unbiased Predictor," *PLoS Genetics*, Vol. 9, No. 7 (2013), p. e1003608; R. Makowsky et al., "Beyond Missing Heritability;" J. Yang, B. Benyamin, B. P. McEvoy, S. Gordon, A. K. Henders, D. R. Nyholt, P. A. Madden, A. C. Heath, N. G. Martin, G. W. Montgomery, M. E. Goddard, P. M. Visscher, "Common SNPs explain a large proportion of the heritability for human height," *Nature Genetics*, Vol. 42 (2010), pp. 565-569.
44 See D. F. Gudbjartsson, G. B. Walters, G. Thorleifsson, H. Stefansson, B. V. Halldorsson, P. Zusmanovich, P. Sulem, S. Thorlacius, A. Gylfason, S. Steinberg, A. Helgadottir, A.

by applying more refined methods and explaining 45% of variance.[45] However, there are also other factors that should be taken into account, such as insufficient sample sizes,[46] rare (low frequency) variants or undetected structural variation, including copy number variants[47] over-estimated heritability or epistasis (interactions among loci)[48]. The problem is described very well by A. J. Clarke and D. N. Cooper:

> So, where is this "missing heritability"? We respond to this question in two different ways. First, we believe that complex disorders are indeed complex and that genetic studies of complex disorders in humans face a number of challenges including gene-gene and gene-environment interactions and epigenetic modification of the genome. Second, we shall argue that high estimates of heritability have been misinterpreted as showing that a predisposition to such a condition (one with high heritability) must have been transmitted through the family from parent to child. The complexity of these

---

Ingason, V. Steinthorsdottir, E. J. Olafsdottir, G. H. Olafsdottir, T. Jonsson, K. Borch-Johnsen, T. Hansen, G. Andersen, T. Jorgensen, O. Pedersen, K. K. Aben, J. A. Witjes, D. W. Swinkels, M. den Heijer, B. Franke, A. L. Verbeek, D. M. Becker, L. R. Yanek, L. C. Becker, L. Tryggvadottir, T. Rafnar, J. Gulcher, L. A. Kiemeney, A. Kong, U. Thorsteinsdottir, K. Stefansson, "Many sequence variants affecting diversity of adult human height," *Nature Genetics*, Vol. 40 (2008), pp. 609-615; G. Lettre, A. U. Jackson, C. Gieger, F. R. Schumacher, S. I. Berndt, S. Sanna, S. Eyheramendy, B. F. Voight, J. L. Butler, C. Guiducci, T. Illig, R. Hackett, I. M. Heid, K. B. Jacobs, V. Lyssenko, M. Uda et al., "Identification of ten loci associated with height highlights new biological pathways in human growth," *Nature Genetics*, Vol. 40 (2008), pp. 584-591; Maher, "Personal genomes", T. A. Manolio et al., "Finding the missing heritability of complex diseases;" M. N. Weedon, H. Lango, C. M. Lindgren, C. Wallace, D. M. Evans, M. Mangino, R. M. Freathy, J. R. Perry, S. Stevens, A. S. Hall, N. J. Samani, B. Shields, I. Prokopenko, M. Farrall, A. Dominiczak et al., "Genome-wide association analysis identifies 20 loci that influence adult height," *Nature Genetics*, Vol. 40 (2008), pp. 575-583.

45 See J. Yang et al., "Common SNPs explain a large proportion of the heritability for human height."
46 See J. H. Park, S. Wacholder, M. H. Gail, U. Peters, K. B. Jacobs, S. J. Chanock, N. Chatterjee, "Estimation of effect size distribution from genome-wide association studies and implications for future discoveries," *Nature Genetics*, Vol. 42 (2010), pp. 570-575.
47 See Manolio et al., "Finding the missing heritability of complex diseases."
48 See Maher, Manolio et al., "Finding the missing heritability of complex diseases;" O. Zuk, E. Hechter, S. R. Sunyaev, E. S. Lander, "The mystery of missing heritability: Genetic interactions create phantom heritability,"*Proceedings of the National Academy of Sciences of the United States of America*, Vol. 109 (2012), pp. 1193-1198.

common conditions is apparent from the range of factors that need to be considered as potentially contributing to the "missing heritability".[49]

Epigenetic mechanisms may be yet another possible solution to the mystery of the "missing heritability" and may account for its significant fraction. It is extremely important that epigenetic modifications, although they do not change the genomic sequences, can alter gene expression in a heritable manner. These changes may be triggered by environmental factors and affect an individual during his/her lifetime or may be involved in a real transgenerational inheritance. The main mechanisms of epigenetic changes include DNA methylation, histone modifications, chromatin remodeling (e.g., activity of chaperones), and "RNA epigenetics". DNA methylation is carried out by DNA methyltransferases and is crucial for gene regulation – gene expression dynamics depend on a balance between DNA methylation/demethylation processes.[50] Histone posttranslational modifications (mainly methylation and acetylation) and changes in chromatin organization are also important for the regulation of gene transcription.[51] There is also a growing body of evidence concerning the role of various types of RNA (e.g., long non-coding RNAs (ncRNAs), specific mRNAs and siRNAs/miRNAs) in epigenetic regulation.[52]

Epigenetic changes may be affected by numerous factors, such as stress, diet and nutrition status, pharmacological treatment, and exposure to toxic compounds. The changes accumulate during lifetime, so the number of differences increases even for monozygotic twins.[53] Epigenetic mechanisms may

---

49 See A. J. Clarke, D. N. Cooper, "GWAS: heritability missing in action?," *European Journal of Human Genetics*, Vol. 18 (2010), pp. 859–861.
50 See M. Ehrlich, M. Lacey, "DNA methylation and differentiation: silencing, upregulation and modulation of gene expression," *Epigenomics*, Vol. 5 (2013), pp. 553–568; C. M. Rivera, B. Ren, "Mapping human epigenomes," *Cell*, Vol. 155 (2013), pp. 39–55.
51 See Rivera, Ren, "Mapping human epigenomes;" N. R. Rose, R. J. Klose, "Understanding the relationship between DNA methylation and histone lysine methylation," *Biochimica et Biophysica Acta*, Vol. 1839, No. 12 (2014), pp. 1362–1372.
52 See N. Liu, T. Pan, "RNA epigenetics," *Translational Research: The Journal of Laboratory and Clinical Medicine*, Vol. 165, No. 1 (2015), pp. 28–35; G. Zheng, J. A. Dahl, Y. Niu, Y. Fu, A. Klungland, Y. G. Yang, C. He, "Sprouts of RNA epigenetics," *RNA Biology*, Vol. 10, No. 6 (2013), pp. 915–918.
53 See M. F. Fraga, E. Ballesta, M. F. Paz, S. Ropero, F. Setien, M. L. Ballestar, D. Heine-Suñer, J. C. Cigudosa, M. Urioste, J. Benitez, M. Boix-Chornet, A. Sanchez-Aguilera, C. Ling, E. Carlsson, P. Poulsen, A. Vaag, Z. Stephan, T. D. Spector, Y. Z. Wu, C. Plass, M. Esteller, "Epigenetic differences arise during the lifetime of monozygotic twins," *Proceedings of the National Academy of Sciences of the United States of America*, Vol.

modify various sorts of phenotypic traits (molecular, cellular, morphological, and physiological, but also behavioral) and increase variation in the human population.[54] A lot of research implicated such epigenetic processes as e.g., in brain development and plasticity[55], diseases, such as diabetes and obesity[56], cancer[57], syndromes involving chromosomal instabilities and mental retardation[58], also in neurological diseases and aging.[59]

All the aspects and intricacies of the relations between genome and phenotypic traits demonstrate clearly how complex and unpredictable these links are. I have already described some aspects of population biology (genetic and phenotypic diversity), but it is also necessary to discuss the concept of normality and introduce an evolutionary perspective in order to fully understand the problem.

5 Populations and evolution

The concept of normality, which is also connected with definitions of "health" and "disease", is extremely important in biology and medicine. The problem has

---

102, No. 30 (2005), pp. 10604–10609; Trerotola et al., "Epigenetic inheritance and the missing heritability."

54 See W. W. Burggren, D. Crews, "Epigenetics in comparative biology: why we should pay attention," *Integrative and Comparative Biology*, Vol. 54, No.1 (2014), pp. 7–20; M. Fagiolini, C. L. Jensen, F. A. Champagne, "Epigenetic influences on brain development and plasticity," *Current Opinion in Neurobiology*, Vol. 19, No. 2 (2009), pp. 207–212.

55 See A. Aksoy-Aksel, F. Zampa, G. Schratt, "MicroRNAs and synaptic plasticity--a mutual relationship," *Philosophical Transactions of the Royal Society of London. Series B, Biological Sciences*, Vol. 369, No. 1652 (2014), p. 20130515; I. A. Qureshi, M. F. Mehler, "An evolving view of epigenetic complexity in the brain," *Philosophical Transactions of the Royal Society of London. Series B, Biological Sciences*, Vol. 369, No. 1652 (2014), p. 20130506.

56 See A. W. Drong, C. M. Lindgren, M. I. McCarthy, "The genetic and epigenetic basis of type 2 diabetes and obesity,"*Clinical Pharmacology and Therapeutics*, Vol 92, No. 6 (2012), pp. 707–715.

57 See M. A. Dawson T. Kouzarides, "Cancer epigenetics: from mechanism to therapy," *Cell*, Vol. 150, No. 1 (2012), pp. 12–27.

58 See G. Egger, G. Liang, A. Aparicio, P. A. Jones, "Epigenetics in human disease and prospects for epigenetic therapy," *Nature*, Vol. 429, No. 6990 (2004), pp. 457–463.

59 See R. Cacabelos, C. Torrellas, "Epigenetics of Aging and Alzheimer's Disease: Implications for Pharmacogenomics and Drug Response," *International Journal of Molecular Sciences*, Vol. 16, No. 12 (2015), pp. 30483–30543; T. C. Roberts, K. V. Morris, M. J. Wood, "The role of long non-coding RNAs in neurodevelopment, brain function and neurological disease," *Philosophical Transactions of the Royal Society of London. Series B, Biological Sciences*, Vol. 369, No. 1652 (2014), p. 20130507.

been analyzed from the philosophical perspective by many authors, but I will discuss their ideas only in brief. For example, R. Wachbroit argues that the concept of normality is unique for biology and cannot be found in physics or chemistry.[60] It must be noted that normality can be defined only in relation to a specified group of individuals (a reference class), e.g., an age group, individuals of a particular sex, or the whole population. There are also various ways of understanding normality, mainly as an evaluative (related to some norms, e.g., cultural or conventional), theoretical (referring to the essential, typical or exemplary state), or statistical concept. Statistical normality means the numerical average (as the mean, median etc.) state of individuals in a reference class, but a portion of a distribution should also be taken into account.[61]

Most analyses focus on the definitions of "health" and "disease", presenting various approaches – mainly naturalist[62], normativist[63] or hybrid theories[64]. The naturalist value-free solution is based on scientific data, normativist perspective

---

60 "Thus far I have tried to identify a biological sense of 'normality' that is distinct from 'normality' as either a statistical term or as a value term. Normality in this sense is allied with the contrast between function and malfunction. I have also tried to show that the standard accounts of biological functions – goal theories or etiological theories – cannot explain normality because they presuppose it. If what I have said so far is correct, the concept of normality is at least as central in biology as is the concept of function (...) while the concept of normality is important to the biological sciences, nothing like it can be found in the nonbiological sciences such as physics or chemistry. Physics may use the concept of statistical normality, but that (...) is a different concept (See R. Wachbroit, "Normality as a Biological Concept," *Philosophy of Science*, Vol. 61 (1994), pp. 579–591 [p. 587]).

61 See Wachbroit, "Normality as a Biological Concept"; M. Ereshefsky, "Defining 'health' and 'disease'," *Studies in History and Philosophy of Science. PartC. Studies in History and Philosophy of Biological and Biomedical Sciences*, Vol. 40 (2009), pp. 221–227.

62 See C. Boorse, "On the Distinction between Disease and Illness," *Philosophy & Public Affairs*, Vol. 5, No. 1 (1975), pp. 49–68; C. Boorse, "Health as a theoretical concept," *Philosophy of Science*, Vol. 44 (1977), pp. 542–573; C. Boorse, "A rebuttal on health," in: *What is disease?*, ed. J. Humber and R. Almeder(Totowa, NJ: Humana Press, 1997), pp. 1–134; J. Scadding, "The semantic problem of psychiatry," *Psychological Medicine*, Vol. 20 (1990), pp. 243–248.

63 See Engelhardt, "Ideology and etiology;" W. Goosens, "Values, health, and medicine," *Philosophy of Science*, Vol. 47 (1980), pp. 100–115; J. Margolis, "The concept of disease," *The Journal of Medicine and Philosophy*, Vol. 1 (1976), pp. 238–255.

64 See L. Reznek, *The nature of disease* (London: Routledge & Kegan Paul, 1987); J. Wakefield, "The concept of mental disorder: on the boundary between biological facts and social values," *American Psychologist*, Vol. 47 (1992), pp. 373–388.

reflects value judgments,[65] while hybrid theories try to merge both approaches. An alternative approach focuses on the distinction between state description and normative claims that should be taken into consideration in medical discussions.[66]

The most prominent and influential definition of "health" has been created by Christopher Boorse[67], but it has also been the subject of an extensive critique.[68] The concept is known as the "biostatistical theory" (BST), "a name emphasizing that the analysis rests on the concepts of biological function and statistical normality".[69] In biology, the employed statistical concept of normality also accounts for the genetic and phenotypic diversity in populations. Qualitative trait categories that occur most frequently in the population are considered normal and are presented as fractions or percentages of the pool. For quantitative (metric) traits, normality is understood as the numerical mean of the trait measures together with its standard deviation.[70] For example, the average human brain volume is

---

65 "Disease does not reflect a natural standard or norm, because nature does nothing – nature does not care for excellence, nor is it concerned with the fate of individuals qua individuals (…) Health (…) must involve judgments as to what members of that species should be able to do – that is, must involve our esteeming a particular type of function" (See Engelhardt, "Ideology and etiology," p. 266).
66 See Ereshefsky, "Defining 'health' and 'disease'."
67 See Boorse "On the Distinction between Disease and Illness," "Health as a theoretical concept," "A rebuttal on health."
68 See R. Amundson, "Against normal function," *Studies in History and Philosophy of Biological and Biomedical Sciences*, Vol. 31 (2000), pp. 33–53; R. Cooper, "Disease," *Studies in History and Philosophy of Biological and Biomedical Sciences*, (2002), Vol. 33, pp. 263–282; J. D. Guerrero, "On a naturalist theory of health: a critique," *Studies in History and Philosophy of Science. Part C. Studies in History and Philosophy of Biological and Biomedical Sciences*, Vol. 41 (2010), pp. 272–278; P. H. Schwartz, "Defining Dysfunction: Natural Selection, Design, and Drawing a Line," *Philosophy of Science*, Vol. 74 (2007), pp. 364–385.
69 See Boorse, "A rebuttal on health", p. 4. Boorse summarizes his theory in the following way: "1. The *reference class* is a natural class of organisms of uniform functional design; specifically, an age group of a sex of a species. 2. A *normal function* of a part or process within members of the reference class is a statistically typical contribution by it to their individual survival and reproduction. 3. A *disease* is a type of internal state which is either an impairment of normal functional ability, i.e., a reduction of one or more functional abilities belowtypical efficiency, or a limitation on functional ability caused by environmental agents. 4. *Health* is *the* absence of disease." (Boorse, "A rebuttal on health," pp. 7–8).
70 Standard deviation (SD) is a measure of variability of the set of observations, showing how individual values (numbers) cluster round the mean x. "A small standard

1500 cm³, but within the range of 750 cm³ – 2400 cm³ it is considered to be normal.[71]

Genetic and phenotypic diversity within a species is extremely important in the process of evolution, as it allows to adapt to various environmental conditions. I have already mentioned two well-known examples of such adaptation – lactase persistence and copy number variants of amylase gene. I will not discuss biological mechanisms of evolution in detail, as it is beyond the scope of my analysis. It must be stressed, however, that there are many misconceptions concerning evolutionary processes. It is a fairly common belief that evolution leads to the formation of new organisms (or whole species) of increasing perfection, a notion expressed also in the concept of the "ladder of life" or *scala naturae*. This way of thinking is no longer valid in biology and furthermore, it was never employed in Darwin's theory of natural selection.[72] The latter, however, is extremely difficult to rectify, which can be seen even in some translations of the title of Darwin's most influential book. The full title *On the Origin of Species by Means of Natural Selection, or the Preservation of Favoured Races in the Struggle for Life* conveys the basis of the theory but it does not reflect the notion

---

deviation tells us that the observations cluster closely round their mean, while a large standard deviation says that the observations are much more scattered" (See G. M. Clarke, D. Cooke, *A basic course in statistics*, Third edition (London, Melbourne and Auckland: Edward Arnold, A division of Hodder & Stoughton Ltd., 1992), p. 44). Normality in a population is usually understood as x ± SD.

71 Real diversity in a population surpasses the normality range. The brain volume lower than 750cm³ (200–750cm³) is characteristic for microcephaly, a disorder connected with over 400 syndromes of various etiology (See S. Mahmood, W. Ahmad, M. J. Hassan, "Autosomal Recessive Primary Microcephaly (MCPH): clinical manifestations, genetic heterogeneity and mutation continuum," *Orphanet Journal of Rare Diseases*, Vol. 6 (2011), p. 39; E Roberts., D. J. Hampshire, L. Pattison, K. Springell, H. Jafri, P. Corry, J. Mannon, Y. Rashid, Y. Crow, J. Bond, C. G. Woods, "Autosomal recessive primary microcephaly: an analysis of locus heterogeneity and phenotypic variation," *Journal of Medical Genetics*, Vol. 39 (2002), pp. 718–721; C. G. Woods, J. Bond, W. Enard, "Autosomal recessive primary microcephaly (MCPH): a review of clinical, molecular, and evolutionary findings,"*American Journal of Human Genetics*, Vol. 76 (2005), pp. 717–728).

72 It is possible that Charles Darwin was, to some extent, influenced by the notion of the "ladder of life" as suggested by some authors (e.g., U. Kutschera, "From the scala naturae to the symbiogenetic and dynamic tree of life," *Biology Direct*, Vol. 6 (2011), p. 33), but it did not play a substantial role in his theory.

of increasing perfection of organisms.[73] Darwin indeed uses the term "perfect" in his book multiple times, but in the sense of being "perfect for something" or being "perfectly adapted" to some conditions. While it is worth mentioning that the basic aspects of Darwin's theory have withstood the challenges brought about by the development of molecular biology and genetics, the theory did go through some changes after incorporating new scientific data and transformed into the neo-Darwinian theory of evolution.[74] In the recent years, as a result of a better understanding of the influence of the environment on the genome, the theory was challenged again which, in consequence, led to the unifying of the neo-Darwinian and the neo-Lamarckian concepts into a new theory.[75]

---

73 For example, the full title of Darwin's book in Polish is *O powstawaniu gatunków drogą naturalnego doboru czyli o utrzymywaniu się doskonalszych ras w walce o byt*. The term "doskonalsze rasy", used for "favoured races", can be literally re-translated as "more perfect races" or "more excellent races". Clearly, the phrase "more perfect" could be understood as "more perfectly adapted", but still the term is philosophically more ambiguous than the one originally used by Darwin.

74 The changes in Darwin's theory are well described by R. S. Singh: "The modern theory of evolution has enjoyed success by bringing together Mendelian genetics and Darwinian evolution; although the theory has faced many challenges from different corners, it has survived them all (...) Many developmental biologists felt Darwin's natural selection theory was inadequate to explain the diversity of developmental complexity observed among organisms. Yet 150 years after Darwin, although our notions about the fine details of mutation, gene regulation, and selection mechanisms may have changed, no new forces of evolution have been added to the important forces of evolution, i.e., mutation and selection. Other similar but more technical challenges have come in the form of non-Darwinian evolution from molecular biologists (...), neutral evolution from molecular population geneticists (...), and from punctuated evolution or 'evolution by burst' (...) As these challenges have broadened our horizons and have enriched Darwinian evolution, especially in terms of the ever-unfolding dynamics of mutational and genomic variation (...), these were not challenges to the theory itself but only to the details of the evolutionary mechanics, i.e., about role of selection and rate of change" (See R. S. Singh, "Darwin's legacy: why biology is not physics, or why evolution has not become a common sense," *Genome*, Vol. 54, No. 10 (2011), pp. 868–873 [p. 870]).

75 See W. W. Burggren, "Epigenetics as a source of variation in comparative animal physiology - or - Lamarck is lookin' pretty good these days," *The Journal of Experimental Biology*, Vol. 217, No. Pt 5 (2014), pp. 682–689; E. Jablonka, M. J. Lamb, E. Avital, "Lamarckian mechanisms in darwinian evolution," *Trends in Ecology & Evolution*, Vol. 13 (1998), pp. 206–210; K. Laland, T. Uller, M. Feldman, K. Sterelny, G. B. Müller, A. Moczek, E. Jablonka, J. Odling-Smee, G. A. Wray, H. E. Hoekstra, D. J. Futuyma, R. E. Lenski, T. F. Mackay, D. Schluter, J. E. Strassmann, "Does evolutionary theory need a rethink?," *Nature*, Vol. 514, No. 7521 (2014), pp. 161–164; R. C. Lewontin, *The triple*

There is no excellence in evolution: there are only individuals who are better adapted to a particular array of factors. There is also a certain relativity in the concept of "trait excellence" due to the fact that even some traits connected with diseases may be advantageous in a different environment. The most famous example is the resistance to malaria caused by hemoglobin-inherited disorders – mainly sickle cell disease, but also thalassemia, hemoglobin HbC and HbE.[76] Malaria is one of the biggest global health problems caused by the *Plasmodium*

---

helix: gene, organism, and environment (Cambridge, Mass.: Harvard University Press, 2000); M. Pigliucci, "Do we need an extended evolutionary synthesis?," *Evolution*, Vol. 61 (2007), pp. 2743–2749; R. S. Singh, "Darwin's legacy II: why biology is not physics, or why it has taken a century to see the dependence of genes on the environment," *Genome*, Vol. 58, No. 1 (2015), pp. 55–62; M. K. Skinner, "Environmental Epigenetics and a Unified Theory of the Molecular Aspects of Evolution: A Neo-Lamarckian Concept that Facilitates Neo-Darwinian Evolution,"*Genome Biology and Evolution*, Vol. 7, No. 5 (2015), pp. 1296–1302. R. C. Lewontin describes this step in an interesting way: "Darwin's alienation of the outside from the inside was an absolutely essential step in the development of modern biology (…) But the conditions that are necessary for progress at one stage in history become bars to further progress at another. The time has come when further progress in our understanding of nature requires that we reconsider the relationship between the outside and the inside, between organism and environment." (R. C. Lewontin,, *The triple helix*)

76 Sickle cell disease is caused by a mutation of the β-globin gene, resulting in the formation of hemoglobin HbS. In the homozygous condition it causes sickle cell anemia which is often lethal, and is called a sickle cell trait in the heterozygous condition (milder symptoms). The global number of children born every year with HbS has been estimated to be 275,000 and will probably reach the number of 400,000 by 2050 according to some projections. Substantial protection against malaria, apart from hemoglobin-inherited disorders, is observed also for erythrocyte polymorphisms (ovalocytosis and Duffy blood group), some immunogenetic variants (e.g., HLA alleles, complement receptor 1, NOS2), and enzymopathies (G6PD deficiency and PK deficiency). Mechanism of protection against *Plasmodium* parasites has also been described (See H. F. Bunn, "The triumph of good overevil: protection by the sickle gene against malaria," *Blood*, Vol. 121, No. 1 (2013), pp. 20–25; E Elguero, L. M. Délicat-Loembet, V. Rougeron, C. Arnathau, B. Roche, P. Becquart, J. P. Gonzalez, D. Nkoghe, L. Sica, E. M. Leroy, P. Durand, F. J. Ayala, B. Ollomo, F. Renaud, F. Prugnolle, "Malaria continues to select for sickle cell trait in Central Africa,"*Proceedings of the National Academy of Sciences of the United States of America*, Vol. 112, No. 22 (2015), pp. 7051–7054; B. P. Gonçalves, S. Gupta, B. S. Penman, "Sickle haemoglobin, haemoglobin C and malaria mortality feedbacks," *Malaria Journal*, Vol. 15, No. 1 (2016), p. 26; C. López, C. Saravia, V. Gomez, J. Hoebeke, M. A. Patarroyo, "Mechanisms of genetically-based resistance to malaria," *Gene*, Vol. 467, Nos. 1–2 (2010), pp. 1–12).

falciparum and *Plasmodium vivax* infections, with about 250 million clinical episodes every year (more than 1 million of those infected die). Despite the fact that sickle cell anemia is a severe disease, the HbS allele maintains a surprisingly high frequency in human populations in areas affected by malaria. The similarity of geographical distribution of both diseases led to the hypothesis that heterozygotes have a selective advantage. It has been proven that malaria is indeed a selective factor in some human populations.[77]

Excellence of various kinds is a goal of transhumanists who want to replace evolutionary processes. This idea has been explicitly expressed by Max More in *A Letter to Mother Nature* in which he declares: "We have decided that it is time to amend the human constitution. We do not do this lightly, carelessly, or disrespectfully, but cautiously, intelligently, and in pursuit of excellence. (…) Rather than seeking a state of final perfection, we will continue to pursue new forms of excellence according to our own values, and as technology allows".[78] More proposes seven amendments to the human constitution, one of which concerns human genetics. In the light of our previous analyses, More's declaration seems to be extremely optimistic (or even fantastic). In the fifth amendment he declares:

> We will no longer be slaves to our genes. We will take charge over our genetic programming and achieve mastery over our biological, and neurological processes. We will fix all individual and species defects left over from evolution by natural selection. Not content with that, we will seek complete choice of our bodily form and function, refining and augmenting our physical and intellectual abilities beyond those of any human in history.[79]

6 Closing remarks

The concept of excellence can act as a "regulatory idea" and give us the motivation necessary to promote our personal development. It can also be applied to groups of people, different human populations, or the whole species. The concept, however, is ambiguous and may remain in conflict with our notions of equality and justice. We must be aware that it may be extremely difficult or even impossible to simultaneously execute both excellence and equality. We should also choose

---

77 See E. Elguero et al., "Malaria continues to select for sickle cell trait in Central Africa"; C. López et al., "Mechanisms of genetically-based resistance to malaria.".
78 See M. More, "A Letter to Mother Nature," in: *The Transhumanist Reader. Classical and Contemporary Essays on the Science, Technology, and Philosophy of the Human Future*; ed. More M. & Vita-More N. (Chichester: Wiley-Blackwell, 2013), pp. 456–457.
79 More, "A Letter to Mother Nature," pp. 456–457.

between the norm-referenced and the criterion-referenced notions, but the definition is more ambiguous if applied to phenotypic traits. Despite the staggering developments in molecular biology and genetics, in recent decades we still do not comprehend the complexity of our phenotypic variation and evolutionary processes. Moreover, it is probable that there are whole realms of knowledge that are now unavailable to us, which will be revealed by future discoveries. We experienced such shocks many times in the past, with the most notable examples of the importance of DNA, previously believed to be "non-coding" (also called "junk DNA"), and epigenetics. Both phenomena, which initially seemed to be unimportant or marginal, turned out to be the paradigm-shifting "tip of an iceberg". Thus, we should be aware of the lack of our knowledge about complex traits and evolutionary processes and be very cautious with our aspirations. We must also remember that biological diversity is essential for populations as it allows adaptation to the changing environment and often ensures their survival. Erika Check Hayden pointed to it in a humorous way: "Life is complicated. The more biologists look, the more complexity there seems to be".[80] This does not mean we should not improve ourselves or fight diseases even with our limited knowledge. It seems reasonable, however, not to correct "Mother Nature" too much before we sufficiently understand her; otherwise we might be "throwing the baby out with the bath water".

References

Aksoy-Aksel, A., F. Zampa and G. Schratt. "MicroRNAs and synaptic plasticity–a mutual relationship." *Philosophical Transactions of the Royal Society of London. Series B, Biological Sciences*, Vol. 369, No. 1652, 2014, p. 20130515, doi: 10.1098/rstb.2013.0515.

Alexander, R. P., G. Fang, J. Rozowsky, M. Snyder and M. B. Gerstein. "Annotating non-coding regions of the genome." *Nature Reviews. Genetics*, Vol. 11, No. 8, 2010, pp. 559–571.

Amundson, R. "Against normal function." *Studies in History and Philosophy of Biological and Biomedical Sciences*, Vol. 31, 2000, pp. 33–53.

Berger, S., P. Pérez-Rodríguez, Y. Veturi, H. Simianer and G. de los Campos. "Effectiveness of shrinkage and variable selection methods for the prediction of complex human traits using data from distantly related individuals." *Annals of Human Genetics*, Vol. 79, No. 2, 2015, pp. 122–135.

---

80 See E. Check Hayden, "Human genome at ten: Life is complicated," *Nature* Vol. 464, No. 7289 (2010), pp. 664–667.

Bersaglieri, T., P. C. Sabeti, N. Patterson, T. Vanderploeg, S. F. Schaffner, J. A. Drake, M. Rhodes, D. E. Reich and J. N. Hirschhorn. "Genetic signatures of strong recent positive selection at the lactase gene." *American Journal of Human Genetics*, Vol. 74, No. 6, 2004, pp. 1111–1120.

Blaxter, M. "Genetics. Revealing the dark matter of the genome."*Science*, Vol. 330, No. 6012, 2010, pp. 1758–1759.

Boorse, C. "On the distinction between disease and illness." *Philosophy & Public Affairs*, Vol. 5, No. 1, 1975, pp. 49–68.

Boorse, C. "Health as a theoretical concept." *Philosophy of Science*, Vol. 44, 1977, pp. 542–573.

Boorse, C. "A rebuttal on health." in: *What is disease?*, eds. J. Humber and R. Almeder. Totowa, New Jersey: Humana Press, 1997, pp. 1–134.

Botting, F. "Culture and excellence."*Cultural Values*, Vol. 1, No. 2, 1997, p. 139.

Brusoni, M., R. Damian, J. R. Sauri, S. Jackson, H. Komurcugil, M. Malmedy, O. Matveeva, G. Motova, S. Pisarz, P. Pol, A. Rostlund, E. Soboleva, O. Tavares and L. Zobel, *The Concept of Excellence in Higher Education*, European Association for Quality Assurance in Higher Education AISBL, Brussels, Belgium, 2014; http://www.enqa.eu/index.php/publications/papers-reports/occasional-papers/

Bunn, H. F. "The triumph of good over evil: protection by the sickle gene against malaria." *Blood*, Vol. 121, No. 1, 2013, pp. 20–25.

Burggren, W. W. "Epigenetics as a source of variation in comparative animal physiology – or – Lamarck is lookin' pretty good these days." *The Journal of Experimental Biology*, Vol. 217, No. Pt 5, 2014, pp. 682–689.

Burggren, W. W. and D. Crews. "Epigenetics in comparative biology: why we should pay attention." *Integrative and Comparative Biology*, Vol. 54, No.1, 2014, pp. 7–20.

Cacabelos, R. and C. Torrellas. "Epigenetics of aging and Alzheimer's disease: implications for pharmacogenomics and drug response." *International Journal of Molecular Sciences*, Vol. 16, No. 12, 2015, pp. 30483–30543.

Campbell, A. R. "Building a culture of excellence from the ground up." *Radiology Management*, Vol. 35, No. 3, 2013, pp. 14–18; quiz 19–20.

Cargill, M., D. Altshuler, J. Ireland, P. Sklar, K. Ardlie, N. Patil, N. Shaw, C. R. Lane, E. P. Lim, N. Kalyanaraman, J. Nemesh, L. Ziaugra, L. Friedland, A. Rolfe, J. Warrington, R. Lipshutz, G. Q. Daley and E. S. Lander. "Characterization of single-nucleotide polymorphisms in coding regions of human genes." *Nature Genetics*, Vol. 22, No. 3, 1999, pp. 231–238. Erratum in: *Nature Genetics*, Vol. 23, No. 3, 1999, p. 373.

Carter, N. P. "As normal as normal can be?." *Nature Genetics*, Vol. 36, No. 9, 2004, pp. 931–932.

Chamary, J. V., J. R. Parmley and L. D. Hurst. "Hearing silence: non-neutral evolution at synonymous sites in mammals." *Nature Reviews. Genetics*, Vol. 7, 2006, pp. 98–108.

Check Hayden, E. "Human genome at ten: life is complicated." *Nature*, Vol. 464, No. 7289, 2010, pp. 664–667.

Clarke, G. M. and D.Cooke. *A basic course in statistics*. Third edition. London, Melbourne and Auckland: Edward Arnold, A division of Hodder & Stoughton Ltd., 1992.

Clarke, A. J. and D. N. Cooper. "GWAS: heritability missing in action?." *European Journal of Human Genetics*, Vol. 18, 2010, pp. 859–861.

Conrad, D. F., D. Pinto, R. Redon, L. Feuk, O. Gokcumen, Y. Zhang, J. Aerts, T. D. Andrews, C. Barnes, P. Campbell, T. Fitzgerald, M. Hu, C. H. Ihm, K. Kristiansson, D. G. Macarthur, J. R. Macdonald, I. Onyiah, A. W. Pang, S. Robson, K. Stirrups, A. Valsesia, K. Walter, J. Wei, Wellcome Trust Case Control Consortium, C. Tyler-Smith, N. P. Carter, C. Lee, S. W. Scherer and M. E. Hurles. "Origins and functional impact of copy number variation in the human genome." *Nature*, Vol. 464, No. 7289, 2010, pp. 704–712.

Cooper, R. "Disease." *Studies in History and Philosophy of Biological and Biomedical Sciences*, Vol. 33, 2002, pp. 263–282.

Coulon, L., M. Mok, K. L. Krause and M. Anderson. "The pursuit of excellence in nursing care: what does it mean?." *Journal of Advanced Nursing*, Vol. 42, No. 4, 1996, pp. 817–826.

Dana, A. and T. Tuller. "Mean of the typical decoding rates: a new translation efficiency index based on the analysis of ribosome profiling data." *G3: Genes – Genomes – Genetics*, Vol. 5, No. 1, 2014, pp. 73–80.

Dawson, M. A. and T.Kouzarides. "Cancer epigenetics: from mechanism to therapy." *Cell*, Vol. 150, No. 1, 2012, pp. 12–27.

de los Campos, G., A. I. Vazquez, R. Fernando, Y. C. Klimentidis and D. Sorensen. "Prediction of complex human traits using the genomic best linear unbiased predictor." *PLoS Genetics*, Vol. 9, No. 7, 2013, p. e1003608, doi:10.1371/journal.pgen.1003608.

De Moor, M. H., T. D. Spector, L. F. Cherkas, M. Falchi, J. J. Hottenga, D. I. Boomsma and E. J. De Geus. "Genome-wide linkage scan for athlete status in 700 British female DZ twin pairs." *Twin Research and Human Genetics*, Vol. 10, No. 6, 2007, pp. 812–820.

Douglass, J. A. "PROFILING THE FLAGSHIP UNIVERSITY MODEL: An exploratory proposal for changing the paradigm from ranking to relevancy."

ROPS.CSHE.5.14 (Research and Occasional Papers Series, Center for Studies in Higher Education, University of California, Berkeley), 2014:http://www.cshe.berkeley.edu/publications/profiling-flagship-university-model-exploratory-proposal-changing-paradigm-ranking.

Drong, A. W., C. M. Lindgren and M. I. McCarthy. "The genetic and epigenetic basis of type 2 diabetes and obesity." *Clinical Pharmacology and Therapeutics*, Vol. 92, No. 6, 2012, pp. 707–715.

Egger, G., G. Liang, A. Aparicio and P. A. Jones. "Epigenetics in human disease and prospects for epigenetic therapy." *Nature*, Vol. 429, No. 6990, 2004, pp. 457–463.

Ehrlich, M. and M. Lacey. "DNA methylation and differentiation: silencing, upregulation and modulation of gene expression." *Epigenomics*, Vol. 5, 2013, pp. 553–568.

Elguero, E., L. M. Délicat-Loembet, V. Rougeron, C. Arnathau, B. Roche, P. Becquart, J. P. Gonzalez, D. Nkoghe, L. Sica, E. M. Leroy, P. Durand, F. J. Ayala, B. Ollomo, F. Renaud, F. Prugnolle. "Malaria continues to select for sickle cell trait in Central Africa." *Proceedings of the National Academy of Sciences of the United States of America*, Vol. 112, No. 22, 2015, pp. 7051–7054.

Enattah, N. S., T. Sahi, E. Savilahti, J. D. Terwilliger, L. Peltonen and I. Järvelä. "Identification of a variant associated with adult-type hypolactasia." *Nature Genetics*, Vol. 30, No. 2, 2002, pp. 233–237.

Enattah, N. S., T. G. Jensen, M. Nielsen, R. Lewinski, M. Kuokkanen, H. Rasinpera, H. El-Shanti, J. K. Seo, M. Alifrangis, I. F. Khalil, A. Natah, A. Ali, S. Natah, D. Comas, S. Q. Mehdi, L. Groop, E. M. Vestergaard, F. Imtiaz, M. S. Rashed, B. Meyer, J. Troelsen and L. Peltonen. "Independent introduction of two lactase-persistence alleles into human populations reflects different history of adaptation to milk culture." *American Journal of Human Genetics*, Vol. 82, No. 1, 2008, pp. 57–72.

Engelhardt, T. "Ideology and etiology." *Journal of Medicine and Philosophy*, Vol. 1, 1976, pp. 256–268.

English, A. C., W. J. Salerno, O. A. Hampton, C. Gonzaga-Jauregui, S. Ambreth, D. I. Ritter, C. R. Beck, C. F. Davis, M. Dahdouli, S. Ma, A. Carroll, N. Veeraraghavan, J. Bruestle, B. Drees, A. Hastie, E. T. Lam, S. White, P. Mishra, M. Wang, Y. Han, F. Zhang, P. Stankiewicz, D. A. Wheeler, J. G. Reid, D. M. Muzny, J. Rogers, A. Sabo, K. C. Worley, J. R. Lupski, E. Boerwinkle and R. A. Gibbs. "Assessing structural variation in a personal genome-towards a human reference diploid genome." *BMC Genomics*, Vol. 16, 2015, p. 286.

Ereshefsky, M. "Defining 'health' and 'disease'." *Studies in History and Philosophy of Science. Part C. Studies in History and Philosophy of Biological and Biomedical Sciences*, Vol. 40, 2009, pp. 221–227.

Eynon, N., J. R. Ruiz, J. Oliveira, J. A. Duarte, R. Birk and A. Lucia. "Genes and elite athletes: a roadmap for future research." *The Journal of Physiology*, Vol. 589, No. 13, 2011, pp. 3063–3070.

Eynon, N., E. D. Hanson, A. Lucia, P. J. Houweling, F. Garton, K. N. North and D. J. Bishop. "Genes for elite power and sprint performance: ACTN3 leads the way." *Sports Medicine*, Vol. 43, No. 9, 2013, pp. 803–817.

Fagiolini, M., C. L. Jensen and F. A. Champagne. "Epigenetic influences on brain development and plasticity." *Current Opinion in Neurobiology*, Vol. 19, No. 2, 2009, pp. 207–212.

Feuk, L., A. R. Carson and S. W. Scherer. "Structural variation in the human genome." *Nature Reviews. Genetics*, Vol. 7, No. 2, 2006, pp. 85–97.

Fisher, S. E. "Tangled webs: tracing the connections between genes and cognition." *Cognition*, Vol. 101, No. 2, 2006, pp. 270–297.

Fraga, M. F., E. Ballesta, M. F. Paz, S. Ropero, F. Setien, M. L. Ballestar, D. Heine-Suñer, J. C. Cigudosa, M. Urioste, J. Benitez, M. Boix-Chornet, A. Sanchez-Aguilera, C. Ling, E. Carlsson, P. Poulsen, A. Vaag, Z. Stephan, T. D. Spector, Y. Z. Wu, C. Plass and M. Esteller. "Epigenetic differences arise during the lifetime of monozygotic twins." *Proceedings of the National Academy of Sciences of the United States of America*, Vol. 102, No. 30, 2005, pp. 10604–10609.

Fung, K. L. and M. M. Gottesman. "A synonymous polymorphism in a common MDR1 (ABCB1) haplotype shapes protein function." *Biochimica et Biophysica Acta*, Vol. 1794, 2009, pp. 860–871.

Gamazon, E. R. and B. E. Stranger. "The impact of human copy number variation on gene expression." *Briefings in Functional Genomics*, Vol. 14, No. 5, 2015, pp. 352–357.

Gerstein, M. B., C. Bruce, J.S. Rozowsky, D. Zheng, J. Du, J. O. Korbel, O. Emanuelsson, Z. D. Zhang, S. Weissman and M. Snyder, "What is a gene, post-ENCODE? History and updated definition." *Genome Research*, Vol. 17, No. 6, 2007, pp. 669–681.

Ginevičienė, V., A. Jakaitienė, A. Pranculis, K.Milašius, L. Tubelis and A. Utkus. "AMPD1 rs17602729 is associated with physical performance of sprint and power in elite Lithuanian athletes." *BMC Genetics*, Vol. 15, 2014, p. 58.

Glinsky, G. V. "Phenotype-defining functions of multiple non-coding RNA pathways." *Cell Cycle*, Vol. 7, No. 11, 2008, pp. 1630–1639.

Gonçalves, B. P., S. Gupta and B. S. Penman. "Sickle haemoglobin, haemoglobin C and malaria mortality feedbacks." *Malaria Journal*, Vol. 15, No. 1, 2016, p. 26, doi: 10.1186/s12936-015-1077-5.

Goosens, W. "Values, health, and medicine." *Philosophy of Science*, Vol. 47, 1980, pp. 100–115.

Gudbjartsson, D. F., G. B. Walters, G. Thorleifsson, H. Stefansson, B. V. Halldorsson, P. Zusmanovich, P. Sulem, S. Thorlacius, A. Gylfason, S. Steinberg, A. Helgadottir, A. Ingason, V. Steinthorsdottir, E. J. Olafsdottir, G. H. Olafsdottir, T. Jonsson, K. Borch-Johnsen, T. Hansen, G. Andersen, T. Jorgensen, O. Pedersen, K. K. Aben, J. A. Witjes, D. W. Swinkels, M. den Heijer, B. Franke, A. L. Verbeek, D. M. Becker, L. R. Yanek, L. C. Becker, L. Tryggvadottir, T. Rafnar, J. Gulcher, L. A. Kiemeney, A. Kong, U. Thorsteinsdottir and K. Stefansson. "Many sequence variants affecting diversity of adult human height." *Nature Genetics*, Vol. 40, 2008, pp. 609–615.

Guerrero, J. D. "On a naturalist theory of health: a critique." *Studies in History and Philosophy of Science. Part C. Studies in History and Philosophy of Biological and Biomedical Sciences*, Vol. 41, 2010, pp. 272–278.

Haraksingh, R. R. and M. P. Snyder. "Impacts of variation in the human genome on gene regulation." *Journal of Molecular Biology*, Vol. 425, No. 21, 2013, pp. 3970–3977.

Harris, E. E. and D. Meyer, The molecular signature of selection underlying human adaptations, *American Journal of Physical Anthropology*, Vol. Suppl 43, 2006, pp. 89–130.

Hickey, P. A. "Building a culture of excellence in Boston and beyond." *World Journal for Pediatric & Congenital Heart Surgery*, Vol. 1, No. 3, 2010, pp. 314–320.

Hindorff, L. A., P. Sethupathy, H. A. Junkins, E. M. Ramos, J. P. Mehta, F. S. Collins and T. A. Manolio. "Potential etiologic and functional implications of genome-wide association loci for human diseases and traits." *Proceedings of the National Academy of Sciences of the United States of America*, Vol. 106, No. 23, 2009, pp. 9362–9367.

Iafrate, A. J., L. Feuk, M. N. Rivera, M. L. Listewnik, P. K. Donahoe, Y. Qi, S. W. Scherer and C Lee. "Detection of large-scale variation in the human genome." *Nature Genetics*, Vol. 36, No. 9, 2004, pp. 949–951.

Jablonka, E., M. J. Lamb and E. Avital, "Lamarckian mechanisms in darwinian evolution." *Trends in Ecology & Evolution*, Vol. 13, 1998, pp. 206–210.

Jia, Y., L. Chen, Y. B. Ma, J. Zhang, N. Xu and D.J. Liao. "To know how a gene works, we need to redefine it first but then, more importantly, to let the cell itself decide how to transcribe and process its RNAs." *International Journal of Biological Sciences*, Vol. 11, No. 12, 2015, pp. 1413-1423.

Kabcenell, A. and K. Luther. "Creating a culture of excellence. It's not as difficult as you might think." *Healthcare Executive*, Vol. 27, No. 4, 2012, pp. 68-71.

Kahrer-Sawatzki, H. "What a difference copy number variation makes." *Bioessays*, Vol. 29, No. 4, 2007, pp. 311-313.

Kingsley, D. M. "From atoms to traits." *Scientific American*, Vol. 300, No. 1, 2009, pp. 52-59

Kudla, G., A. W. Murray, D. Tollervey and J. B. Plotkin. "Coding-sequence determinants of gene expression in *Escherichia coli*." *Science*, Vol. 324, No. 5924, 2009, pp. 255-258.

Kuokkanen, M., J. Kokkonen, N. S. Enattah, T. Ylisaukko-oja, H. Komu, T. Varilo, L. Peltonen, E. Savilahti and I. Jarvela. "Mutations in the translated region of the lactase gene (LCT) underlie congenital lactase deficiency." *American Journal of Human Genetics*, Vol. 78, No. 2, 2006, pp. 339-344.

Kutschera, U. "From the scala naturae to the symbiogenetic and dynamic tree of life." *Biology Direct*, Vol. 6, 2011, p. 33, doi: 10.1186/1745-6150-6-33.

Laland, K., T. Uller, M. Feldman, K. Sterelny, G. B. Müller, A. Moczek, E. Jablonka, J. Odling-Smee, G. A. Wray, H. E. Hoekstra, D. J. Futuyma, R. E. Lenski, T. F. Mackay, D. Schluter and J. E. Strassmann. "Does evolutionary theory need a rethink?" *Nature*, Vol. 514, No. 7521, 2014, pp. 161-164.

Lander, E. S. "Initial impact of the sequencing of the human genome." *Nature*, Vol. 470, No. 7333, 2011, pp. 187-197.

Lander, E. S., L. M. Linton, B. Birren, C. Nusbaum, M. C. Zody, J. Baldwin, K. Devon, K. Dewar, M. Doyle, W. FitzHugh, R. Funke, D. Gage, K. Harris, A. Heaford, J. Howland, L. Kann, J. Lehoczky, R. LeVine, P. McEwan, K. McKernan. et al., International Human Genome Sequencing Consortium. "Initial sequencing and analysis of the human genome." *Nature*, Vol. 409, No. 6822, 2001, pp. 860-921; Erratum in: *Nature*, Vol. 411, No. 6838, 2001, p.720.

Laserre, C. "Fostering a culture of service excellence." *The Journal of Medical Practice Management*, Vol. 26, No. 3, 2010, pp. 166-169.

Lettre, G., A. U. Jackson, C. Gieger, F. R. Schumacher, S. I. Berndt, S. Sanna, S. Eyheramendy, B. F. Voight, J. L. Butler, C. Guiducci, T. Illig, R. Hackett, I. M. Heid, K. B. Jacobs, V. Lyssenko, M. Uda; Diabetes Genetics Initiative; FUSION; KORA; Prostate, Lung Colorectal and Ovarian Cancer Screening Trial; Nurses' Health Study; SardiNIA, M. Boehnke, S. J. Chanock, L. C. Groop, F. B. Hu, B. Isomaa, P. Kraft, L. Peltonen, V. Salomaa, D. Schlessinger, D. J. Hunter, R. B. Hayes, G. R. Abecasis, H. E. Wichmann, K. L. Mohlke and J. N. Hirschhorn. "Identification of ten loci associated with height highlights new biological pathways in human growth." *Nature Genetics*, Vol. 40, 2008, pp. 584–591.

Levy, Y. and R. P. Ebstein. "Research review: crossing syndrome boundaries in the search for brain endophenotypes." *Journal of Child Psychology and Psychiatry, and Allied Disciplines*, Vol. 50, No. 6, 2009, pp. 657–668.

Lewontin, R. C. *The triple helix: gene, organism, and environment.* Cambridge, Massachusetts: Harvard University Press, 2000.

Liu, N. and T. Pan. "RNA epigenetics." *Translational Research: The Journal of Laboratory and Clinical Medicine*, Vol. 165, No. 1, 2015, pp. 28–35.

López, C., C. Saravia, A. Gomez, J. Hoebeke and M. A. Patarroyo. "Mechanisms of genetically-based resistance to malaria." *Gene*, Vol. 467, No. 1–2, 2010, pp. 1–12.

MacNamara, A., D. Collins. "Development and initial validation of the psychological characteristics of developing excellence questionnaire." *Journal of Sports Sciences*, Vol. 29, No. 12, 2011, pp. 1273–1286.

MacNamara, A. and D. Collins. "Do mental skills make champions? Examining the discriminant function of the psychological characteristics of developing excellence questionnaire." *Journal of Sports Sciences*, Vol. 31, No. 7, 2013, pp. 736–744.

MacNamara A., P. Holmes and D. Collins. "Negotiating transitions in musical development: the role of psychological characteristics of developing excellence." *Psychology of Music*, Vol. 36, No. 3, 2008, pp. 335–352.

Maher, B. "Personal genomes: The case of the missing heritability." *Nature*, Vol. 456, No. 7218, 2008, pp. 18–21.

Mahmood, S., W. Ahmad and M. J. Hassan. "Autosomal recessive primary Microcephaly (MCPH): clinical manifestations, genetic heterogeneity and mutation continuum." *Orphanet Journal of Rare Diseases*, Vol. 6, 2011, p. 39, doi: 10.1186/1750-1172-6-39.

Makowsky, R., N. M. Pajewski, Y. C. Klimentidis, A. I. Vazquez, C. W. Duarte, D. B. Allison and G. de los Campos. "Beyond missing heritability: prediction of complex traits." *PLoS Genetics*, Vol. 7, 2011, p. e1002051.

Manolio, T. A., F. S. Collins, N. J. Cox, D. B. Goldstein, L. A. Hindorff, D. J. Hunter, M. I. McCarthy, E. M. Ramos, L. R. Cardon, A. Chakravarti, J. H. Cho, A. E. Guttmacher, A. Kong, L. Kruglyak, E. Mardis, C. N. Rotimi, M. Slatkin, D. Valle, A. S. Whittemore, M. Boehnke, A. G. Clark, E. E. Eichler, G. Gibson, J. L. Haines, T. F. Mackay, S. A. McCarroll and P. M. Visscher. "Finding the missing heritability of complex diseases." *Nature*, Vol. 461, 2009, pp. 747–753.

Marcus, G. F. and S. E. Fisher. "FOXP2 in focus: what can genes tell us about speech and language?." *Trends in Cognitive Science*, Vol. 7, No. 6, 2003, pp. 257–262.

Margolis, J. "The concept of disease." *The Journal of Medicine and Philosophy*, Vol. 1, 1976, pp. 238–255.

Marian, A. J. "Molecular genetic studies of complex phenotypes." *Translational Research: The Journal of Laboratory and Clinical Medicine*, Vol. 159, No. 2, 2012, pp. 64–79.

More, M. A Letter to Mother Nature, in: *The Transhumanist Reader. Classical and Contemporary Essays on the Science, Technology, and Philosophy of the Human Future*, eds. M. More and N. Vita-More. Chichester: Wiley-Blackwell, 2013, pp. 456–457.

Park, J. H., S. Wacholder, M. H. Gail, U. Peters, K. B. Jacobs, S. J. Chanock and N. Chatterjee. "Estimation of effect size distribution from genome-wide association studies and implications for future discoveries." *Nature Genetics*, Vol. 42, 2010, pp. 570–575.

Pearson, H. "Genetics: what is a gene?." *Nature*, Vol. 441, No. 7092, 2006, pp. 398–401.

Pennisi, E. "Genomics. ENCODE project writes eulogy for junk DNA." *Science*, Vol. 337, No. 6099, 2012, pp. 1159–1161.

Perry, G. H., N. J. Dominy, K. G. Claw, A. S. Lee, H. Fiegler, R. Redon, J. Werner, F. A. Villanea, J. L. Mountain, R. Misra, N. P. Carter, C. Lee and A. C. Stone. "Diet and the evolution of human amylase gene copy number variation." *Nature Genetics*, Vol. 39, No. 10, 2007, pp. 1256–1260.

Pigliucci, M. "Do we need an extended evolutionary synthesis?." *Evolution*, Vol. 61, 2007, pp. 2743–2749.

Przeworski, M., R. R. Hudson and A. Di Rienzo. "Adjusting the focus on human variation." *Trends in Genetics*, Vol. 16, No. 7, 2000, pp. 296–302.

Qureshi, I. A. and M. F. Mehler. "An evolving view of epigenetic complexity in the brain." *Philosophical Transactions of the Royal Society of London. Series B, Biological Sciences*, Vol. 369, No. 1652, 2014, p. 20130506, doi: 10.1098/rstb.2013.0506.

Rao, M. R. "International centers of excellence for Malaria research." *The American Journal of Tropical Medicine and Hygiene*, Vol. 93, No. 3 Suppl, 2015, pp. 1–4.

Redon, R., S. Ishikawa, K. R. Fitch, L. Feuk, G. H. Perry, T. D. Andrews, H. Fiegler, M. H. Shapero, A. R. Carson, W. Chen, E. K.Cho, S. Dallaire, J. L. Freeman, J. R. González, M. Gratacòs, J. Huang, D. Kalaitzopoulos, D. Komura, J. R. MacDonald, C. R. Marshall, R. Mei, L. Montgomery, K.

Nishimura, K. Okamura, F. Shen, M. J. Somerville, J. Tchinda, A. Valsesia, C. Woodwark, F. Yang, J. Zhang, T. Zerjal, J. Zhang, L. Armengol, D. F. Conrad, X. Estivill, C. Tyler-Smith, N. P. Carter, H. Aburatani, C. Lee, K. W. Jones, S. W. Schere and M. E. Hurles. "Global variation in copy number in the human genome." *Nature*, Vol. 444, No. 7118, 2006, pp. 444–454.

Reich, D. E., S. F. Schaffner, M. J. Daly, G. McVean, J. C. Mullikin, J. M. Higgins, D. J. Richter, E. S. Lander and D. Altshuler. "Human genome sequence variation and the influence of gene history, mutation and recombination." *Nature Genetics*, Vol. 32, No. 1, 2002, pp. 135–142.

Reznek, L. *The nature of disease.* London: Routledge & Kegan Paul, 1987.

Rivera, C. M. and B. Ren. "Mapping human epigenomes." *Cell*, Vol. 155, 2013, pp. 39–55.

Roberts, E., D. J. Hampshire, L. Pattison, K. Springell, H. Jafri, P. Corry, J. Mannon, Y. Rashid, Y. Crow, J. Bond and C. G. Woods. "Autosomal recessive primary microcephaly: an analysis of locus heterogeneity and phenotypic variation." *Journal of Medical Genetics*, Vol. 39, 2002, pp. 718–721.

Roberts, T. C., K. V. Morris and M. J. Wood. "The role of long non-coding RNAs in neurodevelopment, brain function and neurological disease." *Philosophical Transactions of the Royal Society of London. Series B, Biological Sciences*, Vol. 369, No. 1652, 2014, p. 20130507, doi: 10.1098/rstb.2013.0507.

Rose, N. R. and R. J. Klose. "Understanding the relationship between DNA methylation and histone lysine methylation." *Biochimica et Biophysica Acta*, Vol. 1839, No. 12, 2014, pp. 1362–1372.

Rutter, M., T. E. Moffitt and A. Caspi. "Gene-environment interplay and psychopathology: multiple varieties but real effects." *Journal of Child Psychology and Psychiatry, and Allied Disciplines*, Vol. 47, No. 3-4, 2006, pp. 226–261.

Sauna, Z. E. and C. Kimchi-Sarfaty. "Understanding the contribution of synonymous mutations to human disease." *Nature Reviews. Genetics*, Vol. 12, No. 10, 2011, pp. 683–691.

Sawczuk, M., L. K. Banting, P. Cięszczyk, A. Maciejewska-Karłowska, A. Zarębska, A. Leońska-Duniec, Z. Jastrzębski, D. J. Bishop and N. Eynon. "MCT1 A1470T: a novel polymorphism for sprint performance?." *Journal of Science and Medicine in Sport*, Vol. 8, No. 1, 2015, pp. 114–118.

Scadding, J. "The semantic problem of psychiatry." *Psychological Medicine*, Vol. 20, 1990, pp. 243–248.

Schwartz, P. H. "Defining dysfunction: natural selection, design, and drawing a line." *Philosophy of Science*, Vol. 74, 2007, pp. 364–385.

Sebat, J., B. Lakshmi, J. Troge, J. Alexander, J. Young, P. Lundin, S. Månér, H. Massa, M. Walker, M. Chi, N. Navin, R. Lucito, J. Healy, J. Hicks, K. Ye, A. Reiner, T. C. Gilliam, B. Trask, N. Patterson, A. Zetterberg and M. Wigler. "Large-scale copy number polymorphism in the human genome." *Science*, Vol. 305, No. 5683, 2004, pp. 525–528.

Sherry, S. T., M. H. Ward, M. Kholodov, J. Baker, L. Phan, E. M. Smigielski and K. Sirotkin. "dbSNP: the NCBI database of genetic variation." *Nucleic Acid Research*, Vol. 29, 2001, pp. 308–311.

Singh, R. S. "Darwin's legacy: why biology is not physics, or why evolution has not become a common sense." *Genome*, Vol. 54, No. 10, 2011, pp. 868–873.

Singh, R. S. "Darwin's legacy II: why biology is not physics, or why it has taken a century to see the dependence of genes on the environment." *Genome*, Vol. 58, No. 1, 2015, pp. 55–62.

Skinner, M. K. "Environmental epigenetics and a unified theory of the molecular aspects of evolution: a neo-lamarckian concept that facilitates neo-darwinian evolution." *Genome Biology and Evolution*, Vol. 7, No. 5, 2015, pp. 1296–1302.

Smith, W. J. and C. Lusthaus. "The nexus of equality and quality in education: a framework for debate." *Canadian Journal of Education*, Vol. 20, No. 3, 1995 pp. 378–391.

Strachan, T. and A. P. Read, *Human Molecular Genetics*, Second edition. Oxford: BIOS Scientific Publishers Ltd., 1999.

Strike, K. A. "Is there a conflict between equity and excellence?." *Educational Evaluation and Policy Analysis*, Vol. 7, No. 4, 1985, pp. 409–416.

Supek, F., B. Miñana, J. Valcárcel, T. Gabaldón and B. Lehner. "Synonymous mutations frequently act as driver mutations in human cancers." *Cell*, Vol. 156, No. 6, 2014, pp. 1324–1335.

Trerotola, M., V. Relli, P. Simeone and S. Alberti. "Epigenetic inheritance and the missing heritability." *Human Genomics*, Vol. 9, 2015, p. 17, doi: 10.1186/s40246-015-0041-3.

Tropello, P. G. "Magnet status as a competitive strategy of hospital organizations: marketing a culture of excellence in nursing services." *Journal of Hospital Marketing & Public Relations*, Vol. 14, No. 2, 2003, pp. 53–57.

Tsai, Ch.-J., Z. E. Sauna, Ch. Kimchi-Sarfaty, S. V. Ambudkar, M. M. Gottesman and R. Nussinov. "Synonymous mutations and ribosome stalling can lead to altered folding pathways and distinct minima." *Journal of Molecular Biology*, Vol. 383 2008, pp. 281–291.

Usher, C. L. and S. A. McCarroll. "Complex and multi-allelic copy number variation in human disease." *Briefings in Functional Genomics*, Vol. 14, No. 5, 2015, pp. 329–338.

Venter, J. C., M. D. Adams, E. W. Myers, P. W. Li, R. J. Mural, G. G. Sutton, H. O. Smith, M. Yandell, C. A. Evans, R. A. Holt, J. D. Gocayne, P. Amanatides, R. M. Ballew, D. H. Huson, J. R. Wortman, Q. Zhang, C. D. Kodira, X. H. Zheng, L. Chen, M. Skupski et al. "The sequence of the human genome." *Science*, Vol. 291, No. 5507, 2001, pp. 1304–1351. Erratum in: *Science*, Vol. 292, No. 5523, 2001, p. 1838.

Voisin, S., P. Cieszczyk, V. P. Pushkarev, D. A. Dyatlov, B. F. Vashlyayev, V. A. Shumaylov, A. Maciejewska-Karlowska, M. Sawczuk, L. Skuza, Z. Jastrzebski, D. J. Bishop and N. Eynon. "EPAS1 gene variants are associated with sprint/power athletic performance in two cohorts of European athletes." *BMC Genomics*, Vol. 5, 2014, p. 382.

Wachbroit, R. "Normality as a biological concept." *Philosophy of Science*, Vol. 61, 1994, pp. 579–591.

Wakefield, J. "The concept of mental disorder: on the boundary between biological facts and social values." *American Psychologist*, Vol. 47, 1992, pp. 373–388.

Webborn, N., A. Williams, M. McNamee, C. Bouchard, Y. Pitsiladis, I. Ahmetov, E. Ashley, N. Byrne, S. Camporesi, M. Collins, P. Dijkstra, N. Eynon, N. Fuku, F. C. Garton, N. Hoppe, S. Holm, J. Kaye, V. Klissouras, A. Lucia, K. Maase, C. Moran, K. N. North, F. Pigozzi and G. Wang. "Direct-to-consumer genetic testing for predicting sports performance and talent identification: consensus statement." *British Journal of Sports Medicine*, Vol. 49, No. 23, 2015, pp. 1486–1491.

Weedon, M. N., H. Lango, C. M. Lindgren, C. Wallace, D. M. Evans, M. Mangino, R. M. Freathy, J. R. Perry, S. Stevens, A. S. Hall, N. J. Samani, B. Shields, I. Prokopenko, M. Farrall, A. Dominiczak; Diabetes Genetics Initiative; Wellcome Trust Case Control Consortium, T. Johnson, S. Bergmann, J. S. Beckmann, P. Vollenweider, D. M. Waterworth, V. Mooser, C. N. Palmer, A. D. Morris, W. H. Ouwehand; Cambridge GEM Consortium, J. H. Zhao, S. Li, R. J. Loos, I. Barroso, P. Deloukas, M. S. Sandhu, E. Wheeler, N. Soranzo, M. Inouye, N. J. Wareham, M. Caulfield, P. B. Munroe, A. T. Hattersley, M. I. McCarthy and T. M. Frayling. "Genome-wide association analysis identifies 20 loci that influence adult height." *Nature Genetics*, Vol. 40, 2008, pp. 575–583.

Welter, D., J. MacArthur, J. Morales, T. Burdett, P. Hall, H. Junkins, A. Klemm, P. Flicek, T. Manolio, L. Hindorff and H. Parkinson. "The NHGRI GWAS catalog, a curated resource of SNP-trait associations." *Nucleic Acids Research*, Vol. 42 (Database issue), 2014, p. D1001–D1006.

West-Eberhard, M. J. "Are genes good markers of biological traits?" in: *Biosocial Surveys*, National Research Council (US Committee on Advances in Collecting and Utilizing Biological Indicators and Genetic Information in Social Science Surveys), ed. M. Weinstein, J. W. Vaupel and K. W. Wachter. Washington, DC: The National Academies Press, 2008, pp. 175–193.

Williams, A.G. and J. P. Folland. "Similarity of polygenic profiles limits the potential for elite human physical performance." *The Journal of Physiology*, Vol. 586, No. 1, 2008, pp. 113–121.

Woods, C. G., J. Bond and W. Enard. "Autosomal recessive primary microcephaly (MCPH): a review of clinical, molecular, and evolutionary findings." *American Journal of Human Genetics*, Vol. 76, 2005, pp. 717–728.

Yang, J., B. Benyamin, B. P. McEvoy, S. Gordon, A. K. Henders, D. R. Nyholt, P. A. Madden, A. C. Heath, N. G. Martin, G. W. Montgomery, M. E. Goddard and P. M. Visscher. "Common SNPs explain a large proportion of the heritability for human height." *Nature Genetics*, Vol. 42, 2010, pp. 565–569.

Zhang, F. and J. R. Lupski "Non-coding genetic variants in human disease." *Human Molecular Genetics*, Vol. 24, No. R1, 2015, pp. 102–110.

Zheng, G., J. A. Dahl, Y. Niu, Y. Fu, A. Klungland, Y. G. Yang and C. He. "Sprouts of RNA epigenetics." *RNA Biology*, Vol. 10, No. 6, 2013, pp. 915–918.

Zheng, S., H. Kim and R. G.Verhaak. "Silent mutations make some noise." *Cell*, Vol. 156, No. 6, 2014, pp. 1129–1131.

Zuk, O., E. Hechter, S. R. Sunyaev and E. S. Lander. "The mystery of missing heritability: Genetic interactions create phantom heritability." *Proceedings of the National Academy of Sciences of the United States of America*, Vol. 109, 2012, pp. 1193–1198.

Robert Poczobut

Transhumanism and Cognitive Science[1]

Abstract: Works endorsing transhumanist ideas often appeal to various scientific disciplines, including cognitive science. The key transhumanist concept, being the concept of cognitive enhancement, is closely linked to the research conducted in such fields of cognitive science as artificial intelligence, cognitive robotics, and the evolution of cognitive systems. This provokes the following questions: Do the research results of cognitive science really support the transhumanist vision of the future of mankind? What methodological differences and similarities are there between transhumanism and cognitive science? Is the conception of mind (cognition, intelligence) emerging from contemporary research in cognitive science compatible with the notion of the mind adopted by transhumanists? The present paper provides answers to these questions. In the first part of the text, I present the main transhumanist ideas associated with research in cognitive science. In the second part, I discuss the main currents (paradigms) of research in cognitive science, focusing on the problems concerning transhumanism.

Keywords: Cognitive science, Transhumanism, Mind, Cognition, Cognitive enhancement, Computational theory of mind

1 Introduction

Works advocating transhumanism appeal to various scientific disciplines, including cognitive science. One of the key notions of transhumanism – the notion of cognitive enhancement – is closely associated with research conducted in several fields of cognitive science, such as artificial intelligence, cognitive robotics, and the evolution of cognitive systems.[2] This raises a number of questions: Do research results in cognitive science really support the transhumanist vision of

---

1   This article is a revised version of "Transhumanizm a kognitywistyka" [in Polish], *Ethos*, Vol. 28, No. 3, Issue 11 (2015), pp. 233–251.
2   Nick Bostrom and Anders Sandberg give the following definition of cognitive enhancement: "*Cognitive enhancement* may be defined as the amplification or extension of core capacities of the mind through improvement or augmentation of internal or external information processing systems" (N. Bostrom, A. Sandberg, "Cognitive Enhancement: Method, Ethics, Regulatory Challenges", *Science and Engineering Ethics*, Vol. 15 (2009), p. 311). They believe that the greatest progress, in terms of our cognitive capacities, has been achieved through the implementation of new computational and information technologies.

the future of mankind? What are the methodological differences and similarities between transhumanism and cognitive science? Is the account of the mind (cognition) provided by cognitive science compatible with that advocated by transhumanists?

The paper addresses these questions in the next two sections. In the following section, I describe the main transhumanist ideas associated with research conducted in cognitive science; in the subsequent one, I discuss the most important research programs in cognitive science, focusing on problems relevant to transhumanism, and formulate conclusions about the relation between transhumanism and cognitive science. There are, of course, immensely important ethical, social and legal questions connected with the transhumanist vision of the future of mankind and the place of mind in the physical world, but they lie beyond the scope of this paper and will not be discussed herein.

2 Transhumanism – Science, philosophy, a worldview, futurology, science fiction, a naturalized religion, or utopia?

According to the manifesto proclaimed by Nick Bostrom and colleagues, transhumanism is a *way of thinking about the future* based on the assumption that the human species is neither final, ultimate, nor perfect in its form, but represents merely a phase of a wider evolutionary process that will soon give rise to cognitive systems surpassing man in intelligence and other cognitive capacities.[3] Sometimes transhumanism is also characterized as "an intellectual and cultural movement" that calls attention to the possibility of – and the need for – transforming the human condition by means of advanced technologies. Transhumanists invite us to reflect on the promises and dangers of applying science and technology to transcend existing human limitations in order to improve our physical, intellectual, and psychological traits.[4]

---

3   Cf. N. Bostrom, *The Transhumanist FAQ – A General Introduction (Version 2.1.)*, 2003, p. 4: 21 Feb. 2016, http://www.nickbostrom.com/, henceforth abbreviated as TFAQ. This is a revised and extended version of a 1999 manifesto, which has been reworked with the contribution of about a hundred people, whose names are listed on p. 55.

4   Cf. again Bostrom, TFAQ, p. 4. The term "transhumanism" was coined by the biologist Julian Huxley in 1957. According to Huxley, transhumanism is *a secular belief* that, by realizing the potentialities inherent in his nature, man can overcome his limitations. This will be possible when decisions about the future stages of evolution will be made by human intelligence or its transformations rather than by dint of random mutations and natural selection. The prefix trans- in "transhumanism" indicates the possibility of "transcending itself" by the human race. In contrast to later posthumanists, Huxley

Man is defined not only by the features that have already been realized but, above all, by the potential for development that suggests what man can become in the future. Technological progress carries with it not only the hope for improving human living conditions, but also for improving and enhancing the human organism itself (especially the brain) so as to move beyond our species' physical, psychological, and cognitive limitations. One of the basic differences between classical humanism and contemporary transhumanism is that the latter licenses profound interventions into human nature by means of all manner of available technologies – in the name of progress, growth, self-improvement, the betterment of the world and the human condition, exploration of the unknown, and the hope of building a "brave new world".[5]

Many transhumanists maintain that it is already possible for people to become transhuman, or to achieve the intermediate form between man and a future kind of entity that will emerge as a result of a transformation process initiated by humans and continued by some posthuman cognitive systems. This intermediate (transhuman) phase is said to be characterized by: the enhancement of the capacities of the human body by means of prosthetics, neuroprosthetics and cognitive implants; an extended lifespan; asexual reproduction; the ability to prevent disease; the construction of man-computer interfaces; and the development of artificial intelligence, genetic engineering, and neuroengineering, etc.[6]

---

believed that, after the transformation, "man will remain man, though transcending himself, by realizing new possibilities of and for his human nature" (J. Huxley, "Transhumanism," *Journal of Humanistic Psychology*, Vol. 8, No. 1 (1968), p. 76). In Huxley's construal, transhumanism is an enlightened humanism, or "an evolutionary humanism", as he called it.

5   The brothers Aldous and Julian Huxley, grandsons of the eminent biologist and Darwinian, Thomas Huxley, differed quite radically in their views on the predicted results of interventions into evolution and human nature. Julian was one of the first proponents and a staunch advocate of transhumanism, whereas Aldous, author of the famous anti-utopia *Brave New World*, saw transhumanism as a serious threat to the future of mankind. According to Aldous Huxley, an attempt to realize the transhumanist vision would lead the unintended degradation and impoverishment of human nature, and not its improvement, resulting in the extinction of the human race. Transhumanists' reply to such claims is that our species is bound to disappear anyway if we do not transform it by means of modern technology.

6   The term "transhuman" is shorthand for "transitory human", which refers to the intermediate form of man discussed above. Transhumans, according to the futurist F. M. Esfandiary, who proclaimed himself to be an *Übermensch* as early as 1966 and changed his name to FM-2030, are transitional beings, standing between man and a radically new form of intelligence called the posthuman. For a history of transhumanism, see N.

The transhuman which, according to transhumanists, has already entered the evolutionary scene is merely a stepping stone for the posthuman, a creature whose capacities will be so vastly superior to those of man that it will belong to a different species. In many respects, the posthuman will no longer be human. Just as a chimpanzee cannot comprehend *what it is like to be human*, so people cannot really conceive what posthumans will be like. Nick Bostrom has proposed that we apply the name "posthuman" to any creature that has at least one posthuman trait, i.e., a trait that transcends the normal properties associated with the human species (the character of the trait in question may be physical, psychological, intellectual, or otherwise).[7]

According to the transhumanist manifesto, posthumans may be a form of completely synthetic artificial intelligence, enhanced human minds uploaded into electronic machines (so-called enhanced uploads), or the result of multiple incremental changes made to human biology. The latter possibility would probably involve a thorough-going reprogramming of the human body by means of nanotechnology or through a combination of technologies such as genetic engineering, psychopharmacology, anti-aging therapies, neuronal interfaces, advanced information management tools, memory enhancement with promnestic medication, computers integrated into the body, and others.[8]

The most important means of achieving the posthumanist vision are said to include molecular nanotechnology, genetic engineering, artificial intelligence (artificial intelligence systems may become the first posthumans), neurological interfaces, quantum computers, and the most advanced information technologies. Thanks to molecular nanotechnology, people will acquire the ability to

---

Bostrom, "A History of Transhumanist Thought," *Journal of Evolution and Technology*, Vol. 14, No. 1 (2005), pp. 72–101, an electronic version available at the author's website (retrieved 21.02.2016).

7   Cf. N. Bostrom, *Why I Want to be a Posthuman When I Grow Up*, in: *Medical Enhancement and Posthumanity*, ed. G. Gordijn and R. Chadwick (Berlin: Springer, 2008), p. 108. Bostrom allows that, given a very liberal definition of the word "human", some posthumans may be classified as "modified humans".

8   Cf. Bostrom, TFAQ, pp. 5–6. For a discussion of the connection between the idea of the posthuman and Nietzsche's notion of the *Übermensch*, see M. More, "The Overhuman in the Transhuman," *Journal of Evolution and Technology*, Vol. 21, No. 1 (2010), pp. 1–4. There is no agreement among transhumanists about the degree of convergence between Nietzsche's intuitions and the contemporary explications of the notions of the transhuman and posthuman. For example, Nick Bostrom, in his work on the history of transhumanism, is critical of the alleged connection between transhumanism and Nietzsche's philosophy (see note 8).

construct any possible structures they desire (technological progress is limited only by the laws of nature). Technologies operating at the scale of below a micron will allow us to build machines capable of repairing the cells in our bodies and manufacturing systems and devices of the highest possible quality and functionality, with the production process being controlled down to individual atoms.[9]

Transhumanists are convinced that a soon-to-emerge superintelligence (sometimes called an ultraintelligence) will surpass human cognitive capacities in practically every field, including scientific creativity, its technological and medical applications, and social skills (the superintelligence will accommodate various constituents of multiple intelligences, in Howard Gardner's sense, i.e., interpersonal, intrapersonal, verbal-linguistic, musical-rhythmic, etc.). A *weak superintelligence* is characterized by an increased speed of information processing in the brain (thanks to brain-computer interfaces, cognitive implants and other modifications of the brain's hardware), whereas a *strong superintelligence* involves the creation of an intelligence surpassing the kind of intelligence associated with humans not only in terms of speed, but also quality. One cannot create a new kind of intelligence only by increasing the speed of information processing in the brain of a dog or a man. The idea of a strong superintelligence is motivated by the hope of creating a superintelligence that will be completely unattainable for ordinary as well as cognitively enhanced humans.[10]

The emergence of a superintelligence will be possible if Moore's law, according to which processor speed doubles every 18 months, continues to hold in the future. In such a case, artificial intelligence would equate human intelligence within a few decades and then begin to transcend it. A superintelligence of the future, which is now difficult to imagine, need not have much in common with the intelligence of the human brain. Similar breakthroughs have often occurred

---

9   Thanks to molecular robots we would be able to: successfully fight viruses and bacteria, kill cancer cells, control blood vessels, produce indefinite quantities of cheap food, build almost any physically possible machines (especially atom-sized supercomputers), reconstruct damaged organs and tissues. The work on nanotechnology that fans of transhumanism cite especially often is E. Drexler, *Nanosystems: Molecular Machinery, Manufacturing, and Computation* (New York: John Wiley & Sons Inc., 1992).

10 Cf. Bostrom, TFAQ, pp. 12–15. Thanks to neuroenhancement, we will be able to introduce, without surgical intervention, nanomachines into the bloodstream, whence they will travel to the brain through the cardiovascular system and eventually join in the work of natural neurons. The sending of nanorobots to the brain will also enable us to generate full virtual reality. Yet the most significant consequence of these processes is that they will enhance human intelligence.

in the course of evolution. If we compare the brains and cognitive abilities of animals of different species (the fruit fly, the rat, the dog, the chimpanzee, man), we see, on the one hand, the cognitive limitations of a given species, but, on the other hand, we also notice the evolutionary progress which leads to the emergence of ever-higher forms of intelligence.

Transhumanists believe that the same also applies to the human form of intelligence, which is: a) subject to species-specific limitations (defined by the capacities of the human brain, whose general architecture is determined genetically) and, at the same time, b) capable of being improved, enhanced, and even radically transcended (if new hardware and software are produced superior to the human brain and mind). The authors of the transhumanist manifesto stress the philosophical and ethical consequences of the emergence of superintelligence claiming that the appearance of a superintelligence will deal a serious blow to the anthropocentric worldview. Superintelligence may turn out to be the last human invention, for it will be able to conduct scientific investigations on its own, and do so much more efficiently than humans ever could. Biological humans will therefore no longer be the most intelligent life form on earth.[11]

The development of higher forms of intelligence will coincide with the development of virtual and artificial realities. The term "virtual reality" refers to any environment we experience without being physically located in it. Foreshadowed by theater, cinema, and television, virtual reality can simulate, to an extent, what is physically real or create completely new realities, called artificial worlds. The degree of immersion into a virtual environment varies depending on the kind of devices used to generate the virtual reality. Primitive forms of virtual and artificial reality have been in existence for some time now and include flight simulators, computer games and – to mention a more recent gadget – intelligent glasses which, without impairing the visual field of the person wearing them, display virtually generated images and information directly on the retina of the eye, thus creating, as it is called, an augmented reality. Advanced forms of virtual reality clearly require huge computational power. Transhumanists expect that, as computational power increases, new technologies of deep, interactive, and multimodal simulation will be further developed, becoming phenomenally indistinguishable from physical reality (with respect to resolution, the intensity of experience, and the quality of interaction with virtual objects). As a result,

---

11 Cf. Bostrom, TFAQ, p. 13. For a discussion of problems associated with the notion of superintelligence, see N. Bostrom, *Superintelligence: Paths, Dangers, Strategies* (Oxford: Oxford University Press, 2014).

Transhumanism and Cognitive Science 283

there will be artificial virtual worlds with arbitrary laws of nature and principles of organization (the laws of physics as we know them need not hold there). Trans- and posthumans will visit these virtual environments and spend there a considerable part of their lives socializing, playing, or learning, regarding such experiences as better, more interesting, and intense than those in the physical world.[12]

According to the transhumanist vision of progress, there will come a time when the curve of technological progress becomes almost vertical, at which point science and technology will have entered the state called the singularity (the term "singularity" was coined by Vernon Vinge in the early 1990s and later popularized by Ray Kurzweil).[13] At the outset of the singularity, it will become possible for us to upload our minds into computers. Thanks to a thorough scan of the brain's synaptic structure, it will be possible to transfer information patterns to an electronic medium. A variety of hypothetical methods for obtaining such a scan are being considered, including an atom-for-atom examination of the brain performed by nanorobots and a slice-by-slice analysis of the brain in an electron microscope.[14]

Destructive uploading, in which the original vehicle (the brain) is destroyed together with its information patterns (the mind), is distinguished from *non-destructive uploading*, in which the original brain is preserved and can exist

---

12 Bostrom presents interesting and widely discussed arguments for the claim that we might all be a part of a computer simulation generated by an intelligence from outside of our (simulated) universe. Cf. N. Bostrom, "Are You Living in a Computer Simulation?," *Philosophical Quarterly*, Vol. 53, No. 211 (2003), pp. 243–255, available at the author's website (retrieved: 21.02.2016).
13 Kurzweil uses the term "singularity" in the sense of breaking the continuity of human history as a result of a rapid acceleration of scientific and technological growth. He stresses that there is a distant analogy between his use of the word and what the term means in physics (a cosmological singularity is a place where space-time breaks down, with all the matter and energy flowing into a single point of infinite density and energy). In *The Age of Intelligent Machines* (Cambridge, Mass.: MIT Press., 1990), Kurzweil considers the prospect of constructing machines that would be equally intelligent as humans. In *The Age of Spiritual Machines* (New York: Penguin, 1998), he asks what human life would be like if machines significantly outstripped humans in intelligence. In 2009, in cooperation with NASA and Google, Kurzwail founded Singularity University in the Silicon Valley in California; the university promotes the idea of interdisciplinary research and tries to prepare humanity for the radical technological change associated with the appearance of the singularity.
14 Cf. Bostrom, TFAQ, pp. 17–20.

alongside its copy. It is a matter of debate whether a mind upload preserves the personal identity of its subject. Some writers believe that a person exists as long as the information patterns that code his/her memory, basic values, worldview and personality traits remain intact and it is irrelevant whether the information is stored in an electronic device or a natural neural network of the human brain. However, the problem of personal identity becomes more complicated if we assume the possibility of generating several copies of a single mind. In such a case, an individual person would exist in numerically distinct, but mutually indistinguishable copies, or there would be several qualitatively indistinguishable persons. This would give rise to serious social and legal problems. Transhumanists also contemplate mind uploads of cryonics patients, provided that their brains turn out to be sufficiently well preserved. Copies of human minds could then live in a virtual environment and, equipped with virtual bodies, function in the virtual (simulated) world phenomenally indistinguishable from the physical one.[15]

Transhumanism is not a homogenous movement. From the methodological point of view, none of its versions is a scientific theory or discipline – neither formal, empirical, social, humanist, nor applied. Although transhumanists readily appeal to various scientific results, they do not usually claim transhumanism to be scientific. Bostrom characterizes transhumanism as "a philosophical and cultural movement concerned with promoting responsible ways of using technology to enhance human capacities and to increase the scope of human flourishing".[16] The situation is further confounded by the fact that transhumanism encompasses many different currents, including estropianism, democratic transhumanism, posthumanism, prometheanism, singularitarianism, transhumanist socialism, transtopianism, etc.[17]

Various currents of transhumanism contain elements that could be described as scientific, philosophical, futuristic, science-fiction, utopian and even quasi-religious. When considering the relation of transhumanism to religion, Bostrom notes that transhumanism may play the parts previously played by religion: it gives a sense of meaning and purpose, a hope of achieving more than one has accomplished so far, faith in eradicating disease and imperfection, and even a promise of physical immortality. At the same time, it does not appeal to supernatural beings and divine intervention and relies solely on potentialities

---

15 Cf. Bostrom, TFAQ, pp. 15–19.
16 Bostrom, TFAQ, p. 45.
17 For a characterization of these currents, see Bostrom, TFAQ, pp. 44–45.

inherent in nature, which can be realized by means discovered by science and technology.[18]

Transhumanists put a lot of effort into making sure that their predictions square with scientific knowledge (or at least so they claim). They appeal to the traditions of empiricism and critical rationalism, and declare themselves to be free of dogmatism and open to criticism. According to Bostrom, transhumanism is not a closed set of dogmas: it is an evolving worldview, or rather a set of evolving worldviews. The philosophy of transhumanism is still in its initial stages of development and attempts to adjust to new data and challenges. Transhumanists want to know which of their claims are wrong and use this knowledge to revise their position.[19]

3 Cognitive science as a source of inspiration for transhumanism

Cognitive science is a multidisciplinary scientific field that investigates mental and cognitive processes such as perception, memory, reasoning, language, problem solving, intelligence, consciousness, decision-making, the evolution of cognition and communication, cognitive representations, and many others. Cognitive scientists aim to provide an exact description of the structure and functions of the said processes and explain how they arise; transhumanists, by contrast, seek to improve, extend, and enhance these processes.[20] Cognitive scientists conduct painstaking empirical research, construct theoretical models and, at least sometimes, apply the models in medicine; transhumanists, engage in loose speculation concerning the possible trajectories of the evolution of human cognitive abilities, often alluding to research in cognitive science, especially in

---

18 It seems that transhumanism is consistent with pantheism, although I have found no mention of it in the works of transhumanists. Pantheism can be interpreted as a naturalized account of God (in keeping with Spinoza's *Deus sive Natura*).
19 Cf. Bostrom, TFAQ, p. 46. See also papers collected in M. More, N. Vita-More, ed., *The Transhumanist Reader: Classical and Contemporary Essays on the Science, Technology, and the Philosophy of the Human Future* (Chichester: Wiley-Blackwell, 2013).
20 According to Bostrom and Sandberg, cognitive enhancement is a kind of intervention into a cognitive system that is not intended to be therapeutic or to restore a lost function. Rather, it is designed to significantly improve a natural cognitive ability (perception, memory, consciousness, attention, reasoning, etc.). Cf. Bostrom and Sandberg, "Cognitive Enhancement," p. 312.

fields such as artificial intelligence, cognitive robotics, neurocybernetics, cognitive implantology, and the theory of the evolution of cognitive systems.

Cognitive science investigates natural cognitive systems of humans and other animals, as well as the artificial ones, used in electronic devices designed to process information or enhance natural cognitive processes. Given the breadth of its scope and the interest it displays in improving natural cognitive capacities by using advanced technologies, it is hardly surprising that cognitive science attracts the attention of transhumanists. Yet while cognitive scientists use the method of successive approximations to describe and explain cognitive processes occurring in people, other animals, machines, and hybrid cognitive systems, transhumanists prefer dealing in bold extrapolations and futurist hypothesis that are either extremely difficult – or downright impossible – to test. Transhumanism is therefore a collection of highly speculative views, whereas cognitive science, by its very nature, keeps to the scientific method.

Cognitive science studies cognition using the methods and theoretical resources of many disciplines, including humanities, social, formal, and natural sciences, engineering, and medicine, which makes it a very interdisciplinary research field. In a sense, transhumanism is interdisciplinary too, but its interdisciplinarity is of a different kind and serves a different purpose. Transhumanists use the results of selected scientific disciplines, including cognitive science, as a foundation for their program of transforming and improving human nature, especially with respect to intelligence and cognition. The visions of progress they describe go far beyond what science can predict, and base on the kind of speculation that is characteristic of futurology rather than an understanding of the laws and mechanisms of nature.

In cognitive science, interdisciplinarity stems from the complexities of its subject matter and the fact that the theoretical resources of isolated theories and disciplines are insufficient to build multi-level models of cognitive processes. If cognitive scientists formulate predictions about the future development of cognitive abilities, they do so cautiously and by appeal to what has already been achieved in artificial intelligence, robotics, and cognitive implantology. Transhumanists, by contrast, appeal to *pure technological possibilities*. What is more, it is not entirely obvious whether the things they talk about (such as mind uploads and physical immortality) really are technologically possible. Providing a consistent description of a device, such as a perpetual motion machine, is therefore not the same thing as establishing its physical or technological possibility. The perpetual motion machine has, in fact, *been proven* to be impossible which discourages dreamers and visionaries from making any more attempts at constructing it. In some domains, however, science does not allow us to determine with any degree of certainty whether an object described in futurist speculations can actually work, because we do not know all the consequences of the laws of nature (and, plausibly, neither do we know all the laws). Science fiction writers have often made correct predictions as to the future of science and technology. Transhumanists hope that their predictions will prove correct too, but this does not make them scientific.

Initially, cognitive science was dominated by the computational paradigm, which characterized cognitive processes in terms of receiving, storing, and processing information, and usually ignored the physical make-up of the system and the wider context in which the system was embedded. The mind was supposed to be a kind of software, and its function was to monitor, guide and manage the behavior of the hardware. Thus, a system of abstract representations that enables an organism to solve problems and navigate the environment was thought to be the key to understanding the mind. Most transhumanist theories are based on this computational account of the mind, which, in its extreme version, has the distinct features of the mentioned hardware/software dualism.[21]

According to a theory developed by H. Putnam at the end of the 1950's (called "machine functionalism"), mental states can be thought of as analogous to the functional states of the Turing machine. Two physical systems instantiate the same mental state if they follow the same set of instructions (functions, commands, algorithms, programs). An abstract set of instructions governing human mental processes could also be realized by brainless physical structures, which bars the possibility of identifying mental states with brain states. An account of the mind based on the Turing machine analogy (whether deterministic or indeterministic) presupposes that the essence of the mind consists solely in an abstract system of functional relations that can be implemented by a variety of physical systems. However, a system of functional relations is not identical with its physical substrate (realization base).[22]

Putnam did not preclude the possibility of a non-physical realization of the mind, though he never said what precisely such a non-physical substrate might be like. What attests to the radical nature of Putnam's position is his remark that the functional-state hypothesis, though inspired by mechanicism, is not

---

21 Bostrom and Sandberg characterize cognitive processes as processes of receiving information (perception), processing it (reasoning), storing (memory), and using it to guide behavior. The essence of cognition consists in the managing of information by a biological organism or some other kind of structure. Cf. Bostrom and Sandberg, "Cognitive Enhancement," p. 312.
22 Cf. L. A. Shapiro, *The Mind Incarnate* (Cambridge, Mass.: MIT Press, 2005), p. 16.

incompatible with dualism: a system consisting of a body and a soul might very well be a probabilistic automaton.[23] However, regardless of whether every realization of mental states must be physical in nature (only appropriate physical systems can have mental states, as physicalists maintain), two claims still hold: a) a functionally (computationally) conceived mind is not identical with its realizer and b) functional (computational) mental states are multiply realizable.

If mental states are thought of in terms of Turing machines – as computational states, characterized solely in terms of syntax – then we should not impose any substantive constraints on the properties of their physical realizers. On this construal, mental states and mental properties are characterized in a very abstract way, as purely formal states and properties (without recourse to the physical and neurobiological properties of their realizers). In consequence, psychological concepts capture certain formal (structural, computational) patters that can be investigated without appealing to the details of their material constitution. It is important that the notion of function in machine functionalism is characterized differently than in the so-called functionalism of causal roles. In the former case, function is defined in purely mathematical terms (by analogy with the functional states of a Turing machine); in the latter, the notion of function is richer in content, for it carries with it an appeal to the causal roles performed by a system's constituent parts and the way they contribute to fulfilling the system's tasks (or goals).

According to Putnam, machine functionalism is an empirical hypothesis, for it predicts that if we ever built a machine whose functional profile were the same as that of a given kind of mind (i.e., running its characteristic set of algorithms) then, by the same token, we would have replicated that mind. Every structure functionally (computationally) isomorphic to a mental/cognitive system of kind X would have the same mental/cognitive capacities as X. It has been remarked, however, that even if this is an empirical hypothesis it is extremely difficult to test. We might never be able to create a system functionally isomorphic to the mental/cognitive system of man, and building such a system is part of the transhumanist vision. It is also known that, in the case of systems that are simpler than minds, creating their functionally isomorphic simulations is not the same as recreating them. According to Shapiro, constructing a system that is functionally isomorphic to a mind is not the same as recreating the mind: systems which have been built and are functionally isomorphic to hurricanes and airplanes are

---

23 Cf. H. Putnam, "Psychological Predicates," in: *Art, Mind and Religion*, ed. W. H. Capitan and D. D. Merrill (Pittsburgh: University of Pittsburgh Press, 1967), p. 438.

not hurricanes or airplanes themselves. What, then, is so special about the mind that it should be replicated merely by virtue of functional isomorphism, which usually provides us with nothing more than a simulation? It is no more plausible that a system functionally isomorphic to a mind will think than that a system functionally isomorphic to a hurricane will bend palm trees and destroy trailer parks.[24]

Machine functionalism became later rejected by Putnam, as he began to emphasize the necessary connection between the mind, on the one side, and the neurobiological properties of the brain, the body, and the socio-cultural environment, on the other. In this new perspective, the mind, understood as embodied and situated, cannot be abstracted away from the brain, the body, and the wider context, and then transferred to a supercomputer. The problem is not merely technical – it is a consequence of what an embodied and situated mind is. But not everyone has accepted Putnam's new position. The prominent cognitive scientist Philip Johnson-Laird provides, at the end of his book *The Computer and the Mind*, a description of mind transfers that might just as well have been written by a transhumanist prophet. He believes that there is a distant possibility of preserving the computations of a human brain in a non-biological medium. In such an event, a copy of a man's personality could be stored in a computer program.[25]

The recent years have seen the development of several new theoretical perspectives in cognitive science. The computational paradigm has been significantly enriched by approaches that aim to explain the mind and cognition by stressing the constitutive role of the brain, the body (connecting the brain with the environment), evolution, and various developmental and environmental (socio-cultural) factors. New concepts of embodied, embedded, and extended cognition have been proposed, which changed (enriched) our understanding of the mind and cognition. According to the theory of situated cognition, thinking does not occur only in the brain but in the whole body, and the brain needs the outside environment as well as the body to function properly. Peter Gärdenfors believes that there are many cases where the line between the senses and the world is blurred. The captain of a boat sees through a periscope rather than with the naked eye. A blind man touches with his cane rather than his hand. Similarly, we think with signs, calendars, and pocket calculators. There is no way to draw

---

24 Shapiro, *The Mind Incarnate*, p. 19.
25 Cf. Ph.J ohnson-Laird, *The Computer and the Mind. An Introduction to Cognitive Science* (Cambridge, Mass.: Harvard University Press, 1989), p. 430.

a sharp distinction between the thinking that goes on inside the head and the thinking that goes on outside it.[26]

Transhumanists rarely, if ever, discuss the conceptions of embodied and situated cognition (mind) because, at least at first blush, these conceptions seem to be incompatible with the possibility of mind uploads. Nevertheless, they are inspired by theories of extended cognition and augmented reality. In the former case, the idea is to create cognitive systems that are partly constituted by elements of the outside world (as elements of the system's exostructure). Indeed, we can achieve cognitive enhancement by using a variety of tools: an ordinary pencil, ruler, and compass in the hands of a mathematician make a world of difference, as far as her abilities to construct mathematical theories are concerned. Many such theories would have been very difficult to construct without those simple cognitive implements. The same goes for notebooks (external memory), calculators, perception amplifiers, and computers. Using them radically expands human cognitive horizons.[27]

The case of augmented reality is similar. Augmented reality technologies display additional information by integrating it, in a synchronized and interactive way, with data from the physical environment. Devices such as Google Glass enable us to supplement images of the physical world with computer generated images, maps, translations, guides, instruction manuals, etc. – used in entertainment, traveling, a variety of everyday activities as well as medicine (during surgery). The possibility of controlling intelligent glasses by means of brainwaves is being investigated, which would allow the glasses to be used by the disabled. Another idea that is being explored is that of reverse control, i.e., using such a device to influence particular modules of the brain and thereby modify a person's behavior (present science and technology do not allow us to do that).

Transhumanists usually start by describing particular accomplishments in fields such as augmented reality or brain-computer interfaces, which enable humans to communicate directly with a computer. Their next step, however, involves a sort of imaginary enhancement and produces science fiction narratives that assume the possibility of generating complete virtual reality which is phenomenally indistinguishable from the physical world and which

---

26 Cf. P. Gärdenfors, *How Homo Became Sapiens: On the Evolution of Thinking* (Oxford: Oxford University Press, 2003), pp. 15–16.
27 For a discussion of extended cognition, see A. Clark, *Supersizing the Mind: Embodiment, Action, and Cognitive Extension* (Oxford: Oxford University Press, 2008); D. R. Rupert, *Cognitive Systems and the Extended Mind* (Oxford: Oxford University Press, 2009). It is worth stressing that extended-mind theorists make no references to transhumanism.

can be inhabited by humans themselves or copies of human minds in much the same way as physical reality. So far, we have at our disposal two kinds of brain-computer interfaces: a) invasive interfaces, which are devices implanted into the brain (e.g., cognitive implants replacing damaged brain regions responsible for perception or motion), and b) non-invasive ones, which are brainwave detectors mounted on the head that register and influence brain activity with a high degree of accuracy.[28] These achievements, however, are still a far cry from Kurzweil's fantastic vision of billions of nanorobots sent to the brain, assuming full control, generating virtual reality and, at the same time, repairing and regenerating the host's body so as to guarantee his/her physical immortality.

4 Conclusion

According to Michael Gazzaniga, contemporary people are, without a doubt, *fyborgs* (a term coined by the MIT programmer, Alexander Chislenko), or functional cyborgs – biological organisms equipped with technologies that enhance their natural abilities.[29] Indeed: we wear shoes, clothes, eyeglasses, contact lenses, watches, carry phones, use computers, perception amplifiers, and other devices that enhance our physical and cognitive. Owing to an integration of biological and non-biological structures, we are also gradually becoming cyborgs, or cybernetic organisms, whose lives and cognitive processes are supported or enhanced by technical devices. In the widest sense, all humans that have any sort of mechanic or electronic implant (such as a pacemaker or a cochlear implant) are cyborgs. In the narrow sense, however, a cyborg is a human whose nervous system is surgically connected with mechanical or electronic devices, e.g., a computer, in such a way that the computer and the nervous system interact. A "real cyborg" should also possess at least one additional ability that people do not naturally have, such as an infrared vision, superhuman strength, unnaturally developed analytical skills, or supersensitive senses.[30]

---

28 The first electronic brain implant was constructed by José Delgado in 1963. Delgado placed electrical stimulators in the brains of various animals and used them to induce different reactions depending on where in the brain the stimulator was located. In one of the experiments, Delgado used a button to turn off aggression in a charging bull – the bull stopped in his tracks, instantly becoming calm. Cf. M. Gazzaniga, *Human: The Science Behind What Makes Us Unique* (London: HarperCollins, 2008), p. 334.
29 Cf. Gazzaniga, *Human*, p. 325.
30 Cf. Gazzaniga, *Human*, p. 326.

One way of gradually turning people into cyborgs is by using neuroprosthetics – devices implanted into the brain (or connected to the nervous system) which can detect and process nervous impulses reaching the brain restoring a lost or damaged neurological function. The most successful example of a neuroprosthetic to date is the cochlear implant. The device, directly connected to the brain, takes over one of its functions and makes the computer program decide what the patient hears.[31]

Perhaps, as Gazzaniga speculates, in the future, there will be cochlear implants giving us supernatural hearing. Similarly, Rodney Brooks suggests that in the future people will be able to use artificial retinas, which will give them night, infrared, and ultraviolet vision. Such, neuroprosthetics supplemented with enhancement applications will be used for therapeutic purposes, but it will also be possible to replace with them a healthy eyeball, thus giving one supernatural visual abilities. Similar speculations have been made in regards to the artificial hippocampus (aiding and, in future, enhancing memory), genetically modified organisms (modifying the DNA of an embryo influences the DNA of all body cells – in the brain, in the eyeball, in the sex organs, etc.), and artificial chromosomes.

It is worth noting that Gazzaniga, undoubtedly one of the most prominent contemporary neurocognitive scientists, does not engage in speculation or play with pure possibilities, but considers particular results in cognitive implantology, artificial intelligence, and genetic engineering, and enquires about the positive and negative effects of future progress in these fields. He expresses certain misgivings and objections, but also points to the possibilities being opened up by modern science. Gazzaniga believes these possibilities reach indeed very deep – to our mental and cognitive abilities, which constitute the very core of humanity.[32]

References

Anderson, J.R. *How Can the Human Mind Occur in the Physical Universe?* Oxford: Oxford University Press, 2007.

Bechtel, W. *Mental Mechanisms. Philosophical Perspectiveson Cognitive Neuroscience*. New York and London: Routledge, 2008.

---

31 For a detailed description of the way the cochlear implant works, see Gazzaniga, *Human*, p. 335.
32 Cf. Gazzaniga, *Human*, pp. 382–385.

Bechtel, W., A. Abrahamsen and G. Graham. "The Life of Cognitive Science." In: *A Companion to Cognitive Science*, eds. W. Bechtel and G. Graham. Oxford: Blackwell, 1998, pp. 1–104.

Bostrom, N. "Are You Living in a Computer Simulation?" *Philosophical Quarterly*, Vol. 53, No. 211, 2003, pp. 243–255.

Bostrom, N. "A History of Transhumanist Thought." *Journal of Evolution and Technology*, Vol. 14, No.1, 2005, pp. 72–101.

Bostrom, N. "Why I Want to be a Posthuman When I Grow Up." In: *Medical Enhancement and Posthumanity*, eds. G. Gordijn and R. Chadwick. Berlin: Springer, 2008, pp. 107–136.

Bostrom, N. *Superintelligence: Paths, Dangers, Strategies*. Oxford: Oxford University Press, 2014.

Bostrom, N. and A. Sandberg "Cognitive Enhancement: Method, Ethics, Regulatory Challenges." *Science and Engineering Ethics*, Vol. 15, 2009, pp. 311–341.

Drexler, E. *Nanosystems: Molecular Machinery, Manufacturing, and Computation*. New York: John Wiley & Sons Inc, 1992.

Eckardt, B. von *What is Cognitive Science?* Cambridge, Massachusetts: MIT Press, 1992.

Edelman, G. *Second Nature. Brain Sciences and the Human Knowledge*. New Haven and London: Yale University Press, 2006.

Gazzaniga, M., ed. *The Cognitive Neuroscience III*. Cambridge, Massachusetts: MIT Press, 2004.

Gazzaniga, M., R.B. Ivry and G.R. Mangun. *Cognitive Neuroscience. The Biology of the Mind*. New York: Norton, 1998.

Gärdenfors, P. *How Homo Became Sapiens: On the Evolution of Thinking*. Oxford: Oxford University Press, 2003.

Huxley, J. "Transhumanism." *Journal of Humanistic Psychology*, Vol. 8, No.1, 1968, pp. 73–76.

Johnson-Laird, Ph. *The Computer and the Mind. An Introduction to Cognitive Science*. Cambridge, Massachusetts: Harvard University Press, 1989.

Kurzweil, R. *The Age of Intelligent Machines*. Cambridge, Massachusetts: MIT Press, 1990.

Kurzweil, R. *The Age of Spiritual Machines. When Computers Exceed Human Intelligence?* New York: Penguin, 1998.

More, M. "The Overhuman in the Transhuman." *Journal of Evolution and Technology*, Vol. 21, No.1, 2010, pp. 1–4.

More, M. and N. Vita-More, eds. *The Transhumanist Reader: Classical and Contemporary Essays on the Science, Technology, and the Philosophy of the Human Future*. Chichester: Wiley-Blackwell, 2013.

Putnam, H. "Psychological Predicates." In: *Art, Mind and Religion*, ed. W.H. Capitan and D.D. Merrill. Pittsburgh: University of Pittsburgh Press, 1967, pp. 37–48.

Shapiro, L. A. *The Mind Incarnate*. Cambridge, Massachusetts: MIT Press, 2005.

Marta Soniewicka

Human Enhancement: The Question of Fairness and Virtues in Sport Activities[1]

Abstract: The paper poses the question as to what kind of enhancement constitutes an unfair advantage. To illustrate this problem, the cases of Pistorius, Tiger Woods, Kłobukowska, and the PGA Tour v. Martin case are discussed. It is argued that the gist of the matter does not lie in the questionable distinction between therapy and enhancement, or between different kinds of enhancement. In fact, even if a clear distinction between enhancement and therapy could be determined, it would not provide an answer to the question of whether a particular enhancement or therapy in sport is fair or not. In order to answer the posed question, it is necessary to analyze the criteria of procedural fairness, as well as the goal of the sport activity which they serve. The point is that we value the talents and efforts which constitute the internal good of the sport, i.e., excellence. For this reason, one may argue that, in most cases, enhancement should be permissible in sport on the grounds of procedural fairness, i.e., if everybody is equally enhanced. In some of the exceptional cases which are considered in this paper, one may also argue for the allowability of those corrective enhancements which are necessary to allow an individual to participate in a sport activity in the first place, but only if they promote skills and do not replace or overshadow natural abilities, and also if they do not fundamentally alter the character of the discipline, but promote its aim.

Keywords: Enhancement, Sport, Fair play, Unfair advantage, Discrimination

1 Introduction: The *Pistorius Case*

Oscar Pistorius, the South-African athlete also known as "Blade Runner", was called the fastest man on Earth with no legs. He was born without calf bones (fibular hemimelia). Shortly after birth his parents decided on the amputation of both of his legs following the advice of physicians. The legs were amputated below the knee, enabling him to use prosthetic legs. When he was around 2 years old he was able to walk with his prosthetic legs, and from early childhood Pistorius was very keen on sport; at school he used to play cricket, rugby, water polo, tennis and do other sports, such as boxing or wrestling. He began running at the age

---

1 The writing of this article was funded by the Polish National Science Centre (2012/07/D/ HS1/01099). I owe special thanks to Prof. Robert Audi for reading this paper and for his useful comments and remarks which improved the text.

of 16 in order to regain his fitness after a small knee injury, and upon discovering that it was the sport he was best at, it became his profession. Pistorius' running blades were designed and made by the company named Össur which manufactures the so called Flex-Foot Cheetahs, i.e., a carbon-fiber limbs, which are very flexible and made for sprinting and jumping. Soon Pistorius became a champion sprinter among below-knee amputees (both single leg amputees and double leg amputees). He frequently won gold medals in Paralympic Championships and World Cups, setting world records for the 100, 200 and 400-meters. Yet he was more ambitious than that and soon started to compete with able-bodied sprinters. He was successful in this field too, finishing sixth in the South African Championships in 2005 and continuing to participate in international able-bodied events in subsequent years. His goal was to compete at the Olympics, and in 2008 he had his first opportunity to participate in the Summer Olympics in Beijing in China. His idea attracted the attention of the International Association of Athletics Federation who in March 2007 passed competition rules, according to which during competitions it was not allowed to use "any technical device that incorporates springs, wheels or any other element that the user with an advantage over another athlete not using such a device."[2] Thus, Pistorius' participation in the Olympics depended on the decision of the IAAF as to whether his blades could give him unfair advantage. Pistorius' performances were monitored by the IAAF, and he was also invited to perform a series of tests at Cologne Sports University under the supervision of professors in biomechanics. The test results were disseminated in the form of a report, which claimed that the limbs of Pistorius consumed 25% less energy than the natural legs of able-bodied runners running at the same speed.[3] With reference to these findings, the IAAF found Pistorius' Cheetahs ineligible for use in sport competitions under their rules, including the 2008 Summer Olympics, since they were determined to be a technical device which gave him an advantage over other athletes running with natural legs. In other words, the use of the more efficient Cheetahs was a violation of the afore-mentioned rule 144.2(e). Consequently, Pistorius was prohibited from competing against able-bodied runners in the Olympics, but he challenged the ruling and appealed against the decision to the Court of Arbitration for Sport (CAS) in Switzerland. The CAS panel questioned the evidence for the advantage of Pistorius over able-bodied athletes and found insufficient evidence to prove it

---

2 IAAF Competition Rules 2008, Rule 144.2(e) (Assistance to Athletes), p. 100.
3 IAAF announcement – Oscar Pistorius banned on test results, The Science of Sport: 14 Jan. 2008, http://scienceofsport.blogspot.com/2008/01/oscar-pistorius-announcement-banned_14.html

and upheld his appeal; consequently the IAAF council decision was canceled in May 2008.[4] Pistorius qualified for the 400-metres for the 2012 Summer Olympics in London and was the first amputee athlete to compete at the Olympics (he only qualified for the semi-finals and lost). It was quite remarkable that he was chosen to carry the South African flag for the closing ceremony of the Olympics, as well as the opening ceremony of the 2012 Summer Paralympics which took place in London during the same month.

Pistorius' case opened up a heated debate over unfair advantages in sport competition. The question frequently posed in the debate concerned the kind of enhancement which constitutes unfair advantage. The aim of this paper is to address this problem by considering such controversial terms as enhancement and fairness.

2 A definition of enhancement

Current advances in biotechnology and neurosciences have opened up new perspectives and possibilities for technical and genetic interventions that could significantly affect our own understanding and status in the world. The possibilities which are at stake enable the enhancement of human beings in different ways. There are miscellaneous definitions of enhancement. Edmund D. Pellegrino defines enhancement in the following way:

> my operating definition of enhancement will be grounded in its general etymological meaning, i.e., *to increase, intensify, raise up, exalt, heighten, or magnify*. Each of these terms carries the connotation of *going 'beyond' what exists at some moment, whether it is a certain state of affairs, a bodily function or trait, or a general limitation built into human nature* (...) For this discussion, enhancement will signify an intervention that goes beyond the ends of medicine as they traditionally have been held.[5]

To apply this definition, it is necessary to define the ends of medicine first. If one assumed that the *traditional* aim of medicine is to provide treatment, then one would understand as enhancement everything that goes beyond treatment. Yet the distinction between treatment (therapy) and enhancement would remain vague. One might then claim that "treatments are any interventions

---

4 Arbitration CAS 2008/A/1480 Pistorius v/IAAF, award of 16 May 2008, http://jurisprudence.tas-cas.org/sites/CaseLaw/Shared%20Documents/Forms/By%20Year.aspx
5 E. Pellegrino, *Biotechnology, Human Enhancement, and the Ends of Medicine*, The Center for Bioethics and Human Dignity: 30 Nov.2004, www.cbhd.org/resources/biotech/pellegrino_2004-11-30.htm, emphasis MS.

that physicians and their patients agree are useful and proper",[6] in particular those which are aimed at improving health conditions. But such a definition would depend on another unclear notion – health. According to a frequently questioned definition given by the World Health Organization (WHO), health is a state of complete physical, mental and social *well-being* and not merely the absence of disease or infirmity.[7] On the basis of such a broad definition of health, *every* kind of medical intervention aimed at improving well-being would be considered as treatment, and it would also cover the "welfarist definition of human enhancement" which includes "Any change in the biology or psychology of a person which increases the chances of leading a good life in the relevant set of circumstances",[8] thus making the distinction between treatment and enhancement disappear. If we insist on keeping the distinction, we will always have problems with drawing a line between them. For instance, following our linguistic intuitions, enhancement can be defined as an improvement which goes beyond *normal human functioning*, yet it remains controversial as regards how to define "normal". To elucidate the term, one may also distinguish three different kinds of enhancement, namely:

1. Correcting effects of disability or disease, e.g., wearing an artificial leg
2. Increasing natural human potential (natural endowments of capabilities) within the range typical of the species homo sapiens, e.g., raising a person's IQ (The problem is, what would constitute the typical range? Up to what limit would the increased life span or raised intelligence be typical?);
3. Superhuman enhancements (sometimes called posthuman or transhuman): increasing a person's capabilities beyond the range typical for the species homo sapiens, e.g., providing humans with gills (the capacity to breath under water) or a bat sonar.[9]

---

6   E. T. Juengst, "Enhancement Uses of Medical Technology", in: *Encyclopedia of Bioethics*, vol. 2, ed. S. G. Post (New York: MacMillan Reference USA, 2003), p. 753.
7   Preamble to the Constitution of the World Health Organization as adopted by the International Health Conference, New York, 19–22 June, 1946; signed on 22 July 1946 by the representatives of 61 States (Official Records of the World Health Organization, No. 2, p. 100) and entered into force on 7 April 1948.
8   J. Savulescu, A. Sandberg and G. Kahane, "Well-Being and Enhancement," in: *Enhancing Human Capacities*, ed. J. Savulescu, R. ter Meulen and G. Kahane (Oxford: Wiley-Blackweel, 2011), p. 82.
9   Savulescu, Sandberg and Kahane, "Well-Being and Enhancement," pp. 61–62.

What is more, one may differentiate between enhancements *ex-ante* (introduced to children before their birth) and *ex-post* (introduced to existing persons after their birth), which plays a significant role in bioethical discussions.

Considering the case of Pistorius, his artificial legs can be seen both as therapy and as enhancement, depending on our definitions. And if we assumed that Pistorius blades made for running are a kind of enhancement, we would still have a problem with regards to the type of enhancement they belong to since besides correcting the effects of disability they definitely do much more – they increase natural human potential by enabling Pistorius to run both fast and faster. According to the controversial tests performed in Germany, the Cheetahs would also increase Pistorius' capabilities beyond the range typical for human beings since they are more effective than natural legs and use less energy in running.

The same problem can be also shown by means of another example – the laser eye surgery performed by American golf star Tiger Woods[10]. Woods, who suffered from serious problems with vision and would be considered legally blind without glasses or lenses, in 1999 decided to undergo Lasik surgery to correct his sight. The results of the procedure were amazing. One may pose the question whether his surgery was a mere correction of poor vision or rather an enhancement which gave him eagle sight and improved his golf game (he won a series of tournaments just after the surgery). In other words, the question is whether such surgery provided him with a superhuman vision, superior to the normal abilities of other golf players? If this were true, it would constitute an unfair advantage and could be considered grounds for disqualification from further competitions. As can be seen, the problem does not depend on a questionable distinction between therapy and enhancement or a vague distinction between different kinds of enhancement. No matter if we call the blades or laser eye surgery corrective enhancement (or therapy) or superhuman enhancement, it would not solve the problem of whether these kinds of interventions should be eligible in sport or not. To answer this question, it is therefore necessary to analyze the idea of fairness.

3 Justice as personal desert and justice as fairness

Considering the afore-mentioned cases, one can pose the crucial question: what does it mean, "to deserve to win"? Joel Feinberg identifies desert with worthiness

---

10 Cf. M. Sandel, *The Case Against Perfection. Ethics in the Age of Genetic Engineering* (Cambridge, Mass. and London: Harvard University Press, 2007), p. 37 ff.

and distinguishes it from eligibility and entitlement.[11] Eligibility means the satisfaction of necessary conditions (i.e., minimal qualification, e.g., medical certificate of athletics). Entitlement means satisfaction of a sufficient condition (e.g., crossing the finish line before other competitors). Moral desert, on the other hand, means the satisfaction of certain conditions of *worthiness* (i.e., being in virtue of some characteristic or prior activity). It may happen that team A which was much poorer than team B won a football game because of pure luck. We would then say that team A was entitled to win, yet team B deserved to win more than team A which actually won (this would not affect the entitlement).

The grounds for the rational decision that *one deserves something* may be constituted both by the features of an appropriate kind (merits, e.g., natural talents) and the results of actions (deserts *sensu stricto*, e.g., efforts).[12] Our legal systems and political doctrines tend to evolve in the way that we reward and punish for what people have done, rather than for who they are. Yet, the clear separation of actions from their agents is impossible and counterintuitive. We used to value both people's virtues and their activities, in particular when the latter result from the former. The main problem with moral desert is that it requires a concept of the good according to which conditions of worthiness are specified (e.g., an idea of a good life or an idea of a good sportsman). This is the reason why "none of the most prominent contemporary versions of philosophical liberalism assigns a significant role to desert at the level of fundamental principle."[13] Liberals agree on the derivative status of desert, which is either derived from a just institutional scheme of a society (Rawls), social utility (utilitarian liberals), or from natural entitlements (Nozick). They reject the idea of moral desert as an independent standard by reference to which the justice of institutions is to be evaluated.

One may distinguish two levels of justice – the substantive and formal justice, as Chaim Perelman points out.[14] Substantive justice (first-level rules of justice) consists of a set of substantive principles of distribution of rights and duties.

---

11 J. Feinberg, "Justice and Personal Desert," in: *Doing and Deserving. Essays in the Theory of Responsibility* (Princeton, NJ: Princeton University Press, 1970), p. 57.
12 J. Lucas, *Responsibility* (Oxford: Clarendon Press, 1995), p. 124.
13 See S. Scheffler, *Boundaries and Allegiances. Problems of Justice and Responsibility in Liberal Thought* (Oxford: Oxford University Press, 2008), p. 14 ff. Moral desert in its pre-institutional sense plays no role in distributive justice, yet it may still be evoked in the case of retributive justice sometimes, as Rawls suggests. See J. Rawls, *Theory of Justice* (Oxford: Oxford University Press, 1999), pp. 314–315.
14 See Ch. Perelman, *The Idea of Justice and the Problem of Argument*, trans. J. Petrie (London: The Humanities Press, 1963), pp. 36–45.

Formal justice (second-level rules of justice) is concerned with application of principles of justice and can be identified with a rule that equal people should be treated equally. This is the principle of equal measurement since it requires that no one derives any benefits from the mere fact that he or she is a particular person. Formal justice requires the impartial and consistent application of adopted principles of justice, but it does not determine what these principles are. One can imagine situations in which although similar people in similar situations are treated similarly, there is no justice (e.g., a situation in which all the parties are treated equally unjustly). The rule of equal treatment is empty; there is no content since the criteria of differentiation as well as its aims come from the substantive level of justice, which must complement the formal level of justice. Wherever there is no substantive justice, formal justice is worthless. Wherever there is no formal justice, one may suppose that there is no substantive justice either. Both levels of justice are required for realizing a full-fledged theory of justice. The application of principles of justice may be specified within a special procedure. It leads us to another notion – procedural justice – which is a link between the first-level rules of justice and the second-level rules of justice. Procedural justice guarantees passing from the level of the principle of justice to the level of its application in the best possible way.[15] John Rawls distinguishes among pure procedural justice imperfect and perfect procedural justice. The former guarantees just results by obeying the adopted rules (if parties obey the rules, they must accept the result as just, regardless of what they are, e.g., in a lottery). Imperfect and perfect procedural justice is characterized by obeying the rules aimed at bringing the particular just result which is defined separately (e.g., a criminal procedure which is constructed so as to bring about the truth of who is guilty). By fairness Rawls understands guaranteeing equal opportunities to all people, and to realize this aim he argues for the principles of justice that could be chosen by people situated in fair conditions (having no knowledge of their features and social background).

As John Harris argues, "In sport (…) we are talking always about procedural justice or fairness. (…) The key to procedural fairness (…) is that the same enhancement opportunities are, in principle, available to everyone."[16] Considering the case of Pistorius' competitions in which he won against able-bodied athletes, one may say that he was entitled to win, yet one may also wonder if he deserved

---

15 See Rawls, *Theory of Justice*, pp. 74–77.
16 J. Harris & S. Chan, "Enhancement is good for you. Understanding the ethics of genetic enhancement," *Gene Therapy*, Vol. 15 (2008), pp. 338–339.

the prize assuming that his blades gave him an unfair advantage over the other competitors. Thus, the criteria of eligibility for runners should be formulated in such a way that they guarantee equal opportunities, which would result in fair competition. Consequently, athletes may only use such devices which are available to all of them equally, which brings about unification of sport uniforms and shoes etc. According to this line of argumentation, the use of blades seems to be eligible only for amputees since ordinary runners cannot use such a device, as the IAAF argued. One has to emphasize that the problem of the use of blades is not about blades as such but rather about the assumed *competitive advantage* which they could provide to an amputee over an able-bodied runner.

To evaluate competitive advantage we need to be able to compare such runners according to *the same measure of value*. We could not compare the abilities of athletes from different kinds of sports. For instance, it seems unjustified to claim that a football star such as Lewandowski is better than the tennis star Radwańska, just as it would be to say that Shakespeare was better than Bach. Both are excellent in their own disciplines and incomparable – they are like chalk and cheese. The abilities of athletes of different sex or age seem incomparable, too. Thus, the distinction between sport disciplines, separation of sex, as well as differentiation of weight and age in different disciplines are justified "to ensure that athletes are competing on an equal basis, considering their physical status".[17] The question remains whether one may evaluate the results of able-bodied and disabled athletes using the same criteria; put differently, are natural legs really comparable to Cheetahs? The answer to this question is crucial for identifying the justified criteria of differentiation in sport. Unjustified criteria would constitute grounds for discrimination, which I will address in the next section.

4 Fair competition and discrimination in sport

By discrimination one may understand a situation in which individuals who are alike in *every relevant respect* (not in absolutely every respect) are treated differently, or when individuals who are different in *some relevant respect* are treated alike.[18] Direct discrimination takes place when discrimination is based on the unjustified criteria of different treatment. Indirect discrimination takes place when, despite being based on neutral criteria, different treatment brings about

---

17 E. Hay, "Sex determination in putative female athletes," *JAMA*, Vol. 221 (1972), pp. 998–999.
18 J. Feinberg, "Social Philosophy," in: *Foundations of Philosophy Series*, ed. E. Beardsley and M. Beardsley (Englewood Cliffs, NJ: Prentice-Hall, 1973), p. 99.

the unjustified exclusion of particular individuals or groups of people that are unable to meet such criteria. Indirect discrimination concerns all situations in which a definition of an aim of distribution or a definition of a particular social institution is unjustified.

Before I turn to the problem of discrimination against disability in sport, I will introduce some interesting cases concerning gender tests in sport, which could be interpreted as discrimination. The Polish athlete Ewa Kłobukowska set a world record in the women's 100 meters sprint in Prague, in 1965, and won gold and silver medals in sprint in the European Athletics Championships in Budapest in the following year.[19] In 1967 she was subjected to traditional gender tests, which she failed. Consequently, she was banned from professional competitions, and the IAAF erased her achievements from the record books. She was found illegible on the grounds of having too many chromosomes (she had an XXY sex chromosome complement which constitutes a chromosomal abnormality disorder called mosaicism). "Normally, a person's anatomical sex (and thus social and legal sex) is determined by the type of sex chromosome contributed by the father. However, there are a number of genetic disorders which interfere with the process of sex development and lead to paradoxical findings between anatomical sex and chromosomal sex."[20] Such genetic disorders are numerous and differ in their results – some of them result in the ambiguity of the external genitalia, some of them do not; some of them result in the greater production of male hormones, some do not, etc. In other words, in some cases a genetic disorder may give a competitive advantage to a female athlete who because of it becomes taller and stronger than most women. Yet, it was not the case of Kłobukowska, who happened to be Barr positive, meaning that she had two active X chromosomes (as women have), and in 1968 gave birth to her son, which would be not possible otherwise. Had she been tested one year later, she would have been found eligible for competition. Instead, she was not only unfairly disqualified but also suffered humiliation and public embarrassment because of her gender abnormality. "Since then, international sports organizations have taken great pains to maintain absolute secrecy",[21] however, not always successfully, as it turns out given another famous example of ambiguity in gender tests in sport – the case of Caster Semenya, a South African athlete who won gold

---

19 See Wikipedia online: https://en.wikipedia.org/wiki/1966_European_Athletics_Championships_%E2%80%93_Women's_100_metres
20 M. A. Ferguson-Smith and E. A. Ferris, "Gender verification in sport, the need for change?," *British Journal of Sports Medicine*, Vol. 25, No. 1 (1991), p. 18.
21 Ferguson-Smith and Ferris, "Gender verification in sport, the need for change?," p. 19.

in the women's 800 meters at the Berlin World Athletic Championships in 2009 and silver medals at the World Championships and Summer Olympics in 2011. A few hours after she won gold medal in Berlin, her gender was contested by the IAAF, which subjected her to a procedure of gender tests on the grounds of her astonishing breakthroughs in her achievements and on the grounds of the information given by her on her blog. Although the IAAF did not disclose the test results, "the South African Ministry of Sport (...) decreed that in any case she can continue running with women in her own country".[22] The case of Semenya gained a lot of public attention. The female athletes who lost against Semenya in the competition claimed that she should not run with women. The public opinion in South Africa, however, defended Semenya; many commentators were emphasizing that the success of Caster was a result of her hard work in training. Finally, she was cleared by the IAAF and was allowed to return to competition in 2010.

It is worth noting that the IAAF used an unfortunate parallel comparing Semenya's case with the suspicion of drug abuse aimed at improving sport results. If the analogy between gender abnormality and doping were correct, it would mean that having a genetic disorder may be considered as *cheating*. According to dictionaries, cheating can be understood as breaking rules in order to gain an advantage at something or to prevent somebody from getting what he/she deserves by the use of deceit or fraud.[23] Thus, the analogy neglects the fact that genetic disorder is not the result of an intentional free act, but a condition which is independent of our will. Consequently, only situations involving a voluntary change of sex or withholding information about one's sex in order to gain advantage in sport could be considered as cheating.

The main problem with gender test results is that the general rule of differentiating sex into two categories – male and female – does not cover all the miscellaneous exceptions which lie between. Around 1.7% of the human population have different sex disorders which can be classified as sex chromosome abnormalities, gonadal abnormalities, and sex hormone abnormalities.[24] Most of these conditions are not visible, and those who have them do not even know they belong to the "sex grey zone" until in some cases, for instance, they discover their sterility. Therefore, one has to specify what kind of sex disorders disqualify from sport competitions based on sex categories, and which do not. One may claim

---

22 S. Camporesi and P. Maugeri, "Caster Semenya: sport, categories and the creative role of ethics," *Journal of Medical Ethics*, Vol. 36 (2010), p. 378.
23 *Meriam-Webster Dictionary*, http://www.merriam-webster.com/dictionary/cheat
24 Camporesi and Maugeri, „Caster Semenya," pp. 378–379.

that disorders which do not result with any physical advantages derived from an increased level of androgens (e.g., androgen insensitivity, in which people who are genetically XY, i.e., males look like females and have female gender identity) should not disqualify them from competition. But even if a genetic disorder results in the production of a significantly higher level of virilising hormones (which increases muscle bulk, e.g., congenital adrenal hyperplasia), the question whether disqualifying such people is justified still stands. It seems unfair that persons who suffer from this kind of disorder, independent of their free will, should be banned from professional competitions in sport, when there is no category in which they would fit. Specifically, such a situation could be considered as indirect discrimination, since some categories of people would never meet the criteria of eligibility and therefore would be excluded from professional sport competitions. Moreover, assuming such reasoning may allow one to argue that those who undergo a treatment aimed at lowering the level of androgens should be permitted to compete with other females.[25]

What is more, the biological diversity among human beings actually plays a significant role in sport and it seems unfair to pick on one specific disorder to define the terms of eligibility in sport competitions. Perhaps,

> we should look at this biological diversity as an opportunity rather than a threat, After all, every athlete who performs outstandingly is 'exceptional', in one way or another. The National Basketball Association, for instance, has several players with acromegaly, a condition that cases the overproduction of growth hormone and that is usually classified as pathological, but which turns out to be a powerful advantage when playing basketball at professional levels.[26]

To put it differently, in sport activity one may transform one's own abnormality into success, just as Oscar Pistorius transformed his disability into an ability. When we compare our contemporary Olympics with the ancient-Greek Olympics we can easily understand that our idea of sport and our concept of "natural abilities" have changed significantly. It is not a coincidence that Usain Bolt, who set a world record in 100 meters in 2012, would beat the athletes who competed before him. Comparing the time results of the Olympic winners in the last 100 years, we come to the conclusion that the Olympic medalist Francis Lane who was the fastest in 1896 would be 23.6 meters (!) behind Bolt.[27] There is some

---

25 Camporesi and Maugeri, „Caster Semenya," p. 379.
26 Camporesi and Maugeri, „Caster Semenya," p. 379.
27 K. Quely and G. Roberts, "One Race, Every Medalist Ever," *The New York Times*: 5 Aug. 2012, http://www.nytimes.com/interactive/2012/08/05/sports/olympics/the-100-meter-dash-one-race-every-medalist-ever.html?_r=0

mystery about the main factors of such an improvement in the results of runners in the last 100 years. Some argue that today's athletes run faster because of new track design, training technologies, high level equipment (shoes), running outfit, etc. As far as the phenomena of Jamaican athletes like Bolt is concerned, there are some scientists who claim that they are so good at sprinting because they have a high level of serotonin in their bodies, while others claim that their success depends on a homogenous gene pool. The point is that science does not provide any clear answers for our judgments about justice or fairness – these normative notions need normative criteria of differentiating people on which they should be based.

Equality of opportunities requires "leveling the field" for those who are unequal for arbitrary reasons. If one has lost one's legs, would we consider one's carbon-fiber artificial limbs as a kind of compensation which levels the field or rather as something that makes one unfairly advantaged – in other words, is this person disabled, or "too-abled" with prosthetic limbs? Pistorius frequently repeated: "I don't see myself as disabled. (...) There's nothing I can't do that able-bodied athletes can do."[28] He put emphasis on his abilities instead of disabilities, arguing that: "Some people view themselves as disabled because they have one or two disabilities. (...) But what about the millions and millions of abilities they have? So what if you have a leg or two missing?"[29] According to Pistorius, the idea of banning him from the Olympics was based on misjudgments and biases, which resulted in discrimination. A similar problem of discrimination against disability in sport was invoked in the case of an American golf player, Casey Martin.[30] He suffered from a degenerative circulatory disorder, which caused him pain while walking. Long distance walking (which is a part of playing golf) exposed him also to the risk of hemorrhaging and fracture. Due to his condition, Martin tried to obtain permission from the Professional Golfers' Association (PGA) for the use of a golf cart during tournaments. The PGA rejected his request on the grounds that it would contravene one of the rules which prohibits using golf carts in professional tournaments. Martin questioned this decision in

---

28 S. Camporesi, "Oscar Pistorius, enhancement and post-humans," *Journal for Medical Ethics*, Vol. 34, No. 9 (2008), p. 639.
29 R. Davis, "Oscar Pistorius's murder trial: what's disability got to do with it?," *The Guardian*, 8 July 2014, http://www.theguardian.com/world/2014/jul/08/oscar-pistorius-murder-trial-disability
30 See: M. Sandel, *Justice, What's the Right Thing to Do?* (New York: Farrar, Straus and Giroux, 2009), pp. 203–207.

court, arguing that the PGA rules discriminate against people with disabilities and filed the action

> under Title III of the Americans with Disabilities Act, which, among other things, requires an entity operating "public accommodations" in its policies "when (…) necessary to afford such (…) accommodations to individuals with disabilities, *unless the entity can demonstrate that making such modifications would fundamentally alter the nature* of such (…) accommodations".[31]

The case was considered by the US Supreme Court, which ruled (7-2) in favor of Martin, confirming his right to use a golf cart in tournaments. The Court argued that permitting Martin to use a cart would not give him a competitive advantage over other players since walking fatigue is still less than the fatigue which Martin suffers from coping with his disability. The conclusion of the ruling was that since "permitting Martin to use a cart would not 'fundamentally alter' the nature"[32] of golf, than prohibiting him to use a cart was understood as denying him equal access to professional golf competitions on the basis of his disability. In other words, the decision of the PGA was indeed found to be discriminatory. Justice Antonin Scalia strongly dissented, arguing that the ruling of the Court distorted common sense and exercised a benevolent compassion, which is not the aim of law.[33] He disputed the right of the Court to decide "What is Golf?", and argued that the question whether someone riding around a golf course from shot to shot *really* is a golfer is not the kind of a jurisprudential question for which the judges have prepared studying law for so many years.[34] Scalia argued that the Court's decision was based on the wrong assumption that "PGA TOUR golf must be classic 'essential' golf".[35] According to him, there is no "legal obligation to play classic, Platonic golf"; on the contrary, the PGA adopts rules arbitrarily and may promote a new way of playing golf and whoever dislikes these rules does not have to play.[36] In other words, he objected to the idea of essentialism in defining sport activities, claiming that:

> the very nature of a game is to have no object except amusement (that is what distinguishes games from productive activity), it is quite impossible to say that any of

---

31 PGA Tour v. Martin, 532 U.S. (2001), p. 661, https://supreme.justia.com/cases/federal/us/532/661/case.html
32 PGA Tour v. Martin, pp. 661–662.
33 PGA Tour v. Martin, Scalia dissenting, p. 691 ff.
34 PGA Tour v. Martin, Scalia dissenting, p. 700.
35 PGA Tour v. Martin, Scalia dissenting p. 699.
36 PGA Tour v. Martin, Scalia dissenting p. 700.

a game's arbitrary rules is "essential." Eighteen-hole golf courses, 10-foot-high basketball hoops, 90-foot baselines, 100-yard football fields – all are arbitrary and none is essential.[37]

In Scalia's opinion, the rules of sport are supported by tradition and the authority of entities such as the PGA or the IAAF, who are entitled to change them according to their will. The argumentation of Justice Scalia provides us with some insights about the discussed problem of disability and fairness in sport. As Scalia accurately emphasizes, the law should assure that disability will not deny anybody equal chances to participate in competitive sporting events, but it does not have to guarantee an equal chance to win. Talents, just as excellence in sports, are distributed by nature unequally, which is neither just nor unjust. The unequal distribution of excellence in sport "is precisely what determines the winners and losers – and artificially to 'even out' that distribution (…) is to destroy the game."[38] In other words, we reward excellence which results from both talents and efforts, and the former are usually based on a kind of abnormality (the best ones are always *beyond* the average which means they are not "normal" in a descriptive sense). "Leveling the field" cannot be understood as giving privileges to those who are *below* a certain level of excellence. Thus, the idea of fairness seems to be guaranteed in the best way by general uniform rules, as it was said before. Yet, procedural fairness is not the only criteria of our judgments concerning sport competition, as Michael Sandel accurately points out.[39] The court decision was based on the consideration of both criteria: competitive fairness and the *nature* of sport. Scalia questioned the latter by arguing that it would lead to essentialism – an approach which assumes that an appropriate meaning of a social practice (like sport activity) results from its essential nature. He rejected the assumptions that anyone, and judges in particular, could have access to the essence of golf or any kind of "Platonic idea of golf". At the same time, in his defense of arbitrariness of sport rules Scalia invoked "the very nature of sport", understanding by "nature" a meaning which results from a social convention (tradition, etc.). Following Sandel, one may reject both essentialism and conventionalism, and claim that a meaning or a value of a social practice is a matter of constant interpretation based on the shared values and intuitions of those who participate in it. In other words, *equality* of opportunities should be complemented by *recognition* of the

---

37   PGA Tour v. Martin, Scalia dissenting pp. 700–701.
38   PGA Tour v. Martin, Scalia dissenting, p. 703.
39   Sandel, *Justice, What's the Right Thing to Do?*, p. 206.

values that we want to reward. I will address this idea in the following – and final – section.

5 The teleological perspective: definition of social practice

In the line of argumentation presented by Justice Scalia one may find a certain incoherence. On the one hand, he argues for the arbitrariness of sport rules, claiming that the *only* aim of sport games is enjoyment. On the other hand, he admits that *excellence* determines the winners and losers, though we may disagree as to which rules test the excellence of individuals in the best way. "But the rules are the rules", he asserts,[40] and one must either accept them or choose a competition that suits one better. This means that in the considered case, according to Scalia, the PGA has the sole authority to decide whether the rules are justified as relevant to the aim of the game of golf. If we argue that golf is aimed at amusement, which is certainly true, we also have to admit that it is not its only aim since golf, just like every other sport, is also aimed at excellence. Bearing this in mind, one may distinguish between sport rules which are, in fact, arbitrary (like the rules of many games, including lotteries), and competitive rules which are adopted to test individuals' skills. Both of them may be questioned. If the PGA adopted a rule stating that only black people can participate in tournaments, the rule would be considered discriminatory, since the color of skin has nothing to do with the idea of sport as such. We would not accept the line of argumentation that "the rules are the rules", and for this reason white people who disagree with them should organize their own games with different rules. Rules are considered discriminatory if they directly, or indirectly, exclude some categories of people from participation in tournaments on the basis of unjustified criteria like the color of skin, genetic disorder or disability. The rule regarding 100-yard football fields is arbitrary but not discriminatory since it does not make it impossible for some categories of people to play football. The rule, which requires walking the course between holes, made it impossible for Martin to participate in tournaments and therefore was considered unjustified. It was analyzed whether the rule is necessary to test excellence in golf – and the answer was "no", which means that riding a golf cart would not provide Martin with an advantage over other players. In this situation, there were two possible solutions: either to allow all players to ride golf carts (i.e., to change the rules of tournaments) or to make an exception for disabled people such as Martin (i.e., to individualize the

---

40   PGA Tour v. Martin, Scalia dissenting, p. 700.

rules for exceptional reasons). Neither would result with the equalization of the distribution of excellence, as Scalia argued, since the questioned rule was arbitrary and not aimed at testing the excellence of golf players. In other words, the decision of the Court guaranteed equal chances to participate in tournaments, and not equal chances to win them.

The Pistorius case seems more complicated since the IAAF rule 144.2I was a competitive rule aimed at preventing the competitive advantage of some runners. The use of natural legs is central to the considered discipline of sport, in contrast to walking in golf. Thus, the CAS did not analyze the nature of sprinting but addressed only the question of fairness by analyzing the problem of whether prosthetic legs constitute an advantage over natural limbs in running. The decision was negative – there was no convincing evidence for such an advantage. In consequence, the IAAF amended its rule and shifted the burden of evidence to the athlete using assistance:

> For the purpose of this Rule, the following examples shall be considered assistance, and are therefore not allowed: (…) (d) The use of any mechanical aid, unless the athlete can establish on the balance of probabilities that the use of an aid would not provide him with an overall competitive advantage over an athlete not using such aid.[41]

All the considered examples give us some interesting insights into understanding justice. Even if we consider procedural justice in sport, which is aimed at guaranteeing equal opportunities, i.e., fair competition, we cannot neglect the objective of sport activity, which determines the procedural rules. The criteria of differentiating people (the criteria of eligibility) depend on it and if we want to treat people equally, after Aristotle we have to pose the crucial question: equal in what? As Aristotle put it in *Politics*:

> In all arts and sciences the end in view is some good. (…) justice involves two factors – things, and the persons to whom things are assigned – and it considers that persons who are equal should have assigned to them equal things. But here there arises a question which must not be overlooked. Equals and unequals – yes; but equals and unequals *in what*? (…) If you were dealing with a number of flute-players who were equal in their art, you would not assign them flutes on the principle that the better born should have a greater amount. Nobody will play the better for being better born; and it is to those who are better at the job that the better supply of tools should be given. If our point is not yet plain, it can be made so if we push it still further. Let us suppose a man who is superior to others in flute-playing, but far inferior in birth and beauty. Birth and beauty may be greater goods than the ability to play the flute, and those who possess them may, upon balance, surpass the flute-player more in these qualities than he surpasses them

---

41 IAAF Competition Rules (2016–2017), Rule 144.3(d), p. 153.

in his flute-playing; but the fact remains that *he* is the man who ought to get the better supply of flutes. If it is to be recognized in connexion with a given function, superiority in quality such as birth – or for that matter wealth – ought to contribute something to the performance of that function; and here these qualities contribute nothing to such performance.[42]

Applying this analogy to sprinting, one could say that athletes who are superior to others in running deserve to win. It means that the rules of eligibility and the rules of entitlement in this discipline should be designed in such a way that they test excellence in running and prevent any kind of *superiority which does not derive from one's own talent and effort*. To elucidate this problem one may invoke the useful distinction, made by Alasdair MacIntyre, between goods which are external and internal to a practice.[43] External goods are "contingently attached" to a practice by "the accidents of social circumstance".[44] Internal goods, in contrast, can be only specified in terms of a particular practice and "can only be identified and recognized by the experience of participating in the practice in question."[45] MacIntyre illustrates this with the example of playing chess. When one wants to teach a child to play chess and engage the child in playing by offering a candy for winning, the candy are external goods (equivalent to money, fame, power or prestige for adults). The internal goods of playing chess would be for instance strategic imagination, specific analytical skills, competitive intensity, commitment to self-improvement, etc. One may believe that engaging a person in a certain activity, like chess playing, will sooner or later open up his or her interest in internal goods. But since external goods are not specific to that particular activity, a person interested in them may find other alternatives to achieve them – a child interested in getting candy, not in the game itself, may be tempted to cheat in order to win (the excellence is here less important than the reward).

Another significant difference between external and internal goods lies in the fact that, once achieved, external goods become one's possession, making them the object of competition in which there are losers and winners (especially when scarce resources are concerned). Although internal goods may be the result of competition to excel, "their achievement is good for the whole community who

---

42 Aristotle, *The Politics*, ed. and trans. E. Barker (New York: Oxford University Press, 1946), Book III, chap. Xii [1282b].
43 A. MacIntyre, *After Virtue. A Study in Moral Theory* (Notre Dame: University of Notre Dame Press, 2007), pp. 188–191.
44 MacIntyre, *After Virtue*, p. 188.
45 MacIntyre, *After Virtue,* p. 188.

participate in the practice."⁴⁶ The distinction provided by MacIntyre helps us understand the nature of sport, which is aimed at promoting excellence based on our own talent and effort. Excellence belongs to the internal goods of sport, which include in the case of running athletic skills, competitive intensity, commitment to self-improvement, etc. Rewards (such as medals, money, fame) can be defined as external goods to sport activity. Thus, to answer the question posed in the introduction – what kind of enhancement constitutes unfair advantage – one should analyze whether a certain enhancement corrupts the internal goods of sport activity or rather promotes them. In the case of Pistorius, who put so much determination and effort into transforming his disability into ability, one may admit that excellence and self-improvement were definitely his aim. And this is the reason why his achievements were so appreciated by the public and contributed to the whole community.⁴⁷ If Pistorius had aimed only at external goods such as fame and rewards, excellence would have not mattered for him at all, and instead of hard work and training he would have preferred to use artificial legs which enabled him to run fast without effort which should not have been permitted unless everybody could use such legs. Provided that his artificial legs were not more effective than natural limbs, we should accept the CAS decision as being fully justified.

6 Concluding remarks

Even if a clear distinction between enhancement and therapy could be determined, it would not provide an answer to the question whether a particular enhancement or therapy in sport is fair. In order to establish if it so, it is necessary to analyze the criteria of procedural fairness as well as the end of the sport activity which they serve. This brings to the question about the definition of sport – its nature and what we value about it in particular. Following basic intuitions, we can admit that we value talents and efforts, which constitute the internal good of sport, e.g., excellence. Thus, one may argue that, in most cases,

---

46 MacIntyre, *After Virtue*, pp. 190–191. A similar example comes from Sandel, who poses the question of whether parents should pay children to read books if they want to cultivate the reading habit in them; he claims that the problem with doing good things for bad reasons is that the *reasons affect the meaning and the aim of the practice*. M. Sandel, *What Money Can't Buy. The Moral Limits of Markets* (New York: Farrar, Straus and Giroux, 2012), pp. 57–58.

47 It is not a coincidence that both Pistorius and Semenya were chosen by *Time Magazine* as the most influential people in the world.

enhancement should be eligible in sport on the grounds of procedural fairness, i.e., if everybody is equally enhanced (e.g., via standardized equipment). In some exceptional cases, which were considered in this paper, one may also argue for the eligibility of corrective enhancements which are necessary to allow an individual to participate in sport activity, but only if they promote skills and do not replace or overshadow natural abilities (if they do not give this individual an unfair advantage), and if they do not fundamentally alter the character of a given discipline, but promote its aim.

References

Aristotle. *The Politics*. Ed. and Trans. E. Barker. New York: Oxford University Press, 1946.

Camporesi, S. "Oscar Pistorius, enhancement and post-humans." *Journal for Medical Ethics*, Vol. 34, No. 9, 2008, p. 639.

Camporesi, S. and P. Maugeri. "Caster Semenya: sport, categories and the creative role of ethics." *Journal of Medical Ethics*, Vol. 36, 2010, pp. 378–379.

Davis, R. "Oscar Pistorius's murder trial: what's disability got to do with it?" *Th eGuardian*. http://www.theguardian.com/world/2014/jul/08/oscar-pistorius-murder-trial-disability (8 July 2014).

Feinberg, J. "Justice and personal desert." In idem: *Doing and Deserving. Essays in the Theory of Responsibility*. Princeton, New Jersey: Princeton University Press, 1970.

Feinberg, J. *Social Philosophy, Foundations of Philosophy Series*, eds. E. Beardsley and M. Beardsley. Englewood Cliffs, New Jersey: Prentice-Hall, 1973.

Ferguson-Smith, M.A. and E.A. Ferris. "Gender verification in sport, the need for change?" *British Journal of Sports Medicine*, Vol. 25, No. 1, 1991, pp. 17–20.

Harris J. and S. Chan. "Enhancement is good for you. Understanding the ethics of genetic enhancement." *Gene Therapy*, Vol. 15, 2008, pp. 338–339.

Hay, E. "Sex determination in putative female athletes." *JAMA*, Vol. 221, 1972, pp. 998–999.

Juengst, E.T. "Enhancement Uses of Medical Technology." In: *Encyclopedia of Bioethics*. Vol. 2, ed. S.G. Post. New York: MacMillan Reference USA, 2003.

Lucas, J. *Responsibility*. Oxford: Clarendon Press, 1995.

MacIntyre, A. *After Virtue. A Study in Moral Theory*. Notre Dame: University of Notre Dame Press, 2007.

Pellegrino, E. *Biotechnology, Human Enhancement, and the Ends of Medicine.* The Center for Bioethics and Human Dignity. www.cbhd.org/resources/biotech/pellegrino_2004-11-30.htm (30 November 2004).

Perelman, Ch. *The Idea of Justice and the Problem of Argument.* Trans. J. Petrie. London: The Humanities Press, 1963.

Quely, K. and G. Roberts. "One race, everymedalist ever." *TheNewYorkTimes* http://www.nytimes.com/interactive/2012/08/05/sports/olympics/the-100-meter-dash-one-race-every-medalist-ever.html?_r=0(5 August 2012).

Rawls, J.*Theory of Justice.* Oxford: Oxford University Press, 1999.

Sandel, M. *The Case Against Perfection. Ethics in the Age of Genetic Engineering.* Cambridge, Massachusetts and London: Harvard University Press, 2007.

Sandel, M. *Justice, What's the Right Thing to Do?* New York: Farrar, Straus and Giroux, 2009.

Sandel, M. *What Money Can't Buy.The Moral Limits of Markets.*New York: Farrar, Straus and Giroux, 2012.

Savulescu, J.A. Sandberg and G. Kahane. "Well-Being and Enhancement." In: *Enhancing Human Capacities*, ed. J. Savulescu, R. ter Meulen and G. Kahane. Chichester: Wiley-Blackwell, 2011.

Scheffler, S. *Boundaries and Allegiances. Problems of Justice and Responsibility in Liberal Thought.* Oxford: Oxford University Press, 2008.

Filip Kobiela

The Brave New Athlete? The Meaning of Perfection in Contemporary Professional Sport[1]

Abstract: The main purpose of this paper is to provide an analysis and critique of Michael Sandel's argument against human enhancement in sport. In order to examine the problem, I use Bernard Suits' theory of game-playing ("game" is an activity based on voluntary choice of imperfect means), as well as Wolfgang Welsch's account of sport as seen from the aesthetic perspective. The image of sport that emerges from these considerations is as follows: by nature, sport does not aim at perfection understood literally, but it is rather an aesthetic enterprise in which participants are acting out (in a different way than in a theatre) the drama of striving for perfection. In light of this philosophy of sport, all forms of doping, including genetic doping, which is one of the main subject of Sandel's work, seem to be a consequence of a mistaken view of the nature of sport. Thus, contrary to Sandel's view, the acceptance of the Promethean, quasi-mythological view of sport, is not so much an expression of a lack of gratitude, but rather – of a lack of wisdom.

Keywords: Bernard Suits, Wolfgang Welsch, Michael Sandel, Sport, Game, Perfection

1 The three levels of reflection on sport

In 2008, during the Olympic Games in Beijing, I came across a virtually surreal sight: a group of exorbitantly tall, black basketball players from the USA stood among the crowd of much smaller fans, contrasting with them in such a stark way that it made the basketballers look like ancient heroes, half-people, and half-gods. Similar observations of ancient athletes of superhuman size and strength surely had to inspire the creators of one of the myths pointing to the heroic deeds of Heracles as the origin of Olympic Games.[2] There are, certainly, more profound reasons for associating athletes with superhuman beings, related to their strength and skills, including the *perfection* with which they control their bodies. Colin McGinn claims that sport requires two kinds of harmony: the harmony

---

[1] The writing of this article was funded by the Polish National Science Centre (2012/07/D/HS1/01099).
[2] The contemporary ideological superstructure of the Olympic movement refers to these mythological origins as well.

between the body and the mind, and the harmony between different parts of the body. Such a state is experienced by athletes and perceived by spectators as a somewhat godlike condition, because gods are not divided in their being. When a person functions as a single unified entity, working as an organic whole, it might even be seen as a *sui generis* ontological category: embodied agency. According to McGinn,

> When people idolize their athletic heroes as gods, is it this impression of ontological unity that prompts their adulation? The outstanding athlete seems to operate on another plane of coordinated movement, above mere mortals.[3]

At the same time, in contemporary discussions on doping in sport a significant role is played by another superhuman character from Greek mythology – the Titan Prometheus.[4] Mythological thinking – more precisely, the idea of a mythical hero surpassing human limits – has been accompanying sport throughout its history and became an important subject of philosophical reflection on sport. Certainly, this mythological or quasi-religious level of reflection on sport (which could be called the *mythos* of sport) can at least partially explain the phenomenon of the popularity of sport and its powerful impact on the mass imagination.

However – returning to the experience in Beijing – the god-like athletes, despite the spectacular physical contrast between them and other men, are, in the light of philosophical reflection, the same human beings as anyone else. Within a population there are equally significant, although more disguised disproportions (e.g., related to the power of creative imagination, moral sensitivity, speed of counting, persistence of philosophical reflection.), which do not evoke any associations with heroes. Somehow on the antipodes of the above-mentioned tendency to worship sport heroes, there is also the need to identify with athletes treating them as real people and not "gods". As Wolfgang Welsch aptly observes, "we take the athletes' performance to be not totally beyond our scope. We even take it to be ours in a way. There is a feeling of *mea res agitur* – like in the theatre where when we see kings or people of excellence we don't think they are of an ontologically different kind."[5] Moreover, "natural" competitions, such as running

---

3   C. Mc Ginn, *Sport* (Stockfield: Acumen Publishing, 2008), p. 118.
4   See, for example, T. Franssen, "Prometheus on Dope. Natural Aim for Improvement or a Hubristic Drive to Mastery?" in: *Athletic Enhancement, Human Nature and Ethics*, ed. J. Tolleneer, S. Sterckx and P. Bonte (Dordrecht: Springer Netherlands, 2012), pp. 105–123.
5   W. Welsch, "Sport – Viewed Aesthetically or Even as an Art?" in: *The Aesthetics of Everyday Life*, ed. Andrew Light and Jonathan M. Smith (New York: Columbia University Press, 2005), p. 149.

or jumping, test abilities which are not solely characteristic of the human world, but also the animal kingdom. Embodied agency as an ontological category might be particularly intensively manifested in sport, but it does not provide an ontological demarcation line between human and superhuman athletes. The "divinity" ascribed to athletes should rather be understood in a figurative way, as in the expression "divine body", as a figure of a certain level of excellence and not of superhuman status.

The pursuit of better results, so characteristic of contemporary sport, generated the phenomenon of doping (and, recently, also gene doping), which raises ethical concerns, constituting the second level of reflection on sport. Doping is clearly not the only problem in sport ethics, though it is, undoubtedly, a significant issue related to the problem of perfection. This second level of reflections on sport is constituted by questions about the morality of human interactions in sportand can be subsumed under the heading *ethos* of sport. Despite the impression that ethics is the main philosophical issue of sport (which impression is contributed to by the educational aspect of sport, e.g., the fair play rule), I think that the principal philosophical reflection on sport is rather ontological in nature. As noted by Scott Kretchmar, "soft metaphysics is a precursor to good sport ethics."[6] Moreover, considerations sometimes labelled as "sport ethics" are rather of an ontological nature at their core – as are, in fact, some of the analyzes by Michel Sandel included in the subsection entitled "The Essence of the Game" in his work on human enhancement.[7] In this way we are approaching the third, most elementary level of reflection on sport, which concerns the *logos* of sport. My aim in the present article is to analyze this principal, ontological aspect, which, however, generates certain consequences for both the *ethos* and the *mythos* of sport. Above all, however, I will focus on revealing the ludic and aesthetic nature of sport, as they are of key importance for the issue of enhancement.

---

6   R. S. Kretchmar, "Soft Metaphysics: a Precursor to GoodSportsEthics," in: *Ethics and Sport*, ed. M. J. McNamee and S. J. Parry (London and New York: E & Fn Spon, 1998), pp 19–34.
7   M. Sandel, *The Case Against Perfection. Ethics in the Age of Genetic Engineering* (Cambridge, Mass and London, UK: The Belknap Press of Harvard University Press, 2007), pp. 36–44.

2 Sport as a voluntary choice of imperfect means

As noted by Hans Lenk, the term *philosophy* was once introduced (within the so called *panegyric analogy*) in connection with the Olympic Games of Antiquity.[8] However, despite this connection – promising, as it seemed, for the philosophical reflection on sport – a specialized branch of philosophy undertaking the issue of sport has only emerged in the second half of the 20th century. This somewhat delayed,[9] but natural and necessary, emergence of a sub-discipline dedicated to sport is associated with a certain unfortunate phenomenon. Philosophers of sport, focused on developing concepts arising within their tradition, do not always take full advantage of the achievements generated within mainstream philosophy. At the same time, the "mainstream" philosophers, undertaking the problems related to sports do not take full advantage of the contribution of the philosophy of sport, as can be observed in the above mentioned work by Sandel. One of the objectives of this article is to fill such a gap related to the issues raised by Sandel: the *telos* of sport within the context of human enhancement and the achievement of perfection.

A variety of meanings attributed to sport in everyday language exists, at least to some extent, also on the theoretical level. What is more, sport as a cultural phenomenon is subject to changes in its very constitution. It is thus necessary to clarify its meaning at the beginning, as it will constitute the basis for further considerations. I will therefore continue to analyze professional sport only, i.e., activities which are highly specialized, institutionalized, strictly regulated games generally based on physical skills.

When at the 1988 Seoul Olympics Ben Johnson beat in a spectacular way the world record in track and field crown discipline – 100m sprint – to be then stripped of the title for using an illegal doping substance, the commentators observed that he still remained the fastest man on Earth, although his feat was not of sporting nature any more due to his breaking of the doping rules. Through this example, therefore, we can clearly see that the literal interpretation of the

---

8   H. Lenk, *S.O.S. - Save Olympic Spirit: Toward a Social Philosophy of the Olympics (Selected Writings by Hans Lenk)* (Kassel: Agon Sportverlag, 2012), pp. 25–26.

9   Philosophy of sport emerged approximately thirty years after sport began to be taken seriously as a cultural phenomenon by large numbers of people, thus turning from an idle cultural curiosity into an often hotly debated ethical and aesthetic phenomenon. Compare W. Morgan, "Athletic Perfection, Performance-Enhancing Drugs, and the Treatment-Enhancement Distinction," *Journal of the Philosophy of Sport*, Vol. 36 (2009), p. 5.

famous Pierre de Coubertin motto "citius, altius, fortius" (which is, as a matter of fact, quite often recalled in the context of relations between sport and human enhancement) is highly misleading, as sprint is *not* all about being the fastest, leastwise, not the fastest by any means.

The precise meaning of this assertion, which I call the *negative characteristic of sport*, and which has key significance for capturing the essence of human enhancement and perfectionism in sport, was only comprehended with the application of the breakthrough game playing theory developed by Bernard Suits.[10] In this theory, games (sport – with some reservations – is a subset of games[11]) are voluntary activities in which means employed to achieve the goal *are not* the most efficient. The principal type of rules governing games are the so-called constitutive rules, i.e., players voluntarily accept (this acceptance is called the *lusory attitude*) prohibitions against applying the most efficient means. Therefore, in the case of sprint races, the crude fact of crossing the finish line ahead of others – called the *prelusory goal* – is, after all, not the real goal in this discipline; the goal is to cross the finish line ahead of others in accordance with certain rules, which impose some additional limitations. In light of this theory, Ben Johnson's feat was simply completing the *prelusory goal* of a sprint race in a record-breaking time. However, since the proper goal (the so-called *lusory goal*) of this competition is the realization of its *prelusory goal* only according to the rules, Ben Johnson by breaking the rules did not only not *win* the race, but, in a strict sense, he did not even *participate* in it.

This aforementioned prohibition is itself an illustration of the peculiarity of the ludic sphere in culture to which sport belongs. To capture the contrast between broadly understood work and sport we need to consider a hypothetical example of a scientific achievement, such as proving a new mathematical logic theorem. Should it become apparent that the author used in the process a special doping substance for scientists and, moreover, used a properly programmed computer, it would not limit the value of his work or make his proof invalid.

---

10 I present this theory on the basis of the article by B. Suits, "The Elements of Sport," in: *Ethics in Sport*, ed. W.J. Morgan, (Champaign, IL: Human Kinetics, 2007), pp. 9–19, and B. Suits' work, *The Grasshopper: Games, Life and Utopia* (Toronto: University of Toronto Press, 1978).

11 There is a small group of sports, which Bernard Suits calls *performances* or *judged events*, and David Best calls *aesthetic sports*, whose inclusion in the set of games in Suits' sense is a matter of controversy, see. D. Best, "Sport and Art," *Journal of Aesthetic Education*, Vol. 14, No. 2, (1980), pp. 70–72; B. Suits, "Tricky Triad: Games, Play and Sport," *Journal of the Philosophy of Sport* Vol. XV (1988), pp. 1–9.

The maximum achievable efficiency in realizing the cognitive goal (as long as it does not harm others) is the essence of scientific work, and the possible health hazard of the researcher exposed to such doping would be considered heroism – sacrificing one's well-being in the name of the science. One who recognizes this view at the same time recognizes a certain axiological relation between the positive cognitive value of scientific discovery and the negative value of a researcher's impaired health. The mechanism is observable to an even greater extent in the case of rescuers, fire fighters, or soldiers. Having performed this type of extrapolation, I will now transpose this classic discussion on more traditional forms of doping onto the area of partially implemented and partially anticipated fields of genetic human enhancement.

In the case of technical activities (work), the limitations of efficiency are dictated by certain necessities (mainly legal or ethical in nature); in sport, the acceptance of limitations generated by constitutive rules is of an entirely different character.[12] The above mentioned lusory attitude consists in the acceptance of unnecessary difficulties only because they make such activities possible. Constitutive rules – in contrast to the so-called regulative rules – not only regulate a previously existing activity, but also produce – constitute – a new activity (a chess game is created when the rules of a chess game are created). As, potentially, any achievable, specific state of affairs may become a prelusory goal, and there is an infinite number of methods to limit the efficiency of reaching this state of affairs by constitutive rules, the spectrum of possible games is virtually infinite. Because limitations in the form of constitutive rules are settled arbitrarily within the community of players, what we are dealing with here is an extensive area of creative cultural activities, which are in many respects similar to art, and in particular to theatre. The only certainty here is the inefficiency that makes games possible, i.e., their fundamentally negative nature: they are *not* the most efficient (i.e., "perfect") pursuits of specific goals, otherwise they would have lost their identity and become activities similar in nature to work. It should be noted that many sport disciplines (javelin throw may be a model example here) could also be viewed in this context as a symbolic representation of certain historical instrumental activities ousted by more efficient forms of action.[13] Since, however, the empirically existing set of sport competitions is only a small subset of all the possible competitions of this sort, the question is whether it is only a

---

12 On the issue of a more extensive confrontation of work (instrumental activities) and ludic activities, see Suits, *The Grasshopper…*, in particular chapters 3 and 15.
13 There are some interesting hypothetical examples on this subject provided by Suits, *Tricky Triad…*, pp. 3–5.

randomly arranged set or there are some reasons for creating and cultivating certain sport games.

The problem of the positive characteristics of sport – in contrast to its generally accepted negative characteristics in the form of Suits' theory – has not found an acceptable solution (it is one of the reasons for the above mentioned variety of meanings of the term *sport*). The function of the prohibitions generating particular games is to create conditions enabling to perform certain activities otherwise impossible or difficult to carry out.[14]

It should be noted that in so far as the goal of a given game is a certain *state of affairs*, the activities performed in pursuing it belong to a different ontological category of *processes*. Depending on the type of difficulties occurring during the pursuit of the goal, the game writer shall be given the opportunity to model various processes involving different skills to achieve it. Therefore, e.g., the prohibition against the use of hands generates specific abilities proper to football. The creation and acceptance of restrictions constituting the game is actually a pretext to realize a certain potential. But why would anybody arrange conditions to create such – as it seems – absurd activities (and the related skills) as, for instance, pole vault? I think that the famous answer given by George Mallory to the question of why he climbs mountains – "Because they exist" – will provide important guidance. The free will of an entity is a sufficient reason to undertake autotelic activities. If a certain form of expression of human potential – such as sport, but also, for example, playing the saw or swallowing gold fish – has its followers, it is only wise to simply accept this strictly ludic phenomenon. Possible further reasons (health, education, etc.) are of secondary nature, and become an interpretation or ideology imposed from the outside. I, thus, accept that sport is an autotelic form of expression (and celebration) of human physical potential. The primary impulse here is the acceptance of certain possibilities of embodied existence, and especially – a driving force behind its growth – undertaking the development of physical skills as such (and not as a means to achieve a goal).

This portrayal of the idea of sport may seemingly differ considerably from its media image (suggesting "positivist" progress and instrumental efficiency) but, in fact, it makes it possible to see some key dependencies in order to solve the general question concerning human enhancement in sport. To complement the positive characteristics of sport, I will now move on to aesthetic reflections.

---

14 See Suits, *The Elements of Sport*, p. 14.

3 Sport as the drama of striving for perfection

While sport games provide their fans with heartfelt emotions and at the same time attract their attention to the rivalry between the leading players, the fundamental question – "What is sport for?" – remains in the shade. Stephen Mumford accurately pointed out that while *competitors* aim solely at victory, *sport itself* aims at beauty – its constitutive rules set up situations in which beauty is likely to manifest; furthermore – the rule changes themselves are precisely focused on the need to make the sport more attractive to watch.[15] It should be noted that between the goal of an individual player – victory, and the goal of sport itself as a branch of culture – beauty, there is an intermediate link in the form of athletic excellence. Competition is a means of achieving this goal, and the essence of athletic excellence is to show the beauty of the human body in motion. It needs to be observed that if the player's pursuit of victory would lessen in comparison to his pursuit of beauty, not only would his results suffer, but paradoxically, also the beauty of the whole event would deteriorate, in a process similar to the working of the invisible hand of the market.

Welsch presents a discussion of a similar problem in the context of comparing sports to art.[16] He claims that one of the main objections against recognizing sport as art is that sport is merely a profane activity aiming at victory and thus it lacks symbolic meaning as well the quality of being an end in itself. Welsch refutes this objection by different means than the distinction between the participant's goal and the overall aim of the discipline. He indicates certain similarities between sports and performing arts: they both have their ends in themselves and as autotelic activities they do not serve other purposes. The aim of winning cannot be reached *directly* but only *through the sporting performance*. The proper work of the athlete is the performance; it might lead to victory, but this secondary result cannot affect the autotelic nature of sport. Furthermore, Welsch claims that sport's autotelic character is connected with its symbolic meaning. Sport is semantically intense and intrinsically artistic, but this impression is not a result of the fact that sport is explicitly about something (like a play in the theatre). Sport is a drama *without a script*:[17] sporting events act out the most basic features

---

15 This is one of Mumford's theses presented in his defense of the aesthetic way of watching sports, see S. Mumford, *Watching sport: aesthetics, ethics and emotion* (London: Routledge, 2011).
16 Welsch, "Sport...," pp. 143–148.
17 Even though there is no script in a sport event, the rules afford it with a kind of meta-script, a mechanism generating a particular, improvised narration. This is the reason why a close football game can be perceived as an unpredictable, one-time,

of the human condition, and can display all the dramatic traits of human existence. One of these features of the human condition is striving for perfection. McGinn describes it in a similar manner: "Sport lends itself to narrative exposition, because there is an inherent drama to it; this is why sport enjoys the media saturation it does – it's a story of stories."[18]

It is crucial to note that sport is an *artistic* display of a human struggle for perfection. The tricky thing is that this quasi-theatrical meaning is expressed through the efforts undertaken by an athlete who appears to be literally striving for perfection. This structure might make the artistic, metaphorical dimension of the struggle for perfection in sport difficult to grasp. Sport is not so much literally *striving for* perfection, as rather *acting out* (in a different way than in a theatre) *the drama of striving for perfection*. Therefore its literal achievement is not expected, and this aesthetic conclusion complements my former ontological thesis claiming that imperfection is somehow a necessary condition for sport to exist.

4 Sport *versus* human enhancement

There are two reasons why contemporary sport is attractive research material to ponder on the idea of perfection within the context of human enhancement. First of all, the idea of sport includes exposing human physical potential in a manner that is accessible to an audience, owing to which at least some relevant, observable phenomena provide handy illustrative material. In this respect, also the variety of sport events is reminiscent of art works which, in their essence, are things (in light of the institutional definition of art provided by Arthur Danto) *submitted for assessment*. Second of all, sport belongs to the avant-garde areas in which human enhancement is becoming a fact[19] (by the way, these practices have their origins in the phenomenon of doping which is specific to sport), thus enabling us to base our reflections on a real, rather than only hypothetical or anticipated, context (following Habermas – "realistically expected"). Gene doping (gene therapy) is an exceptionally important issue in sport as it could, for example, support muscle growth, facilitate regeneration, and improve overall

---

quasi-theatrical show. This uniqueness and particularity of the specific place and time of a sport event – as opposed to the repeatability of a scientific experiment – makes the nature of sport specifically dramatic (especially in the case of higher ranked events).
18 McGinn, *Sport*, p. 111.
19 See A. Miah, "Be Very Afraid: Cyborg Athletes, Transhuman Ideals & Posthumanity," *Journal of Evolution and Technology*, Vol. 13, No. 2 (2003).

body fitness, thus obviously contributing to the abilities of the player. If we add to this the phenomenon of cyborgization and other relevant technologies, the disturbing (for some) perspective of creating a "superathlete" – whom I suggest to call "the brave new athlete" – becomes quite probable. For these reasons, the discussion on human enhancement in sport, with the important work of Sandel in focus, is progressing rapidly.[20]

In his analysis of enhancement, Sandel draws attention to the fact that the traditional argument against enhancement concerning the violation of an individual's autonomy[21] cannot be applied in sport: "An athlete who genetically enhances his muscles does not confer on his progeny his added speed and strength; he cannot be charged with foisting talents on his children that may push them toward an athletic career."[22] Additionally, Sandel does not accept the argument against enhancing technologies originating in the degradation of human agency in sport. The core of this argument is as follows: because there is an inverse correlation between enhancement and achievement, the use of enhancing technologies lessens our admiration for the achievement, and the athlete's agency and responsibility are thus degraded. However, both the athletes and the audience are aware that athletes do not develop their skills in isolation, and they are never totally responsible for their achievement.[23]

Instead of this Sandel offers another line of argumentation, indicating an opposite problem: not the degradation of agency, but hyper-agency. According to Sandel, enhancing technologies "represent a kind of hyperagency, a Promethean

---

20 Sandel is one of the three – next to Habermas (J. Habermas, *The Future of Human Nature*, transl. H. Beister, M. Pensky and W. Rehg (Cambridge: Polity 2003)) and Fukuyama (*Our Posthuman Future. Consequences of the Biotechnology Revolution* (New York: *Farrar, Straus* and *Giroux*, 2002)) – most renown opponents of human enhancement, however, he is the only one who engaged in the issue of human enhancement in sports in a more systematic way. See also N. Bostrom and J. Savulescu, "Introduction: Human Enhancement Ethics: The State of the Debate," in: *Human Enhancement*, ed. N. Bostrom and J. Savulescu (New York: Oxford University Press, 2009).
21 In my analysis of Sandel's arguments I significantly benefited from the very clear explanations of their premises and conclusions presented in the materials developed by A. Agler, *Sandel on Bionic Athletes*: 20 Nov. 2015, http://www.davidagler.com/teaching/philoftech/enhancement/Handout4_BionicAthletes.pdf
22 Sandel, *The Case* ..., p. 9.
23 The refutation of the argument is not provided by Sandel (he immediately proceeds to the hyper-agency argument); for a more detailed discussion see Agler, *Sandel on Bionic Athletes*, p. 3.

aspiration to remake nature, including human nature, to serve our purposes and satisfy our desires. The problem is (…) the drive to mastery. And what the drive to mastery misses, and may even destroy, is an appreciation of the gifted character of human powers and achievements."[24] The outline of Sandel's argument is as follows: sport has its essence (*telos*), and the *telos* of sport is excellence, which in turn consists of the display of *natural* talents. The athletes who desire to obtain enhanced capacities express an *ingratitude* towards their natural capacities, and thus degrade the *telos* of sport. Such a degraded sport (devoid of its *telos* – in which natural talents are not appreciated) would be a mere spectacle – like WWF's staged wrestling – and not a sport with integrity. The described mechanism of the degradation of sport leads to the conclusion that using enhancement technologies is morally wrong because it corrupts sport (a valuable part of human culture) and should be avoided even if there were no other reasons (like health risks) against its use.

The two most important presuppositions made by Sandel are as follows: (S1) the existence of the *telos* of sport and (S2) the identification of excellence through the display of *natural* talents. In defending the notion of the *telos* of sport Sandel appeals to the philosophically interesting case of Casey Martin, a golfer who, due to a certain kind of disability, made a request to use a golf cart during professional competition. After getting a refusal from the golf association (PGA Tour), Martin pursued a legal case – and as a result the Supreme Court held in his favor on the grounds that walking was inessential to golf. However, Justice Antonin Scalia dissented, arguing that "To say that something is 'essential' is ordinarily to say that it is necessary to the achievement of a certain object. But since it is the very nature of a game to have no object except amusement (that is what distinguishes games from productive activity), it is quite impossible to say that any of a game's arbitrary rules is 'essential.'"[25] Scalia's conclusion is correct, because if we allow some players to move around on a golf cart, we cause a medley of two different games, since the constitutive rules referring to Martin are different (no prohibition on moving on a golf cart) from the rules applied to other players.[26] However, the argumentation and interpretation of the conclusion performed by Sandel is far more important than the conclusion itself. Scalia is correct in juxtaposing sport and "productive activity", but his further analysis

---

24 Sandel, *The Case* …, pp. 26–27.
25 Sandel, *The Case* …, p. 43.
26 Martin's competitors might claim that walking *is*, in fact, essential to golf, since it causes fatigue that might influence the player's ability to make a precise shot. In this respect, there is an analogy between golf and biathlon.

is simplified. His error consists in taking into account only one meaning of the term *essence*. In this ordinary sense, all restrictions imposed by the constitutive rules of the game are inessential, since the game has no external aim. But in another sense, all these restrictions are essential, since they are responsible for the game's identity (essence = what constitutes identity). Finally, in the third sense, some restrictions might more or less contribute to the *telos* of the game (understood as skills centrally tested in the game) and in this sense they are more or less "essential". In this third sense walking is less "essential" to the *telos* of golf than using a club.

Sandel claims that failing to recognize the *telos* of sport, i.e., the fact that sport is designed to call for and celebrate certain talents worth admiring might result in neglecting the meaning of the outcome of the game and thus the degradation of the game into a mere spectacle. Indeed, as follows, the popularity and cultural importance of a game are related to the way in which the game tests some skills that might be appreciated (to illustrate this, we can compare, for example, a sack race and running). However, recognizing the *telos* of the game (S1) correctly by no means allows Sandel to identify tested and admired skills with *natural* skills only (S2). This is an additional presupposition that cannot be justified solely by analyzing the structure of the game. As viewed from Suits' perspective, games consist of restrictions placed upon the means of achieving the prelusory goal. These restrictions might or might not involve using enhancing technologies, and this distinction is different from the distinction between valuable and shallow games.

Additionally, Sandel's minor mistake consists in implicitly suggesting a false dichotomy between a spectacle – a source of amusement, and (proper) sport – a subject of appreciation. But in the case of sport as observed by an educated fan, the source of amusement is exactly the subject of appreciation. This point has been clearly recognized by Suits: "All sports appear to be games of skill rather than games of chance. I suggest that the reason for this is that the main requirement in sports, for participants and spectators alike, is that the participants perform actions that must be admirable in some respect. The exercise of virtually any skill (…) will elicit some degree of admiration."[27] Contrary to Sandel, sport is therefore a kind of spectacle, and its *telos* need not be narrowed down only to manifestations of natural ability. However, despite my disagreement with Sandel's conclusion, I would like to present an argument against enhancing technologies

---

27 Suits, "The Elements of Sport," p. 15.

in sports, which is based, similarly to Sandel's, on the belief that *telos* in sports does exist.

5 The meaning of perfection in sport

As I attempted to show, reflection on sports consists of certain levels, where the principal ontological part impacts the other – ethical and mythological – parts. Researching the structure (logos) of sports, exposed mainly through Suits' work, reveals an interesting dialectic of efficiency and inefficiency, perfection and imperfection, as well as the necessity to take into account certain distinctions relating to the goal of sport. While Sandel presents the *quasi*-theological argument against enhancement in sport, I propose a ludic-aesthetic perspective, which reveals not the unethicality, but rather the meaninglessness of human enhancement in sport. Rather than speaking of *fair-play*, I suggest focusing on the *meaningful-play*. Paradoxically, from this perspective, sport – a leading area in doping use (including gene doping) – is the last area in which such procedures would have any greater sense.

Allowing the enhancement of the cognitive abilities of a scientist, a surgeon, or a plane pilot has reasonable grounds, as it can indicate a possibility of fulfilling a certain goal, appreciated in itself; a goal beyond the activity used to achieve it. But this kind of rationale is missing from activities of sporting nature (enhancing one's predispositions to throw a javelin would have been reasonable in the Bronze Age). When the abilities of players and the speed of balls in table tennis became so perfect that an average spectator was not able to follow the game, new, slower balls were introduced, thus also reintroducing the spectacle aspect of the competition (if the average height of basketball players had reached 3 meters, the baskets would have been raised, too). Similar rule modifications are common, and their message is clear: the abilities tested in sports are involved in the making of a spectacle whose goal is not to generate perfection in the absolute sense.

I will now consider the problem of the meaning of perfection in more detail. As Władysław Tatarkiewicz noted, the coexistence of these two concepts of perfection (precise and colloquial), raises the so-called *paradox of perfection*: "If the world were perfect it could not improve and thus would not possess 'true perfection' which depends on progress. And so 'the world is perfect through its imperfection.'"[28] Tatarkiewicz also points out a similar phenomenon in the domain of technology: in some cases, the imperfection in the structure of matter (e.g., irregularity

---

28 W. Tatarkiewicz, *On Perfection* (Warsaw: Warsaw University Press, Center of Universalism, 1992), p. 18.

in conductor crystals) contributes to the perfection of technology – thus, irregularity can be useful.[29] A solution to this paradox might consist in indicating an equivocation: the confusion of two different concepts of perfection expressed by the same term. The concept of "perfection" is ambiguous: in the strict sense of the term it designates that which possesses all the virtues (*perfectus* = *teleos* = finished = flawless); in the loose (colloquial) sense it designates that which possesses virtue greater than the others.[30] This distinction has its roots in Latin: *perfectio* involves no comparison (if something is perfect, it is such without comparison to other things), whereas *excellentia* is a distinction between many and implies comparison (it is thus a relative term; it designates which is the best, which is better than other things, which stands out among others).[31] The paradox of excellence, which can thus be solved in relation to the world as a whole, can, however, prove to be a strong heuristic tool on a local level. I assert that this is exactly the situation in sports as perceived through Suits' theory. Let us imagine the sterility of sprint, for instance, if all the participants of the race achieve identical results, equal to the limit of human potential. Their perfection would, at the same time, mark the end of this discipline. Sandel remarks in a similar manner: "A game in which genetically altered sluggers routinely hit home runs might be amusing for a time, but it would lack the human drama and complexity of baseball, in which even the greatest hitters fail more often than they succeed."[32] It is through this *imperfection* of the players that sport as a domain of culture bears its specific *perfection*. It should be noted that in so far as breaking records (where it is possible – so in a relatively narrow group of disciplines) does bring a type of progress, perfection of sport itself is not connected to it. Most memorable sport events are more or less evenly spread throughout the history of sport, and the development of enhancing technologies does not contribute, or contributes only to a minimal degree, to the development of sport itself, as seen in its ludic-aesthetic nature.

In the same way that an authentic player maintains his lusory attitude, a mindful sports fan presents his *spectator's ludic attitude*, analogously to an art viewer presenting his aesthetic attitude in his perception of art. The development of the proper attitude requires the spectator to be acquainted with the rules, as well as to have a more profound understanding of the essence of sport competitions. This understanding could be served by a practice of placing a forklift next to a barbell waiting to be lifted by an athlete – the situation would point

---

29 Tatarkiewicz, *On Perfection*, p. 21.
30 Tatarkiewicz, *On Perfection*, p. 18.
31 Tatarkiewicz, *On Perfection*, pp. 15–16.
32 Sandel, *The Case* …, p. 36.

to the conventionality and theatricality of a sporting performance. Its function would not be the moderation of the athletes' pride, but rather the dissipation of the fans' tendency to treat sport too literally, thus misinterpreting the essence of a sport performance.

Some known negative phenomena related to cheering and fan behavior result from attitudes foreign to the ludic approach (e.g., tribal). There is reason to fear that this naive manner of watching sport by literally interpreting the drama of pursuit of perfection, in conditions excluding the possibility of this perfection (constitutive rules), will become a force gradually pushing for the implementation of enhancing technologies into sports. However, whoever rejects this mythological, Promethean vision of sports will observe that "breeding" super-athletes is not so much an expression of a lack of gratitude, but rather of a lack of wisdom. However, Sandel is definitely on the right track when he observes that "there is something unsettling about the specter of genetically altered athletes lifting SUVs or hitting 650-foot home runs or running a three-minute mile."[33] We could also add – a significant fact within the aesthetic framework of sports – that it is rather kitschy, as has been aptly observed (within a broader context of bringing to life the idea of a superhuman) by Bronisław Łagowski.[34] Treating athletes as superhumans is an incorrect way of interpreting sports, while striving to enhance athletes is a wrong direction in the development of sport itself. Both errors result from mistaking the specifics of perfection within the framework of sport.

References

Agler, D. W. *Sandel on Bionic Athletes*. http://www.davidagler.com/tcaching/philoftech/enhancement/Handout4_BionicAthletes.pdf(20 November 2015).

Best, D. "Sport and Art." *Journal of Aesthetic Education*, Vol. 14, No. 2, 1980, pp. 69–80.

Bostrom, N. and J.Savulescu. "Introduction: Human Enhancement Ethics: The State of the Debate." In: *Human Enhancement*, eds. N. Bostrom and J. Savulescu. New York: Oxford University Press, 2009, pp. 1–21.

Franssen, T. "Prometheus on Dope. Natural Aim for Improvement or a Hubristic Drive to Mastery?" In: *Athletic Enhancement, Human Nature and Ethics*, eds. J. Tolleneer, S. Sterckx and P. Bonte. Dordrecht: Springer Netherlands, 2012, pp. 105–123.

---

33 Sandel, *The Case* …, p. 12.
34 B. Łagowski, „Czy monstrualna mutacja?," in: *Łagodny protest obywatelski*, B. Łagowski (Kraków: Księgarnia Akademicka, 2001), pp. 115–117.

Fukuyama, F. *Our Posthuman Future. Consequences of the Biotechnology Revolution*. New York: Farrar, Straus, and Giroux, 2002.

Habermas, J. *The Future of Human Nature*. Trans. H. Beister, M. Pensky, and W. Rehg. Cambridge: Polity, 2003.

Kotarbiński, T. *Praxiology. An Introduction to the Science of Efficient Action*. New York: Pergamon Press, 1965.

Kretchmar, R. S. "Soft Metaphysics: A Precursor to Good Sports Ethics." In: *Ethics and Sport*, ed. M. J. McNamee and S. J. Parry. London and New York: E & Fn Spon, 1998, pp. 19–34.

Lenk, H. *S.O.S. - Save Olympic Spirit: Toward a Social Philosophy of the Olympics (Selected Writings by Hans Lenk)*. Kassel: Agon Sportverlag, 2012.

Łagowski, B. „Czy monstrualna mutacja?" In: *Łagodny protest obywatelski*, ed. B. Łagowski. Kraków: Księgarnia Akademicka, 2001, pp. 115–117.

McGinn, C., *Sport*. Stockfield: Acumen Pub, 2008.

Miah, A. "Be Very Afraid: Cyborg Athletes, Transhuman Ideals & Posthumanity." *Journal of Evolution & Technology*, Vol. 13, No. 2, 2003.

Morgan, W. "Athletic Perfection, Performance-Enhancing Drugs, and the Treatment-Enhancement Distinction." *Journal of the Philosophy of Sport*, Vol. 36, 2009, pp. 162–181.

Mumford, S. *Watching Sport: Aesthetics, Ethics and Emotion*. London: Routledge, 2011.

Sandel, M. *The Case Against Perfection. Ethics in the Age of Genetic Engineering*. Cambridge, Massachusetts, and London, England: The Belknap Press of Harvard University Press, 2007.

Suits, B. *The Grasshopper: Games, Life and Utopia*. Toronto: University of Toronto Press, 1978.

Suits, B. "Tricky Triad: Games, Play and Sport." *Journal of the Philosophy of Sport*, Vol. XV, 1988, pp. 1–9.

Suits, B. "The Elements of Sport." In: *Ethics in Sport*, ed. W. J. Morgan. Champaign, Illinois: Human Kinetics, 2007, pp. 9–19.

Tamburrini C. M. and T. Tännsjö, eds. *Genetic Technology and Sport: Ethical Questions*. London: Routledge, 2005.

Tatarkiewicz, W. *On Perfection*. Warsaw: Warsaw University Press, Center of Universalism, 1992, pp. 11–51.

Welsch, W. "Sport – Viewed Aesthetically, or Even as Art?" In: *The Aesthetics of Everyday Life*, eds. A. Light and J. M. Smith. New York: Columbia University Press, 2005, pp. 135–155.

Notes on Contributors

Robert Audi: John A. O'Brien Professor of Philosophy, University of Notre Dame, USA

Michał Bizoń: MA in Classics and MSc in Physics, PhD in Philosophy, Assistant Professor at the Institute of Philosophy, Jagiellonian University in Krakow, Poland

Thomas Douglas: B.Med.Sc. in Bioethics and M.B.Ch.B. in Medicine, B.A. in Philosophy, Politics & Economics, PhD in Philosophy, Senior Research Fellow in the Oxford Uehiro Centre for Practical Ethics and a Golding Junior Fellow at Brasenose College, Oxford, UK

Tomasz Fiedler: MA in Classics, MA in Philosophy, STL in Theology, Doctoral student at the Pontifical Oriental Institute in Rome, Italy

Jacek Jaśtal: MA and PhD in Philosophy, Habilitation in Philosophy, Associate Professor at the Institute of Economics, Sociology and Philosophy, Cracow University of Technology, Poland

Jan Kiełbasa: MA and PhD in Philosophy, Habilitation in Philosophy, Associate Professor at the Institute of Philosophy, Jagiellonian University in Krakow, Poland

Filip Kobiela: MA and PhD in Philosophy, Assistant Professor at the Department of Humanities, University of Physical Education in Krakow, Poland

Wojciech Lewandowski: MA and PhD in Philosophy, Assistant Professor at the Department of Applied Ethics, Faculty of Philosophy, the John Paul II Catholic University in Lublin, Poland

Beata Płonka: MSc in Biology, Doctor of Natural Sciences (Biophysics), PhD in Philosophy, Senior Lecturer at the Department of Biophysics, Faculty of Biochemistry, Biophysics and Biotechnology, Jagiellonian University in Krakow, Poland

Robert Poczobut: MA and PhD in Philosophy, Habilitation in Philosophy, Associate Professor at the Department of Sociology and Cognitive Science, Faculty of History and Sociology, University of Bialystok, Poland

Marta Soniewicka: MA in Law, PhD in Law and Philosophy, Assistant Professor at the Department of Philosophy of Law and Legal Ethics, Faculty of Law and Administration, Jagiellonian University in Krakow, Poland

Adriana Warmbier: MA in Polish Philology and Philosophy, PhD in Philosophy, Assistant Professor in the Institute of Philosophy at the Faculty of Philosophy of the Jagiellonian University in Krakow, Poland

Karol Wilczyński: MA in Philosophy, Ph.D. student of the Inter-University Program of Interdisciplinary PhD Studies at the »Artes Liberales« Academy and Jagiellonian University in Krakow, Poland

Wojciech Załuski: MA in Law, PhD in Law and Philosophy, Professor at the Department of Philosophy of Law and Legal Ethics, Faculty of Law and Administration, Jagiellonian University in Krakow, Poland

www.ingramcontent.com/pod-product-compliance
Ingram Content Group UK Ltd.
Pitfield, Milton Keynes, MK11 3LW, UK
UKHW041924210426
5322IPUK00002B/44